30th IFHTSE Congress 2025

Suzhou, China
19-22 August 2025

ISBN: 979-8-3313-2986-0

Printed from e-media with permission by:

Curran Associates, Inc.
57 Morehouse Lane
Red Hook, NY 12571

Some format issues inherent in the e-media version may also appear in this print version.

Printed by Curran Associates, Inc. (2025)

International Federation for Heat Treatment & Surface Engineering (IFHTSE)
CH-8404 Winterthur
Sulzer Allee 25, Postfach
c/o Listemann AG
Switzerland

www.ifhtse.org

Additional copies of this publication are available from:

Curran Associates, Inc.
57 Morehouse Lane
Red Hook, NY 12571 USA
Phone: 845-758-0400
Fax: 845-758-2634
Email: curran@proceedings.com
Web: www.proceedings.com

目 次

IFHTSE | CHTS

The 30th World Congress of International Federation for Heat Treatment and Surface Engineering

August 19 to 22, 2025, Suzhou, China

Abstract Collection

Thermally Activated Martensite Formation: An Exception or the Rule?

Matteo Villa[1,2], Marcel A. J. Somers[1*]

[1] Technical University of Denmark, Department of Civil and Mechanical Engineering, DK-2800 Kgs. Lyngby, Denmark

[2] Now with: Polytechnic University of Bari, Department of Mechanics, Mathematics and Management, IT-70125 Bari, Italy

E-mail address: majs@dtu.dk (Marcel A. J. Somers)

Abstract: Generally, the formation of martensite in steel is represented by horizontal lines in the CCT and TTT diagrams implying that martensite formation is athermal, i. e., independent of time. Nevertheless, hardening of precipitation hardening stainless steels that contain a significant fraction of austenite after cooling to room temperature is achieved by isothermal treatment in the sub-zero Celsius temperature regime. In the last 15 years, we investigated the kinetics of martensite formation with in-situ magnetometry, complemented by electron microscopy and X-ray diffraction studies. Our investigations revealed that thermally activated formation of martensite in a wide variety of steels and model alloys is the rule rather than the exception. For all the investigated steels and model alloys it was observed that martensite can form during cooling, isothermally as well as during heating. The kinetics of martensite formation depends on the steel chemistry and relates intimately to martensite's internal structure. In this overview of our work, the data is put into perspective and it is demonstrated that Zhao & Notis' presentation of lath and plate martensite as individual C-shaped curves can be extrapolated to the sub-zero Celsius regime to provide a comprehensive understanding. This paves the way for development of new steel compositions.

Keywords: martensitic transformations; steel; heat-treatment; materials characterization

Current Status and Prospects of Surface Heat Treatment Technologies Utilizing Nitrogen Effectively

Yoichi Watanabe

Nihon Parkerizing Co., Ltd., Technical Division 5-1, 4-Chome Ohkami, Hiratsuka, Kanagawa, 254-0012 JAPAN

E-mail address: watanabe. yoichi@ parker. jp (Yoichi Watanabe)

Abstract: Surface heat treatment technologies that effectively utilize nitrogen, particularly nitriding and nitrocarburizing, have become increasingly important in the context of global efforts toward decarbonization, energy efficiency, and sustainable manufacturing. These processes introduce nitrogen into the surface layer of steels, leading to significant improvements in surface hardness, wear resistance, and fatigue strength. Unlike conventional carburizing processes, nitriding treatments are conducted at relatively low temperatures (typically below 600 ℃), which are without phase transformations and minimize dimensional distortion. This makes them particularly well-suited for precision components used in sectors such as automotive, aerospace, and energy systems. In the automobiles industry, for example, the demand for lightweight and high-strength components is accelerating, especially in line with the shift toward electrification. Nitriding-based treatments help extend the service life of critical parts such as gears and shafts, reduce frictional losses, and enable more compact and efficient mechanical systems. These benefits contribute directly to reducing energy consumption and CO_2 emissions throughout the product lifecycle. Moreover, the fact that these treatments can be performed without molten salts or high carbon atmospheres makes them more environmentally friendly and adaptable to modern production systems. However, despite these advantages, several technical challenges remain. The hardening depth achieved by nitriding is generally shallower compared to carburizing, and the effectiveness of the process can vary significantly depending on the steel's composition, particularly its alloying elements having a strong nitrogen affinity such as Cr, Mo, Al, V. As a result, precise control of steel grade and alloy selection, process parameters, and surface activation becomes essential. In addition, there is a growing need to develop high-speed, high-efficiency nitriding technologies that align with Industry 4.0 principles, including digital process monitoring, reduced cycle times, and better integration with advanced manufacturing lines. This keynote lecture will first provide a comprehensive review of the current status of nitriding-based surface heat treatment technologies, with a focus on conventional nitriding/nitrocarburizing, high-temperature nitriding, and carbonitriding quenching. It will examine the historical development, key mechanisms, and accumulated industrial data supporting their adoption. Furthermore, the presentation will reassess the untapped potential of these treatments in addressing modern engineering challenges, particularly in relation to component multifunctionality, surface fatigue behavior, and microstructural optimization. Finally, we will discuss emerging trends and future prospects for next-generation surface heat treatment technologies. This includes the integration of plasma and gas nitridingsystems with real-time process control, hybrid treatments combining nitriding with advanced coatings, and efforts to improve sustainability and energy efficiency at the system level. Through this exploration, the aim is to clarify the strategic role of nitriding-based heat treatment in future manufacturing, and to identify key research directions and development opportunities for the coming decades.

Keywords: Nitrogen; nitriding; nitrocarburizing; carbonitriding; alloying steel; fatigue strength; wear resistance

Multi-Mechanism Study on the Improvement of Mechanical Property and Shape Memory Effect of NiTiNb Shape Memory Alloys

Liu Shan[1], Wei Huijun[1], Feng Zhihao[1], Guo Rui[1], Li Jianhui[1*], Hu Jianwen[a*]

[1] School of Materials Science and Engineering, Hebei University of Science and Technology, Shijiazhuang 050018, China

E-mail address: lijianhui_97@163.com (Li Jianhui)

E-mail address: hujianwen05@163.com (Hu Jianwen)

Abstract: In order to investigate the microstructure influence mechanism of NiTiNb shape memory alloys on the mechanical property and functional stability, various amounts of rolling deformation (30%, 65%, and 80%) was designed to obtain different microstructures characterized by grain sizes, dislocation density, and precipitates distribution, respectively. Multiple microscopic mechanisms, especially the synergistic relationships that improve the mechanical and functional properties of NiTiNb alloys, were explored via XRD, SEM, EBSD, TEM, STEM and DSC. The results indicate that compared to the as-cast state, the phase transformation temperature hysteresis, mechanical property, and shape memory effect of the hot-rolled NiTiNb alloy are markedly improved with an increase at least 20%, after a severe rolling deformation of 80%. The increase of the wide thermal hysteresis of the NiTiNb alloy is primarily attributed to the elastic strain relaxation of the martensite interface, which is influenced by the β-Nb phase. This relaxation reduces the driving force of the martensite reverse transformation and enhances the martensite stability. The favorable mechanical property is due to the obvious grain refinement and the substantial increase in dislocation density that introduced by rolling deformation. Additionally, the rolling deformation facilitates the precipitation of the β-Nb phases and promotes the fragmentation and spheroidization of the NiTi (Nb) eutectic structure. The notable enhancement of the shape memory effect is mainly derived from the microstructure refinement and β-Nb spheroidization. The above research will lay a scientific foundation for the development of NiTiNb shape memory alloys used as pipe joints in aerospace fields.

Keywords: NiTiNb shape memory alloy; grain refinement; β-Nb phase; phase transformation; mechanical property; shape memory effect

A Study Foroptimized Properties of Methylsilsesquioxane Based Xerogel Monoliths via Ambient Pressure Drying Method

Xiaoyong He[1*], Guangming He[2], Lunkun Niu[3], Chongyi Wu[4]

[1] Nanchang Hangkong University, School of Material Science and Engineering Key Laboratory for Microstructural Control of Metallic Materials of Jiangxi Province (Nanchang Hangkong University) NO. 696, Fenghe South Avenue, Nanchang, Jiangxi, China

[2] Shangrao Vocational Technical College, School of electronics and Automation NO. 358, Xuri North Avenue, Shangrao, Jiangxi, China

[3,4] Nanchang Hangkong University, School of Material Science and Engineering NO. 696, Fenghe South Avenue, Nanchang, Jiangxi, China

E-mail address: xiaoy_hexmu@163.com (Xiaoyong He)

E-mail address: srzyhgm@163.com (Guangming He)

Abstract: A series of Methylsilsesquioxane based silica xerogel monoliths were prepared by using methyltriethoxysilane (MTES) silicon precursor via ambient pressure drying. Through the designed orthogonal tests, four important parameters including the molar ratio of cetyltrimethylammonium bromide (CTAB) and MTES (n(CTAB)/n(MTES)), the molar ratio of urea and MTES (n(Urea)/n(MTES)), the molar ratio of deionized water and MTES(n(Water)/n(MTES)), and the aging time (t_a) were examined with three levels. The results revealed that the influencing significance lessened for the average transmittance according to n(Water)/n(MTES) > n(CTAB)/n(MTES) > n(Urea)/n(MTES) > t_a, while for the thermal conductivity according to t_a > n(CTAB)/n(MTES) > n(Urea)/n(MTES) > n(Water)/n(MTES) and for the water contact angle according to n(Water)/n(MTES) > n(Urea)/n(MTES) > t_a > n(CTAB)/n(MTES). Under the optimized experimental conditions by using n(CTAB)/n(MTES) = 0.080, n(Urea)/n(MTES) = 0.242, n(Water)/n(MTES) = 8, the maximized average transmittance of 92.71% was achieved with the silica xerogel monolith at t_a = 3 d, while the minimized thermal conductivity of 22.54 mW/(m · K)) with the silica xerogel monolith at t_a = 6 d, besides under n(CTAB)/n(MTES) = 0.016, n(Urea)/n(MTES) = 0.726, n(Water)/n(MTES) = 14 and t_a = 6 d, the maximized water contact angle 132.0° could be obtained.

Keywords: silica xerogel; transparency; efficient thermal; hydrophilicity

Microstructural Evolution and Mechanical Properties of Al10(CoFeNi1.5)90 High Entropy Alloy Processed by Cold Rolling and Subsequent Heat Treatment

Mazullah[1,2], Muhammad Ismail[1], Zhang Ke[1], Niu Yuxuan[1], Yin Shipan[1], Hammad Ahmad[3], Xiong Zhiping[1,2]

[1] National Key Laboratory of Science and Technology on Materials under Shock and Impact, School of Materials Science and Engineering, Beijing Institute of Technology, Beijing 100081, China

[2] Tangshan Key Laboratory of High-Performance Metals and Ceramics, Tangshan Research Institute BIT, Tangshan 063000, China

[3] State Key Laboratory of Advanced Machining Technology Lab (AMT), School of Mechanical Engineering, Beijing Institute of Technology, Beijing 100081, China.

Abstract: The automotive, aerospace, and nuclear sectors currently require structural materials with high strength and sufficient ductility. This demand that can be met by high entropy alloys (HEAs) due to their versatile design capabilities. This study synthesized the Al10(CoFeNi1.5)90 (at. %) HEA using a cold crucible suspension furnace under an inert atmosphere. The microstructures and mechanical properties were systematically examined. The as-cast (AC) and cold rolled (CR) microstructures exhibited a fully face centered cubic (FCC) phase. However, after heat treatment at 575 and 600 ℃ for 0.25 to 15 h following CR, the microstructure consisted of lamellar/granular FCC and BCC phases. Aging at 575 and 600 ℃ for 5 h significantly enhanced the UTS (1680 ± 6 and 1419 ± 4 MPa) as compared with AC state; however, the uniform elongation decreased (0.9 ± 0.1, 13.6 ± 3%). The kinetics of phase transformation were evaluated using the JMAK equation, obtaining transformation parameters (n, k) of (1, 1.16) at 575 ℃ and (0.86, 1.035) at 600 ℃, indicating a fast transformation.

Keywords: high entropy alloys; solid solution; lamellar microstructure; heat treatment

Microstructural Stability of γ/γ′ Innovel Co-Based Superalloys Modified by Ti/Nb/Ta

Li Jinlin[1*], Li Zhen[2], Wang Qing[3]

[1] School of Materials Science and Engineering, Engineering Research Center of High Entropy Alloy Materials (Liaoning Province), Dalian University of Technology, Dalian, China

[2] School of Mechanical Engineering, Dalian University of Technology, Dalian, China

[3] School of Materials Science and Engineering, Engineering Research Center of High Entropy Alloy Materials (Liaoning Province), Dalian University of Technology, Dalian, ChinaE-mail address: ljl950627@ mail. dlut. edu. cn (Li Jinlin)

E-mail address: lizhen@ dlut. edu. cn (Li Zhen)

Abstract: Coherent precipitation of cuboidal γ′-Co_3 (Al, W) nanoparticles in face-centered-cubic (FCC)-γ matrix is of great significance for improving high-temperature mechanical properties of Co-based superalloys. The present work developed a series of Co-based superalloys in light of the cluster composition formula of [Al_1-$(Co,Ni)_{12}$]($Al_{0.5}(Ti/Nb/Ta)_{0.5}W_{0.5}Mo_{0.5}Cr_{0.5}Co_{0.5}$), where the addition of Ti, Nb, and Ta is mixed with an equimolar ratio. It is found that these designed alloys possess the coherent microstructure of cuboidal γ′ nanoprecipitates in the γ matrix. The microstructural evolution of coherent γ/γ′ during aging at 1173 K shows that these superalloys exhibit a higher thermal stability. Even after aging for 1000 h, there does not exist any other precipitated phases on grain boundaries, except the coarse γ′ precipitates. Also, the coarsening rate constants of cuboidal γ′ nanoprecipitates in these alloys are relatively-lower ($K = 5.76 \sim 6.03\ nm^3/s$), which is mainly ascribed to a moderate lattice misfit ($\varepsilon = 0.28 \sim 0.45\%$) of γ/γ′. The stable γ/γ′ microstructure renders the alloys with prominent mechanical properties, as evidenced by the high yield strength of $\sigma_{YS} = 438 \sim 445$ MPa at 1143 K.

Keywords: Co-based superalloys, coherent precipitation, microstructural stability, lattice misfit, particle coarsening.

Activation of Edge Dislocation Enables Ductilization of Refractory High-Entropy Alloys

Wang Y. W.

Institute of Materials Modification and Modelling, School of Materials Science and Engineering,

Shanghai Jiao Tong University, Shanghai, 200240, China

Abstract: Refractory high-entropy alloys (RHEAs) demonstrate exceptional high-temperature mechanical properties but frequently exhibit limited room-temperature ductility, constraining their practical implementation. This study examines TiZrHfNbx (x = 0.4 – 1.0) alloys, all retaining single-phase BCC crystal structures. Increasing Nb concentrations markedly enhance fracture strain from 8.76% to 33.21% while maintaining yield strength. Integrated experimental characterization and theoretical modeling reveal distinct dislocation behaviors: $TiZrHfNb_{0.4}$ displays planar-confined, straight screw dislocations that restrict plasticity. Conversely, TiZrHfNb achieves homogeneous dislocation distribution with activated edge dislocations—an unusual phenomenon in BCC systems. This ductility enhancement originates from greater atomic misfit volume and elastic asymmetry within the TiZrHfNb lattice. These findings not only unravel a previously unreported ductilization mechanism but also establish a materials design paradigm for toughening RHEAs through atomic-scale environment engineering.

Keywords: refractory high entropy alloys; ductilization; dislocation configurations; lattice distortion

Pulse Current Induced Damping Enhancement in Micron-Submicron Pure Magnesium

Su Di[1], Zhang Qiang[1], Fan Jianfeng[1]*, Zhang Qiang[1]

[1] College of Materials Science and Engineering, Taiyuan University of Technology, China

E-mail address: sudi1523572@163.com (Su Di)

E-mail address: getotten@gmail.com (Fan Jianfeng)

Abstract: This work presents the concept of uphill process applied in heat treatment of aluminium alloys. This process was developed and applied by Alcoa for the first time more than 50 years ago, with the aim to reduce residual stress mainly after quenching, and can be higher than 80% as has been reported. This gain tends to provide an increase in fatigue life during part or component operation at the same time that is waited a good performance related to corrosion resistance. Starting from the principle that quenching is considered the phase of heat treatment with most residual stress formation tendency, uphill process is applied following this phase and before aging. Thus, this process consists to the immersion of the part in cryogenic environment to obtain proper temperature stabilization. So, following, the part is transferred to a hot steam chamber to generate a temperature gradient to maintain the mechanical properties gained with the heat treatment. This results in low residual stress and better dimensional stability. The aim of this paper is to present results related to tensile test behaviour and to compare the results for some quenching media types with uphill process variations.

Keywords: uphill process; quenching; heat treatment; aluminum alloys; tensile test; mechanical properties

Format Requirements for Extended to the 30th World Congress of International Federation for Heat Treatment and Surface Engineering

Wang Binshan[1], Fan Jianfeng[1*]

[1] Key Laboratory of Interface Science and Engineering in Advanced Materials, Ministry of Education,

Taiyuan University of Technology, Taiyuan 030024, P. R. China

E-mail address: 939849748@qq.com (Wang Binshan)

E-mail address: fanjianfeng@tyut.edu.cn (Fan Jianfeng)

Abstract: This study introduces an electropulse assisted extrusion (EPAE) technique and investigates the effects of eletropulse on the microstructure and mechanical properties of AZ series magnesium alloys during extrusion. The results show that compared with the conventional extrusion technique, EPAE process can promote dynamic recrystallisation (DRX) in the alloy samples during extrusion at low temperatures and low extrusion ratios, resulting in a refinement of the average grain size to submicron levels and a weakening of the basal texture. After EPAE treatment, the AZ31 magnesium alloy achieved a yield strength (YS) of 338 MPa, an ultimate tensile strength (UTS) of 404 MPa and an elongation (TEF) of 16.8%, while the YS, UTS and TEF of AZ91 Mg alloy reached 454 MPa and 539 MPa and 11%, respectively. Under medium-temperature and high-extrusion ratio conditions for complete DRX, the EPAE process significantly reduced the volume fraction of the β-$Mg_{17}Al_{12}$ phase in the AZ91 alloy and promoted its spheroidization, while also increasing grain size and basal texture intensity. Due to the pinning effect and the inhomogeneous distribution of the β-$Mg_{17}Al_{12}$ phase, a bimodal structure was formed in the EPAEed AZ91 alloy. Experimental and theoretical analyses indicated that the increase of Gibbs free energy variation and atomic diffusion flux during extrusion of alloy samples caused by the combined thermal and athermal effects of the pulsed current was the main reason for the experimental phenomena, where the athermal effect playing a dominant role.

Keywords: electric pulse assisted extrusion; mg alloys; dynamic recrystallization; mechanical properties; microstructure

Strategies for Synergistic Enhancement of Strength and Ductility in Medium Manganese Steel

He Binbin[1]

[1] Department of Mechanical and Energy Engineering, Southern University of Science and Technology, Shenzhen, 518055, China

Abstract: In high-end manufacturing sectors such as automotive, aerospace, and defense, metallic materials are required to possess both ultra-high strength and excellent ductility. This is essential to meet the precision demands of forming complex components and to prevent brittle failure during service. However, there is typically a significant trade-off between strength andductility, enhancing one often leads to a reduction in the other. Therefore, achieving a synergistic optimization of these two properties remains a major challenge in materials science and engineering. As the most widely used metallic material, steel offers outstanding industrial manufacturability and cost advantages. With mature metallurgical processes and a high level of production automation, steel continues to hold a strong competitive position among advanced structural materials. This report focuses on medium manganese (Mn) steel, investigating its strengthening and toughening mechanisms. By regulating dislocation behavior, deformation twinning, and martensitic transformation across multiple microstructural scales, it explores pathways to simultaneously improve both strength and ductility. This study provides a new approach to overcoming the strength-ductility trade-off insteel and offers theoretical and technical support for the design and application of high-performance structural materials.

Keywords: Medium manganese steel, Dislocation, Twinning, Martensitic transformation, Strength-ductility.

Sequential γ-ε-α′ Transformation Preserves High Elongation in An Ultrafine-Grained High Mn Steel

Wang Lingyu

State Key Laboratory of Digital Steel, Northeastern University, Shenyang 110819, China

E-mail address: wanglingyu@ral.neu.edu.cn (Wang Lingyu)

Abstract: While grain refinement is a prominent way to improve the yield strength of metastable austenitic steels, it usually leads to a reduction in the uniform elongation as γ-α′ transformation tends to be suppressed. In the current study, it is found that the sequential γ-ε-α′ transformation helps an ultrafine-grained high Mn austenitic steel maintain high uniform elongation while having more than doubled yield strength compared to the coarse-grained counterpart. The transformation kinetics of the ε and α′ phases indicate that grain refinement increases theσ_s, delaying the transformation kinetics for both phases and decreases the maximum volume fraction of ε, but does not affect the final volume fraction of α′, leading to sustained strain hardening and high uniform elongation. The differences in transformation kinetics can be reasonably explained by the intermediate ε phase and the variant selection phenomena in the γ-ε and ε-α′ transformations. Grain refinement enhances variant selection of ε and increases strain energy, making the γ-ε transformation more difficult. Meanwhile, ε is rapidly consumed and transformed into α′. Together, these effects lead to a reduction in the maximum volume fraction of ε. The presence of the intermediate ε phase reduces the driving force, and the coordinated transformation of multiple α′ variants reduces the strain energy, which together drive the sustained transformation of α′. The combination of a higher α′ transformation rate and delayed transformation kinetics enables continued strain hardening, leading to high uniform elongation.

Corrosiveness and Electrochemical Analysis of Bainitic Rails in a Neutral Salt Spray Environment

Wang Genyuan[1*], He Chenggang[1], Liu Jihua[1]

[1] School of Rail Transportation, Wuyi University, Jiangmen, China

E-mail address: 2112418016@wyu.edu.cn (Wang Genyuan)

E-mail address: hechengan@126.com (He Chenggang)

Abstract: As the development of rail transit, rail materials face higher demands for mechanical strength and corrosion resistance. Bainitic rails, with their excellent mechanical properties, hold significant promise for application in the railway industry. However, further research is needed into their corrosion resistance and mechanisms. This study employs U75V rail steel. After austempering to develop a lower bainitic structure, the samples are assessed via neutral salt spray (NSS) testing and electrochemical evaluation. The rust layer characterization involves macroscopic examination using a portable microscope, microscopic and cross-sectional analysis by scanning electron microscopy (SEM), elemental distribution mapping with energy-dispersive X-ray spectroscopy (EDS), and phase identification through X-ray diffraction (XRD). The research reveals that the corrosion of lower bainitic rails occurs in three stages. In the early stage (0-96 h), the corrosion products do not fully cover the substrate. The rust layer changes from orange-yellow to bright yellow. At low magnification, the micro-morphology shows a "cotton-like" structure that gradually becomes denser, and at high magnification, it changes from a "net-like" to "cluster-like" and "disc-like" structure. The rail dissolves upon direct contact with the corrosive medium, with a relatively low average corrosion rate that increases with time. In the middle stage (96-192 h), the corrosion products completely cover the substrate, and the rust layer turns dark yellow. The microstructure of the rust layer becomes more compact than in the early stage. The increase in corrosion rate slows down, and the pore parameter (Q) decreases. This may result from the accumulation of corrosion products, which makes the rust layer more compact and enhances its protective effect. In the late stage (192-240 h), severe local corrosion occurs, and the rust layer turns dark yellow and black. The microstructure of the rust layer becomes more sparse than in the early stage, with a large amount of Fe_3O_4 formed. The pore parameter (Q) increases, the protection index ($\alpha/\gamma*$) decreases, and the stability of the rust layer declines. Energy-dispersive X-ray spectroscopy (EDS) of the rust layer cross-section shows significant chlorine peaks. The average corrosion rate rises sharply, which may be attributed to the catalytic effect of Cl^- ions on the micro-galvanic cells formed by the fine-grained ferrite and cementite in the lower bainite with high dislocation density. In addition, X-ray diffraction (XRD) analysis shows that the main corrosion products of the sample are α-FeOOH, γ-FeOOH, and Fe_3O_4. During the early and middle stages of corrosion, γ-FeOOH is the primary corrosion product. As the corrosion process continues, the amount of γ-FeOOH decreases while the amount of Fe_3O_4 increases. By the late stage of corrosion, Fe_3O_4 becomes the dominant corrosion product. Based on other observations, γ-FeOOH is believed to enhance the protective properties of the rust layer. The corrosion behavior of lower bainitic rails is influenced by multiple factors, including microstructure, rust layer properties, and corrosive media, research shows. Lower bainitic rails exhibit excellent corrosion resistance initially. However, prolonged exposure leads to a significant reduction in their corrosion resistance due to microstructural inhomogeneity and the interaction with corrosive media. This study provides valuable theoretical insights and data support for the marine application and optimization of bainitic rails.

Keywords: lower bainitic rail steel; neutral salt spray test; corrosion mechanism; electrochemistry

Effect of Natural Aging on the Subsequent Heat Treatment in High-Strength Al-Zn-Mg-Cu Alloys

Zhang Zhiwu[1*], Zhou Guowei[2], Li Dayong[1], Peng Yinghong[1]

[1] State Key Laboratory of Mechanical Systems and Vibration, Shanghai Jiao Tong University, Shanghai, China

[2] School of Ocean and Civil Engineering, Shanghai Jiao Tong University, Shanghai, China

E-mail address: zhangzhiwu35@163.com (Zhang Zhiwu)

E-mail address: zgw99@sjtu.edu.cn (Zhou Guowei)

Abstract: Al-Zn-Mg-Cu alloys have become an essential structural material in aerospace, nuclear, and automotive industries, due to their outstanding strength, low density and excellent mechanical properties. Especially, the Al-Zn-Mg-Cu alloy with high Zn content is extensively used in the commercial structures. The increased degree of alloying introduces distinct variations in the precipitation behaviour of strengthening elements compared to conventional 7xxx series alloys, thereby rendering its systematic investigation of considerable scientific and practical significance. In this study, the influence of natural aging on the subsequent heat treatment has been comprehensively investigated on a spray formed Al-Zn-Mg-Cu back-extrude tube. Experimental results demonstrate that natural aging at various stages following thermal deformation leads to a notable deterioration in mechanical properties after identical heat treatment. Specifically, when T6 heat treatment is applied immediately after hot extrusion—including solution treatment at 350 ℃ for 2 h and 475 ℃ for 2 h—the extruded tubes exhibit outstanding tensile strength of approximately 700 MPa. However, following 20 days of natural aging, the application of the same heat treatment yields a reduced tensile strength of 646 MPa, representing a decline of approximately 7.7%. The microstructure after natural aging exhibits a typical fiber texture (<100> and <111>//ED), exerting minimal impact on grain morphology and overall texture evolution. The nanoscale precipitates under different natural aging conditions were also qualitatively studied by transmission electron microscopy (TEM). The results show that the final products indicate a significant reduction in the quantity of precipitates following natural aging, thereby weakening the precipitation strengthening effect. Furthermore, the scanning electron microscopy (SEM) observations show that a larger number of coarse second-phase particles remain in the natural aged samples, which reduces the degree of supersaturation in the matrix and consequently impairs the subsequent artificial aging response. However, by increasing the solution treatment temperature, the mechanical properties can be restored to the level attained in the absence of natural aging. This phenomenon is preliminarily attributed to the high alloying characteristics of high-strength Al-Zn-Mg-Cu alloy, which maintain substantial precipitation driving force even under natural aging conditions. As a result, further growth of existing coarse second-phase particles occurs during the extended natural aging period. Consequently, under the same heat treatment regime, the dissolution efficiency of the coarse second-phase particles are significantly reduced. These findings suggest that in industrial practice, the thermal processing of Al-Zn-Mg-Cu alloy with high Zn content must consider the effects of prolonged natural aging, and that corresponding heat treatment parameters should be adjusted accordingly. This has important implications for optimizing heat treatment strategies in the manufacturing of high-performance Al-Zn-Mg-Cu alloy components, particularly to ensure consistent quality during large-scale production.

Keywords: natural aging; heat treatment; aluminum alloys; tensile test; mechanical properties

Influence of Initial Microstructure on Spheroidizing Annealing Effect in 4Cr5Mo2V Steel and Associated Transformation Mechanisms

Du AoBo[1], Zhu ZhiChao[1], Yao ZhongBo[2], Zhang XinJin[1]

[1] Tianjin Heavy Equipment Engineering Research Co., Ltd., Tianjin, China

[2] Changzhou New Materials Co., Ltd., JiangSu, China

E-mail address: 15512470791@163.com (Du AoBo)

E-mail address: zczhu87@126.com (Zhu ZhiChao)

Abstract: With the rapid development of automotive lightweight design and integrated die-casting technology, the demands for mold materials' high-temperature toughness and thermal fatigue resistance have significantly increased under extreme clamping forces and complex service conditions. Conventional H13 steel (4Cr5MoSiV1) faces limitations in large-scale die-casting mold applications due to its low thermal conductivity and severe high-temperature wear, while the improved DIEVAR steel (4Cr5Mo2V), through optimized C/Si content and enhanced Mo levels combined with remelting process innovations, demonstrates superior crack resistance and ductility, making it an ideal choice for integrated die-casting molds. Although domestic production of 4Cr5Mo2V steel has approached international technical standards in recent years, significant quality inconsistencies persist among manufacturers - particularly for large-scale mold modules exhibiting drastic microstructural disparities between core and surface regions after spheroidization annealing, with NADCA annealing structure ratings spanning from AS4 to AS16, which severely restricts mold size scalability. Current domestic research primarily focuses on quenching-tempering processes and service performance, lacking systematic investigation into annealing technology - especially for large-dimension modules. The critical challenge lies in achieving uniform spheroidized structures throughout massive components, as homogeneous annealing microstructure fundamentally determines subsequent machining performance, heat treatment response, and service reliability. This study systematically investigates the effects of different initial microstructures (martensite, martensite-bainite mixture, and pearlite-austenite mixture) on spheroidization annealing outcomes in 4Cr5Mo2V steel through SEM/EBSD characterization, integrated Thermo-Calc thermodynamic simulations and heat transfer modeling during furnace heating elucidate austenite nucleation/growth mechanisms and carbide dissolution-precipitation kinetics. Experimental results demonstrate that martensitic initial microstructure yields optimal annealed structures (AS1-3), the martensite-bainite mixture produces inferior structures below AS4 rating, while pearlite-austenite combinations result in the worst annealed structures (AS15-16). This discrepancy originates from transformation mechanisms: martensite's diffusionless transformation creates abundant phase boundaries and fine substructures, enabling homogeneous austenite nucleation at lath boundaries during intercritical heating and subsequent uniform carbide spheroidization; As for martensite + bainite organization, bainite is a two-phase mixture of supersaturated ferrite and carbides, which is a coarser organization, and when incompletely austenitized, the new austenite inherits the original coarser bainite orientation and retains it in the isothermal spheroidal annealing process, which results in a spheroidal annealing grade inferior to that of AS4; Pearlite-austenite systems exhibit preferential austenite nucleation at ferrite-cementite interfaces within pearlite colonies, carbon-enriched neo-austenite transforms into densely packed carbide particles during prolonged isothermal treatment, while carbon-depleted original austenite domains develop sparse carbide precipitation preferentially along grain boundaries due to nucleation site scarcity, this heterogeneity in carbide distribution culminates in AS15-16 ratings.

Keywords: 4Cr5Mo2V steel; heat treatment; spheroidal annealing; microstructure

Effect of Chill Grains on High-Temperature Rupture Properties of Gas Turbine Blade Alloy and Their Deformation Damageresponse

Li Min[1,2], Wang Yao[2*], Li Jiasheng[2], Dong Jiasheng[2]

[1] Key Lab for Anisotropy and Texture of Materials, Ministry of Education, Northeastern University, China

[2] Institute of Metal Research, Chinese Academy of Sciences, Shenyang, China

E-mail address: 2300607@ stu. neu. edu. cn (Li Min)

E-mail address: yaowang@ imr. ac. cn (Wang Yao)

Abstract: Gas turbine blade/vane is the core component of modern energy conversion system. These components work under high-temperature, high-pressure and high-speed rotation, which places extremely high demands on the mechanical properties and durability of the material. Technical specification for gas turbine blades/vanes prohibits the presence of chill grains, but major international gas turbine manufacturers often appropriately relax acceptance criteria for chill grains in non-critical zones of blades/vanes, balancing manufacturing cost with service reliability. However, the relevant domestic institutions presently lack the data of technical justification necessary to support the relaxation of chill grain acceptance criteria, which has become an obstacle to the progress of domestic gas turbine component manufacturing technology. In this study, the effect of chill grains on the rupture properties and high-temperature fracture behavior of K452 alloy was investigated. Specimens were extracted from as-cast large-sized vane of heavy turbine to preserve authentic chill grains and equiaxed grains distributions, followed by rupture testing at 900 ℃/120 MPa. Compared to equiaxed grain specimen, an obvious decrease in rupture lifetime of chill grain containing specimen can be found. The fractography and electron backscatter diffraction (EBSD) results demonstrate that cracks preferentially initiated at the interfaces between chill grains and equiaxed grains, subsequently further propagating along these interfaces. Stress concentration at these interfaces between chill grains and equiaxed grains, coupled with localized dislocation activity, is identified as the primary factor of degrading mechanical performance. This study provides preliminary evidential support for assuring the reliable service life of gas turbine blades/vanes.

Keywords: chill grain; cast superalloy; deformation mechanism; gas turbine blade; incoordination deformation

Microstructural Stability and Mechanical Behavior of Cladding Fe-Cr-Al-Mferritic Stainless Steels at High Temperatures

Niu Ben[1], Wang Qing[1*]

[1] School of Materials Science and Engineering, Dalian University of Technology, Dalian, China, 116024

Corresponding author. E-mail addresses: wangq@ dlut. edu. cn (Wang Qing)

Abstract: Fe-Cr-Al-based ferritic stainless steels have great potentials for the application of accident-tolerant-fuel cladding materials due to their excellent mechanical properties and resistances to steam oxidation and corrosion at high temperatures (HTs). The present work investigated the synergetic effect of Nb/Ta/Zr co-alloying on the microstructural stability of precipitates in the body-centered-cubic (BCC) matrix at high temperatures in Fe-13. 5Cr-4. 73Al-2. 07Mo-(0. 34 ~ 0. 5)Nb-(0. 65 ~ 0. 98)Ta-(0 ~ 0. 33)Zr (wt. %) ferritic stainless steels (FSSs). Alloy plates were fabricated by vacuum arc melting and multi-pass hot rolling, then aged at 1073 K for 24 h, and finally re-treated at different temperatures (1273 ~ 1473 K) for 1 h to study the re-dissolution of precipitates. A large amount of Fe_2M (M = Mo, Nb, Zr, Ta) Laves phase particles with a high-entropy effect were dispersed on BCC matrix with a volume fraction of f ~ 4. 6% in 1073 K-aged alloy. These particles were then coarsened and dissolved into matrix obviously after high temperatures re-treatments, as evidenced by the amount of particles with f ~ 0. 7% at 1473 K. More importantly, this value is about ten times higher than that in alloys without Zr addition (f ~ 0. 05%), in which these particles are still distributed both in inner grains and on grain boundaries. The higher microstructural stability at high temperatures is ascribed to the co-precipitation of Fe_2M, $Fe_{23}Zr_6$, and core ($Fe_{23}Zr_6$)-shell(Fe_2M)-structured particles, rather than Laves phase alone. And the formation of core-shell-structured particles at HTs was studied from the viewpoint of solid solubility and diffusion coefficient of M in the matrix. All these aged alloys exhibit prominent mechanical properties at both room and elevated temperatures, showing high yield strength with σ_{YS} = 490 ~ 560 MPa at room-temperature and σ_{YS} = 80 ~ 85 MPa at 1073 K. The strengthening effect was further discussed in light of various strengthening mechanisms, and the calculated strength are in good agreement with the experimental results.

Keywords: Fe-Cr-Al alloy; microstructural stability; phase transformation; strengthening mechanisms

Accelerating Bainite Transformation by Concurrent Pearlite Formation in Medium Mn Steel

Huang L. K. [1]

[1] State Key Laboratory of Solidification Processing, Northwestern Polytechnical University, Xi'an, Shaanxi 710072, P. R. China

E-mail address: hlk@ nwpu. edu. cn (Huang L. K.)

Abstract: Bainite transformation has yet to be utilized and even thoroughly studied in medium Mn steels. Here, we investigate the isothermal bainite transformation in a 10Mn steel at 450 ℃ experimentally and theoretically, focusing on the effect of dislocations introduced by warm deformation. We show that the bainite transformation in the studied medium Mn steel exhibits extremely sluggish kinetics (on a time scale of days), concurrent with the pearlite formation. The introduced dislocations can significantly accelerate bainite transformation kinetics while also facilitating the pearlite reaction. This is likely the first report on the simultaneous occurrence of these two solid-state reactions in medium Mn steels. With respect to the roles of dislocations in the acceleration of bainite transformation observed in this work, we propose a new 'carbon depletion mechanism', in which dislocations-stimulated pearlite formation makes a twofold contribution: facilitating the formation of bainitic ferrite sub-units to further enhance the autocatalytic effect and preventing the carbon enrichment in the remaining austenite. On this basis, a physical model is developed to quantitatively understand the bainite transformation kinetics considering the effect of concurrent pearlite formation, revealing good agreements between model descriptions and experiment results. Our findings, herein, offer fundamental insights into the bainite transformation in medium Mn steels and uncover a previously unidentified role played by introduced dislocations in influencing the kinetics of bainite formation, which may guide its future application in manipulating microstructure for the development of advanced high-strength steels.

Keywords: medium Mn steels; bainite; thermodynamics; kinetics; phase transformation

Near-Theoretical Strength and Large Elasticity of Micron-Sized In-Situ TiB_2 Reinforcements in High Modulus Steels

Li Yizhuang

State Key Laboratory of Digital Steel, Northeastern University, Shenyang 110819, China

Abstract: The presence of hard, brittle second phases in metallic materials often induces stochastic damage that compromises ductility. While particle fracture is commonly attributed to stress concentrations exceeding intrinsic strength, the latter remains poorly characterized. This study quantifies the intrinsic local strength of in-situ formed TiB_2 particles within high-modulus steels, where particles fracture is identified as the dominant damage mechanism. Specifically, in-situ micromechanical testing—performed on micro-cantilever beams and C-shaped specimens fabricated from single-crystal TiB_2 particles via focused ion beam milling—was coupled with finite element simulations to extract anisotropic mechanical properties. Contrary to previous reports of fracture strengths in the gigapascal range, coarse TiB_2 particles in the micrometer scale exhibited strengths up to several tens of gigapascals, accompanied by remarkable elastic deformation (elastic strains up to 6.6%). Control experiments with varied matrix chemical compositions and etching conditions confirmed the robustness of these measurements. Anisotropy assessments across different crystallographic directions revealed strength variations within one order of magnitude, indicating no pronounced directional weakness. The measured particle strengths exceed the matrix's maximum sustainable stress by two orders of magnitude, suggesting that defect-free TiB_2 particles should resist fracture during tensile deformation. Consequently, experimentally observed fractures should originate from pre-existing defects within TiB_2, underscoring defect-free TiB_2 as an ideal reinforcement phase for high-performance metallic systems.

Strong and Ductile High-Mn Cryogenic Steels by Tailoring ε-TRIP

Lai Qingquan

Nanjing Tech University

Abstract: The high-Mn steels are promising candidate for the cryogenic structural applications due to the low cost and good mechanical performance. The high-Mn steels usually involve a relatively low stacking fault energy, triggering additional deformation mechanisms besides dislocation glide (deformation twinning and ε-martensitic transformation). When comparing with the mechanical consequences of deformation twinning, the ε-martensitic transformation is more efficient in enhancing the strain hardening rate, but the drawback is the occurrence of premature failure. Therefore, we have generated the ultrafine-grained high-Mn steel by heavy warm rolling, with the aims of tailoring the ε-TRIP via the grain size effects. The warm-rolled ultrafine-grained high-Mn steel present outstanding cryogenic mechanical properties, including the true tensile strength of 1.8GPa and the true uniform elongation of 0.3. A combination of advanced characterization technique and micromechanical modeling was used for a deeper interpretation of the mechanical properties. In-situ neutron diffraction experiment has been conducted to reveal the micromechanism of plastic deformation, including the progressive formation of ε-martensite by the increase of stacking fault propability and also the intrinsic strain hardening of the ultrafine-grained austenite phase. The micromechanical model demonstrates the salient strain hardening capability of the nanostructured ε-martensite, which is essential for the excellent cryogenic properties of the ultrafine-grained high-Mn steel.

High-Performance Maraging TRIP Steel for Load-Bearing Applications

Liu Xinyu[1*], Chen Zihan[1], He Binbin[1]

[1] Department of Mechanical and Energy Engineering, Southern University of Science and Technology, China

E-mail address: 12332400@mail.sustech.edu.cn (Liu Xinyu)

E-mail address: 12232319@sustech.edu.cn (Chen Zihan)

Abstract: A novel maraging transformation-induced plasticity (TRIP) steel was designed to simultaneously achieve high strength and ductility by integrating precipitation strengthening and the TRIP mechanism. This steel exhibits a tensile strength of 1913 MPa, uniform elongation of 8.0%, and total elongation of 13%. The steel was processed through hot rolling, cryogenic treatment, and aging, resulting in a dual-phase microstructure with the retained metastable austenite distributed as isolated islands among the fine lamellar martensitic laths. Coherent B2-NiAl nanoprecipitates are uniformly distributed within the martensitic matrix, providing significant precipitation strengthening. Electron microscopy revealed heterogeneously distributed austenite grains embedded along martensite block boundaries, with different austenite grain sizes stemming from variations in martensitic transformation kinetics. Synchrotron X-ray diffraction and interrupted tensile tests were combined to investigate the evolution of microstructure including volume fraction of retained austenite and the dislocation density. The transformation of austenite to martensite induced local strain hardening and facilitated coordinated deformation between the newly formed martensite and the surrounding matrix, leading to smoother strain distribution and delaying strain localization. These results demonstrate how TRIP-induced strain hardening and precipitation strengthening contribute to the superior mechanical performance, providing a promising strategy for designing next-generation ultra-high-strength steels with balanced strength and ductility.

Keywords: martensite; metastable austenite; transformation-induced plasticity; precipitates; maraging TRIP steel

Refining Mechanism and Elevated-Temperature Mechanical Properties of Al-Ce Alloys Solidified Under Super Gravity Field

Shen Shuai[1], Wu Chuandong[1*], Gan Zhanghua[1*], Liu Jing[1,2]

[1] State Key Laboratory of Refractories and Metallurgy, Wuhan University of Science and Technology, Wuhan, Hubei, 430081, China

[2] Guangdong Provincial Key Laboratory of New Energy Materials Service Safety, College of Materials Science and Engineering, Shenzhen University, Shenzhen 518060, China

E-mail address: wuchuand@ wust. edu. cn (Wu Chuandong)

E-mail address: ganzhanghua@ wust. edu. cn (Gan Zhanghua)

Abstract: The rapid coarseening of the strengthening phase in industrial heat-resistant aluminum alloy may occur above 300 ℃, resulting in a sharp reduction in high-temperature strength and making it difficult to service at higher temperatures. Therefore, it is urgent to develop a new generation of high-strength heat-resistant aluminum alloy that can service in the 350-400 ℃ range. The Al-Ce alloy, as a new type of eutectic heat-resistant Al alloy, has similar casting properties to Al-Si, a higher eutectic temperature point (640 ℃), extremely low solubility (0. 01 at%), and extremely low diffusion coefficient ($D_{Ce/Al}$ = 4. 7 × 10 − 19 m^2/s at 400 ℃), which enables its eutectic phase to maintain good thermal stability even at 400℃. However, the coarse eutectic phase severely limits its strengthening effect. Currently, the refinement of eutectic $Al_{11}Ce_3$ is detected in super-gravity solidified Al-Ce alloys, which can be ascribe to the enhancement of mass transfer under super-gravity field, affecting the nucleation rate and inhibiting the growth of eutectic phase for this irregular Al-Ce eutectic system. Interestingly, the refined eutectic $Al_{11}Ce_3$ is thermally stable even up to 400 ℃ thermal exposure without obvious coarsening, contributing to the enhancement of elevated-temperature mechanical properties (~ 125. 0 ± 9. 1 MPa at 300 ℃), which is ~74. 8% higher than that solidified under normal gravity field (~71. 5 ±0. 6 MPa). The refinement of eutectic $Al_{11}Ce_3$ with higher number density also contribute to the stress dispersion, the crack prefer to propagate through the eutectic colony, leading to the transformation of fracture mode from brittle to ductile. In order to furhter enhance the mechanical properties of Al-Ce based alloys at elevated temperature, thespatial interconnection structure was constrctured by the incorporation of Ni elements, which is able to form Al-$Al_{11}Ce_3$-Al_3Ni eutectic microstructure with more strengthening phases and stronger pinning effect. Meanwhile, the formation of ultrafine fibrous interconnected $Al_{11}Ce_3$-Al_3Ni eutectic clusters with more "chains" and higher spatial connectivity was detected mainly due to the relatively higher nucleation rate. The high volume fraction and refinement of interconnected $Al_{11}Ce_3$-Al_3Ni eutectic clusters with excellent thermal stability could contribute to enhancement of ductility and tensile strength at elevated-temperature. The tensile strength of Al-Ce-Ni alloy tested at room temperature, 300 ℃, 350 ℃, and 400 ℃ were ~240. 1 ±5. 9 MPa, ~163. 4 ±7. 9 MPa, ~118. 0 ± 6. 9 MPa and ~79. 6 ± 8. 0 MPa, respectively, indicating enhanced strength retention of 68. 1% (300 ℃), 49. 1% (350 ℃) and 33. 2% (400 ℃) at elevated temperatures. Noted that the mechanical properties of the Al-Ce based alloys can be further improved by the regulation of the interconnected structure.

Keywords: heat resistant Al alloys; super-gravity field; eutectic alloy; spatial interconnection structure

Coherent Cr-Rich Nanoprecipitates: Achieving Excellent High-Temperature Strengthin Multi-Principal Element Alloy

Huang Weiming[1], Gan Zhanghua[1*], Wu Chuandong[1*], Liu Jing[1,2]

[1] State Key Laboratory of Refractories and Metallurgy, Wuhan University of Science and Technology, Wuhan, Hubei, 430081, China

[2] Guangdong Provincial Key Laboratory of New Energy Materials Service Safety, College of Materials Science and Engineering, Shenzhen University, Shenzhen 518060, China

E-mail address: ganzhanghua@ wust. edu. cn (Gan Zhanghua)

E-mail address: wuchuand@ wust. edu. cn (Wu Chuandong)

Abstract: Simultaneously enhancing strength and ductility has always been the primary objective in alloy materials research. The rich compositional variations and significant performance potential of multi-principal element alloy (MPEAs) provide important opportunities for the design of advanced materials, particularly face-centered cubic (FCC) MPEAs have attracted widespread attention due to their excellent ductility and high fracture toughness at room and low temperatures. Many attempts have been made to enhance the high-temperature strength of the MPEAs. Reviews of the published papers reveal that the introduction of nano-precipitates ($L1_2$, B2, and σ) into materials is one of the most promising methods for achieving excellent high-temperature strength. The coherent and ordered $L1_2$ phase is an effective strengthening phase in FCC MPEAs. Particularly for the high-entropy alloy strengthened by $L1_2$, the addition of elements such as Co, Cr, and Ti can improve the thermal stability of the $L1_2$ (Ni_3Al) phase and significantly enhance the high-temperature mechanical properties. For instance, Yuan et al. reported that by introducing nanoscale $L1_2$ particles into $Co_{40}Cr_{20}Ni_{30}Mo_2Al_4Ti_4$ MPEAs, the tensile strength of the alloy reached 1072 MPa at 700 ℃. Cao et al. reported $(FeCoNi)_{81}Cr_9Al_8Ti_1Nb_1$ MPEAs with excellent elevated-temperature tensile strength (897 MPa at 700 ℃), due to the pinning effect of the thermal stable precipitates ($L1_2$phase) at grain boundaries. However, in the process aimed at promoting the formation of ordered phases can inadvertently lead to the formation of Cr-rich phases such as σ phases, which damages the performance of the material CoarseCr-rich phase particles are prone to cause local stress concentration and promote crack nucleation, which are the main sources leading to brittleness. Recently, significant efforts have been made to refine the size of the Cr-rich or σ phase. For instance, the σ phase size of $Co_{34.5}Cr_{30}Ni_{26.5}Al_{5.4}W_{3.6}$ MPEA s is 80 ~ 200 nm, $Co_{30}Cr_{30}Fe_{18}Ni_{18}Mo_4$ MPEAs obtained metastable σ phases with a size of 125nm under moderate temperatures and short time annealing treatment Inspection of published papers reveals that that MPEAs with micron-sized σ phases can be obtained through conventional cold rolling and heat treatment, but refining the Cr-rich or σ phase to the nanoscale (< 100 nm) remains limited. It is interesting to refine the size of the precipitates to avoid embrittlement and facilitate its strengthening effect. In this paper, we propose a new method to improve the mechanical properties of MPEAs at room temperature and high-temperature by obtaining ultrafine nano Cr-rich phases through microalloying and annealing design. The results show that the yield strength of the alloy is 1092 MPa (25 ℃) and 550 MPa (700 ℃) respectively, while maintains certain plasticity. The influences of Mo, V and C on the precipitation behavior and spatial distribution of the Cr-rich phase were discussed in detail. This research provides useful information to deepen the understanding of the formation mechanism of nano Cr-rich phase and offer practical information of improving the high-temperature performance of MPEAs by using nanoscale precipitation.

Keywords: multi-principal element alloy; microstructure; mechanical properties; hot-forging; spinodal decomposition

A Study on the Mechanical Properties and Wear Resistance of CADI at Different Isothermal Temperatures

Yang Baoqing, Hu Zhaonan, Wang Hong, Tang Zhiheng, Huang Wenpeng, An Kang, Wang Li

Chongqing Gearbox Co., Ltd., Chongqing, China

Abstract: This paper first demonstrates the difference in friction and wear properties between ADI and CADI under the same heat treatment parameters. Secondly, it studies the mechanical properties and friction and wear performance differences of CADI materials at different isothermal quenching temperatures. By comparing the friction and wear properties of ADI and CADI under the same heat treatment parameters, it is concluded that the friction and wear properties of CADI materials are superior to those of ADI materials. Further analysis shows that CADI materials exhibit the best overall performance in mechanical properties and friction and wear resistance at an isothermal quenching temperature of 270 ℃.

Keywords: CADI; isothermal quenching temperature; mechanical properties; friction and wear properties

Effect of Cryogenic Treatment on Hardness, Wear Resistance and Microstructure of 300M Steel

Zhang Haidong*

Department of Mechanics, Jinzhong University, China

E-mail address: tronmek@outlook.com (Zhang Haidong)

Abstract: This work applies cryogenic treatment to conventional heat treatment of 300M steel to investigate the effect of cryogenic treatment on hardness, wear resistance and microstructure of 300M steel for promisingly improving performance and prolonging service life of conical pick in the future work. In coal mining industry, conical pick is used as a cutter in mining machinery such as shearer and roadheader to directly break and cut coal and rock in coal mining and tunneling operation. In the process of operation, conical pick often suffers from complex and variable loads and other critical working conditions, which leads to wear and failure of conical pick, then resulting in huge consumption of conical picks, increase in production cost and reduction in production efficiency. Therefore, conical pick requires excellent mechanical properties such as hardness and wear resistance, which can be obtained by using proper materials for pick body and improving the treatment process of pick body, so as to reduce the wear failure, prolong the service life and reduce the consumption of conical pick. 300M steel, a modified AISI 4340 steel with addition of vanadium and higher silicon and carbon content, is an ultra-high-strength low-alloy medium-carbon steel in AMS 6417 standard. It has excellent comprehensive mechanical properties such as high strength, good hardenability, good toughness, and high fatigue performance, thus widely used in aerospace, automotive, heavy machinery and other applications, such as aircraft landing gear, fuselage structural components, transmission shafts, high-strength bolts and other key components with strict technical requirements. Cryogenic treatment, as an extension of conventional heat treatment, shown in extensive research, can effectively improve mechanical properties such as hardness and wear resistance of steel by promoting microstructural changes in steel, that are incomplete transformation of residual austenite into martensite, dispersed precipitation of fine carbide particles, and grain refinement. Inpresent research, conventional heat treatment and surface treatment processes are commonly used to investigate the strength, plasticity, toughness, ductility, and other mechanical properties of 300M steel. However, cryogenic treatment technology is rarely applied to investigate hardness and wear resistance of 300M steel. In view of the effect of cryogenic treatment technology on improving the mechanical properties of steel, cryogenic treatment could be an effective and promising technology to supplement conventional heat treatment to improve hardness and wear resistance of 300M steel for promisingly enhancing performance and prolonging service life of conical pick. This work applied cryogenic treatment to the conventional heat treatment of 300M steel, which namely underwent cryogenic treatment preceded by quenching and followed by low-temperature tempering. Afterwards, its hardness and wear resistance were tested, microstructure was observed using scanning electronic microscopy, and the phase composition was analyzed using X-ray diffractometry to investigate the effect of cryogenic treatment on the hardness and wear resistance of 300M steel, as well as the microscopic influencing mechanism of cryogenic treatment. The results showed that the hardness of 300M steel with cryogenic treatment improved slightly, whereas wear resistance improved and the friction coefficient decreased. Furthermore, both the wear resistance and friction coefficient with deep cryogenic treatment were superior to those with shallow cryogenic treatment. This main reason is that cryogenic treatment can promote the segregation of carbon atoms in martensite and then the dispersed precipitation of fine carbides during tempering stage, thereby enhancing both wear resistance and friction coefficient. Hence, this work can provide theoretical and practical references for the application of 300M steel and cryogenic treatment technology to improve the wear resistance of conical pick.

Keywords: 300M steel; cryogenic treatment; hardness; wear resistance; microstructure; mechanism

Role of Heat Treatment in Dislocation Mechanisms in a Magnesium Alloy

Abstract: Magnesium (Mg) alloys have received tremendous attention over the last few decades for improving energy efficiency of various engineering fields. To achieve widespread applications, understanding the mechanical response of the Mg alloysunder different heat treatment conditions becomes a crucial issue for judging their in-service performance. In the present work, the mechanical properties of the Mg alloys were investigated on both solid-solution and aged samples. Microstructure evolution was characterized by the in-situ EBSD technique and the dislocation density, including the fractions of dislocations with different Burgers vectors, was determined based on the synchrotron XRD data. The rate sensitivity of plastic deformation was studied. A marked aging-induced shift in dislocation plasticity is observed and possible mechanisms contributing to this effect are explored.

Effect of Composite Extrusion on Microstructure and Properties of Mg-Gd-Y-Zn-Ca Alloy Containing 14H-LPSO Phase

Chen Hongmei*, Zhang Bin, Huang Tao, Tu Xingjia, Zang Qianhao, Feng Di
School of Materials Science and Engineering, Jiangsu University of Science and Technology, Zhenjiang, 212003, China
E-mail address: hmchen@just.edu.cn (Chen Hongmei)

Abstract: This paper takes Mg6Gd3Y1.5Zn0.4Ca alloy as the research object, and studies the changes in the microstructure of the alloy during the homogenization annealing process. The influence of hot extrusion on the microstructure and properties of the alloy are also analyzed. Compared with the as-cast alloy, more bright white dot-like cluster phases precipitate in the alloy after 3 hours of homogenization treatment, and the volume fraction of blocky LPSO phases at the grain boundaries significantly decreases, while the volume fraction of intragranular lamellar LPSO phases increases. After compound extrusion, the microstructure of the alloy is mainly composed of a small amount of deformed structure and a large number of recrystallized grains. The second phase in the alloy undergoes deformation kinking, and the recrystallized grains in the alloy are mainly distributed around the kinked and deformed second phases. The dynamic recrystallization mechanism is continuous dynamic recrystallization and particle-stimulated nucleation mechanism. After extrusion, the compressive strength of the alloy reaches 465 MPa, and the elongation is 21%. The improvement of mechanical properties is attributed to the dynamic recrystallization phenomenon existing in the alloy and the certain texture strengthening effect exerted by the kinked and deformed lamellar LPSO.

Keywords: rare earth magnesium alloy; hot extrusion; LPSO phase; microstructure

Control of BCC/B2 Coherent Microstructure and Macroscopic Properties in Multi-Principal-Element Alloys

Wang Qing *
Dalian University of Technology, Dalian, China
E-mail address: wangq@ dlut. edu. cn (Wang Qing)

Abstract: The modulation of the BCC/B2 coherent microstructure in conventional alloys remains challenging due to the significant compositional differences between BCC solid solution and its ordered B2 phase, as well as the metastable nature of the B2 phase. These limitations, however, are expected to be overcome in multi-principal-element alloy (MPEA) systems. In this work, we employ a cluster-formula approach to design alloy compositions for precise control of the BCC/B2 coherent microstructure in Al-TM (transition metal) systems. Furthermore, the morphology of coherent nanoparticles and the evolution of BCC/B2 coherent microstructure are systematically investigated through the phase-field simulations and the first-principle calculations. MPEAs across various systems have demonstrated exceptional macroscopic properties, including superior mechanical strength, soft-magnetic behavior, high electrical conductivity, and excellent combustion performance. These findings open new avenues for designing advanced materials with integrated mechanical and functional properties.

Keywords: BCC/B2 coherent microstructure; composition design; simulations; macroscopic properties; multi-principal-element alloys

Study on Annealing Process and Formability of 409L Ultra-Pure FerriticStainless Steel Hot-Rolled Plates

Zhang Jinhui[1*], Piao Rongxun[1], Wang Wensong[1], Wang Junhai[2], Zhao Gang[2]

[1] School of Mechatronics Engineering, Anhui University of Science and Technology, China

[2] Taishan steel, China

E-mail address: 18375372135@163.com (Zhang Jinhui) E-mail address: yhpark617@163.com (Piao Rongxun)

Abstract: 409L ultra-pure ferritic stainless steel is widely used in automotive exhaust systems due to its excellent comprehensive performance and cost advantage. With the development of the domestic automotive industry, the market's demand for lightweight exhaust systems and complex components has upgraded, and higher requirements have been placed on the formability of materials. The continuous annealing process of stainless steel hot-rolled plates is an important link connecting the hot-rolling and cold-rolling processes, and it has a significant impact on the formability of the finished plates. To determine the optimum conditions for the continuous annealing process, the annealing experiments were conducted on the 3mm-thick 409L ultra-pure ferritic stainless steel hot-rolled plates at temperatures ranging from 640 to 950 ℃. The recrystallization behavior of 409L at low temperatures range (640-800 ℃) was investigated, as well as the texture and property changes during the grain growth stage at high temperatures range (800-950 ℃). The research on recrystallization behavior shows that as the annealing temperature increases, the recrystallization of 409L hot-rolled plates becomes more and more complete. Moreover, the tensile strength decreased with temperature, while the average plastic strain ratio $\bar{r}$ increased slightly. The recrystallization temperature was determined to be 745 ℃ by the hardness method. The research on the texture evolution during the grain growth stage indicates that as the annealing temperature increases, the grain size in the microstructure of 409L hot-rolled plates keeps growing, and the average plastic strain ratio $\bar{r}$ shows an increasing tendency and reaches its maximum value at 930℃. With the increase of annealing temperature, the texture of 409L hot-rolled plates changed from the original α-fiber <110>//RD orientation to the γ-fiber <111>//ND orientation, which is beneficial to the formability of stainless steel. At 930 ℃, the texture intensity of γ-fiber reached its maximum value, with its peak point located at {111} <011>. When the annealing temperature is too high, the hot-rolled texture tends to become disperse, leading to a decrease in formability.

Keywords: 409L ultra-pure ferritic stainless steel hot-rolled plates; continuous annealing process; formability; Texture

Research Progress on Surface Strengthening Technology of Heavy-Duty Gears with High Fatigue Resistance

Wang Xiaonan*, Chen Qiyuan, Shen Xinjun, Zhang Qingyu

School of Iron and Steel, Soochow University, China

E-mail address: wxn@suda.edu.cn (Wang Xiaonan)

E-mail address: qychen1992@suda.edu.cn (Chen Qiyuan)

Abstract: High-performance gears demand exceptional resistance to wear and contact fatigue. This study systematically investigates the influence of microstructure, carbon concentration, and hardness distributions on these critical properties in deep-carburized 18Cr2Ni4W gear steel through experiments in conjunction with finite element simulations. Initial experiments focused on conventional carburizing at temperatures of 930 ℃, 950 ℃, and 970 ℃. Results confirm that surface carbon concentration and hardness increase progressively with higher carburizing temperatures. Tensile tests on the carburized surface layers revealed impressive strengths exceeding 2000 MPa across all temperatures; however, a significant reduction in ductility was observed as temperature increased. Sliding wear tests demonstrated that the sample carburized at 970 ℃ exhibited the lowest wear rate compared to those processed at 930 ℃ and 950 ℃. Analysis identified the primary wear mechanisms as adhesive oxidation wear coupled with abrasive wear. Crucially, energy dispersive spectroscopy mapping revealed the most pronounced aggregation of oxygen on the counterface opposing the 930 ℃ carburized sample, indicating the most severe adhesive oxidation wear in this case. Supporting finite element simulations of transient temperature fields during sliding contact predicted a maximum contact point temperature approximately 50 ℃ lower for the 970 ℃ sample compared to the 930 ℃ sample. This significant temperature reduction effectively mitigates oxidation severity. Consequently, the enhanced wear resistance at 970 ℃ is attributed to the synergistic effect of an increased volume fraction of hard carbides within the case and the lower frictional heating suppressing oxidation. Building upon these findings, a novel two-stage heat treatment combining deep carburizing followed by carbonitriding was specifically designed to further elevate the performance limits of 18Cr2Ni4W steel. Microhardness profiling showed that conventional carburizing yielded a surface hardness of 671 HV1 and a core hardness of 465 HV1. In striking contrast, the novel carburizing-carbonitriding process significantly increased the surface peak microhardness to 748 HV1 (representing an ~11% improvement) while slightly reducing the core hardness to 440 HV1 (a ~6% decrease). The wear rate with novel carburizing-carbonitriding heat treatment is the lowest, with the decrease in wear rate from 8.73×10^{-7} $mm^3 \cdot N^{-1} \cdot mm^{-1}$ to 5.89×10^{-7} $mm^3 \cdot N^{-1} \cdot mm^{-1}$, corresponding to a 33% improvement in wear resistance. The cycle time of contact fatigue stress increases from 1.60×10^7 to 3.91×10^7, an increase of 144%. Compared to conventional carburizing process, the carburizing-carbonitriding heat treatment significantly increases carbon concentration within the surface layer. The higher carbon concentration promotes the formation of carbides, and the proportion of large-sized carbides increases. The interstitial nitrogen atoms induce potent solid solution strengthening within the martensitic matrix, significantly increasing its intrinsic microhardness and thereby contributing substantially to the improved resistance against both wear and fatigue crack initiation/propagation. The synergistic combination of refined carbide distribution and nitrogen-strengthened martensite underpins the exceptional gains in both wear resistance and contact fatigue life. This work demonstrates that optimizing carburizing temperature effectively enhances wear resistance by controlling carbide formation and minimizing oxidation through reduced frictional heating. More significantly, the newly developed carburizing-carbonitriding heat treatment offers a highly effective strategy for dramatically improving the surface integrity and service longevity of critical gear components made from 18Cr2Ni4W steel, achieving substantial simultaneous improvements in both wear resistance and contact fatigue performance.

Keywords: heavy-duty gears; carburizing process; carbonitriding process; microstructure; wear resistance, fatigue resistance

Proceedings of IFHTSE 2025

The 30th World Congress of International Federation for Heat Treatment and Surface Engineering
August, 19-22, 2025, Suzhou, China

Design of Low Hardness Abrasion Resistant Steels: Microstructural Tailoring

Xu Xiaojun*, Sheng Liangliang, Zhu Minhao

School of Materials Science and Engineering, SouthwestJiaotong University, Chengdu, 610031, China

E-mail address: xiaojunxu0617@126.com (Xu Xiaojun)

Abstract: Abrasion resistant steels combining high abrasion resistance, good cold formability and superior impact toughness are highly demanded by various applications subjected to abrasive environments. Current developments of high-end abrasion resistant steels are very oriented on the high hardness martensitic concept following the hypothesis that the abrasion resistance holds a monotonous relationship with macro hardness. However, various experimental observations have suggested that the high hardness of martensite does not guarantee a superior abrasion resistance comparing to multiple phase steels with a relative lower hardness owing to the brittle nature of martensite. Therefore, systematic experiments have been conducted to tailor microstructures of advanced steels and targets at a balanced microstructure composing of a soft phase (ferrite) and hard components (martensite), in which the soft phase can provide support and impart compatibility and the hard phase can share the load and resist the destructive action of abrasive particle. In addition to the dual-phase matrix, an introduction of a considerable amount of retained austenite could further enhance the abrasion resistance owing to its significant strain hardening capability via the TRIP effect due to wear. In such a manner, the relative soft matrix offers good toughness and processibility while the self-reinforcing surface involving the TRIP effect provides superior abrasion resistance. Of importance is to unravel the mechanism of stress (abrasion)-induced martensitic transformation and its effect on the reinforcement of abrasion resistance. The dynamic process of microstructural evolutions during abrasion process and the wear mechanism under different loading conditions have been discussed. A promising microstructure with high abrasion resistance and low hardness has been finally obtained.

Keywords: abrasion resistant steels; TRIP effect; low hardness; microstructural tailoring; abrasive wear mechanism

Precipitation Behavior and Mechanical Properties of High-Manganese Steel During Aging

Wu Zhihai

Kunming University of Science and Technology, China

E-mail address: 1771730554@qq.com (Wu Zhihai)

Abstract: This research systematically investigates the correlation between the evolution of precipitation behavior and the resultant alterations in mechanical properties within Fe-17Mn-0.88C high-manganese steel, subjected to varying durations of aging treatment (Aging Time, AT). To comprehensively elucidate the underlying microstructural mechanisms driving these changes, a multi-faceted characterization approach was employed. This integrated methodology combined detailed microstructural examination using scanning electron microscopy (SEM), electron probe micro-analyzer (EPMA), and transmission electron microscopy (TEM) with complementary micromechanical assessment via nanoindentation techniques. The findings reveal a distinct pattern: while micro-sized precipitates exhibited negligible changes in their morphology and population throughout the investigated aging spectrum, a significant and progressive transformation was observed concerning nano-sized V2C precipitates. Specifically, as the aging time extended from 1 hour to 24 hours, the quantity of these nano-scale V2C carbides increased markedly. Furthermore, their spatial distribution within the steel matrix became demonstrably more uniform and finely dispersed. This pronounced evolution in the characteristics of the nano-sized precipitates is conclusively identified as the principal strengthening mechanism responsible for the overall enhancement of the steel's mechanical performance. Quantifying this enhancement, the yield strength of the steel demonstrated a clear upward trajectory corresponding directly with increasing aging time. The yield strength rose progressively from an initial value of 484 MPa after 1 hour of aging to a significantly higher value of 548 MPa following an extended aging period of 24 hours. Analysis attributes this substantial gain in yield strength primarily to the synergistic effects arising from the increased volume fraction of nano-sized V2C precipitates and their enhanced homogeneity of distribution achieved at longer aging durations. Through a holistic evaluation of the mechanical properties across the different aging conditions, the optimal aging time was determined to be 24 hours (designated AT24). At this specific aging duration, the steel simultaneously attained its peak hardness value, measured at 250 HWB, alongside superior impact toughness, quantified as 81 J. Remarkably, the AT24 condition also delivered an impressive yield strength increment of 277 MPa and achieved a peak work-hardening rate reaching 65 GPa, metrics that underscore the potent strengthening and toughening effects conferred by the optimized nano-precipitate structure. To provide a more complete and spatially resolved assessment of the material's mechanical response at the micro and nano scales, the comprehensive mechanical behavior was further characterized using NanoBlitz three-dimensional (3D) mapping technology. This advanced technique yielded detailed maps revealing the distribution of properties. The data confirmed that the AT24 steel exhibits exceptionally high values for both nano-hardness and Young's modulus, specifically exceeding 16.86 GPa and 370 GPa, respectively. These outstanding micromechanical properties fundamentally stem from the potent interactions occurring between the dispersed nano-sized V2C precipitates and dislocations moving through the lattice during deformation. Crucially, this interpretation of the micromechanical data aligns seamlessly and provides strong corroboration for the microstructural observations obtained through SEM, EPMA, and TEM analyses. The consistency across these independent characterization methods confirms that the increased population and highly uniform dispersion of nano-scale V2C precipitates are the dominant microstructural features underpinning the significant improvement in the comprehensive mechanical properties observed in the Fe-17Mn-0.88C high-manganese steel subjected to the 24-hour aging treatment. The precipitation strengthening effect mediated by these nano-particles is thus established as the key factor governing the material's enhanced performance.

Keywords: high-manganese steel; precipitation behavior and mechanical properties; nano-sized V_2C precipitates; nanoBlitz three dimensional (3D)

Transformation Behavior of Additively Manufactured 17-4PH Atsub Zero Temperatures

Ellen K. Troyanosky[1*], Liang Jianyu[2], Thomas L. Christiansen[3]

[1] Worcester Polytechnic Institute Department of Mechanical & Materials Engineering 100 Institute Road, Worcester, MA 01609 Center for Heat Treating Excellence (CHTE)

E-mail address: etroyanosky@ wpi. edu (Ellen K. Troyanosky)

E-mail address: jianyul@ wpi. edu (Liang Jianyu)

Abstract: Additively manufactured (AM) parts fabricated through Laser Powder Bed Fusion (LPBF) often require a modified heat treatment to accommodate for the differences in microstructure and composition as compared to their conventional counterparts. The unconventional microstructure is a consequence of the very fast cooling rate inherent to the LPBF process. 17-4PH stainless steel (17-4PH) is a precipitation hardening steel which hardens through the aging of a martensitic matrix during aging treatment. Stainless steel parts fabricated through LPBF typically result in a relatively high nitrogen content. This is due to i) higher nitrogen content in the N2 gas-atomized powder and ii) additional uptake from reaction with the "inert" N2 cover gas during the LPBF process. An enhanced nitrogen content in 17-4PH will unavoidably lower the martensite start temperature and lead to a higher fraction of retained austenite. Controlled sub-zero treatment can accomplish near complete transformation into martensite; previous ex-situ analysis of sub-zero treatments determined that a combination of boiling nitrogen (−196 ℃) and extended hold in dry ice (−78 ℃) produced the greatest hardness, suggesting a significant degree of isothermal martensite transformation. In this paper, dilatometry is used to track the phase transformations in LPBF and nitrogen-enriched wrought 17-4PH to further examine the martensite formation behavior at sub-zero temperatures. A variety of temperatures and hold times are applied to study the kinetics of the martensite transformation. The extent of martensite transformation is additionally examined through XRD analysis and optical microscopy.

Keywords: Sub-zero treatment; heat treatment; dilatometry; 17-4PH stainless steel; additive manufacturing

Microstructure and Property Evolution of High-Alloy Carburizing Steel Carburized Layer at Different Temperatures

Luo Chunyang[1], Wang Xuefei[1], Wang Haojie[2], Wang Zhaodong[1]

[1] State Key Laboratory of Digital Steel, Northeastern University, Shenyang, China

[2] School of Mechanical Engineering, Shenyang University of Technology, Shenyang, China

E-mail address: luochunyang2020@163.com (Luo Chunyang)

E-mail address: getotten@gmail.com (Wang Xuefei)

Abstract: With the progress of science and technology, the performance requirements of key parts such as bearings and gears in wear resistance, corrosion resistance, fatigue spalling resistance and high temperature strength are getting higher and higher. Through years of iterative development, the content of alloying element in bearing steels has gradually increased, with the total alloy content of third-generation carburized bearing steels exceeds 30 wt. %. However, this kind of high alloy steels faces some problems in carburizing process, including large carbides, reticular carbides and "soft zone" phenomena. In this paper, the effects of quenching process parameters (temperature and time) on the microstructure and properties of the carburized layer are studied. The evolution of microstructure of carburized layer, the dissolution behaviour of carbide and influencing factors of the "soft zone" are analysed, aiming at determining the best process window for refining carbide and eliminating the "soft zone". The design of quenching parameters covers all kinds of industrial production scenes, which provides a reference for actual process optimization.

Keywords: high-alloy carburizing steel; carburized layer; High-temperature quenching; process window; "soft zone" test; carbides

Research Progress of Aluminum Alloy Materials Hot Forming Technology and Equipment

Wang Kaize[1], Zhang Wenliang[1], Luo Ping[1], Li Xianjun[1], Zhang Minghao[1], Wu Xiaolin[1], Hou Junqing[1]

[1] China Academy of Machinery Beijing Research Institute of Mechanical & Electrical Technology Co. , Ltd. ,

Beijing 100083, China

Abstract: Under the background of high-performance and lightweight demands in key fields such as aerospace and automotive industries, aluminum alloys, as a key material, has been applied in an increasingly wide range. With the continuous improvement of the strength of aluminum alloy materials and the increasing complexity of component shapes, traditional cold forming manufacturing technology has become inadequate to meet the processing requirements. Aluminum alloy materials hot forming technology is a method of plastic processing of aluminum alloy materials under high-temperature conditions, which can form thin-walled, complex-shaped components in one step and control the performance of the components. It is a hot topic in academic research and engineering applications. This paper systematically discusses hot forming technology, hot stamping forming processes, domestic and foreign hot stamping forming equipment. It also summarizes the challenges faced by aluminum alloy materials hot forming technology and equipment and looks forward to their future development trends in material innovation, process enhancement, intelligent equipment, and green low-carbon initiatives, aiming to provide references for the development direction of aluminum alloy materials hot forming technology and equipment.

A Special Heat Treatment Process Achieves Ultra-High Room-Temperature Ductility of $Fe_{49}Mn_{33.2}Cr_{9.6}Co_{8.2}$ High-Entropy Alloy

Han Jinke[1,2], Zhou Ge[1,2], Chen Lijia[1,2]

[1] Shenyang University of Technology, 110870, School of Materials Science and Engineering, No. 111, Shenyang Liaoxi Road, Shenyang Economic and Technological Development Zone, China.

[2] Shenyang Key Laboratory of Advanced Structural Materials and Applications, 110870, No. 111, ShenyangLiaoxi Road, Shenyang Economic and Technological Development Zone, China.

E-mail address: Hanjinke16@ 163. com (Han Jinke)

E-mail address: zhouge@ sut. edu. cn (Zhou Ge)

Abstract: Since the discovery of high-entropy alloys (HEAs) in 2004, they have attracted significant attention due to their excellent mechanical properties. However, various HEA systems have experienced a common challenge: when mechanical processing and heat treatment are employed for microstructure control, the strength-ductility balance of the alloy at room temperature is compromised, with a particularly severe loss of ductility. This work introduces the "metastable engineering" concept into HEAs and designs a novel $Fe_{49}Mn_{33.2}Cr_{9.6}Co_{8.2}$ HEA by synergistically regulating stacking fault energy, grain size, phase composition and stability. Following conventional rolling pretreatment, the alloy underwent a heat treatment process involving holding at 950 ℃ for 6 minutes, followed by quenching to prepare test specimens. After a mutual transformation between the FCC phase and the HCP phase occurred during the deformation process at room temperature, a secondary TRIP effect of the FCC occurs at 70% deformation until fracture. This resulted in a five-stage strain hardening behavior. During multiple deformation-induced phase transformations, high-density phase boundaries are generated, continuously refining the microstructure and generating dynamic Hall-Petch effects, giving the alloy a high strain-hardening ability. Additionally, the mutual transformation between the FCC phase and the HCP phase induced by deformation is also beneficial for improving HEA ductility. This is because the movement of dislocations is hindered during the deformation process, leading to dislocation aggregation and the generation of significant local stress fields. This stress can be redistributed within the alloy through a self-regulating mechanism involving deformation-driven phase transition (FCC → HCP → Reversed-FCC → HCP, Triple TRIP effect). This process effectively relieves local stress concentrations and allows the alloy to maintain 91 % ductility at room temperature.

Keywords: high-entropy alloy; transformation-induced plasticity; room-temperature ductility; strain-hardening behavior; stacking-fault energy

Effect of the Solidification-Loading Axis Angle on Hot Ductility and Dynamic Recrystallization Behavior in a 7Mo-0.38N-RE Super Austenitic Stainless Steel

Hu Xin

School of Materials Science and Engineering, Hebei University of Science and Technology, Shijiazhuang, Hebei, 050018, China

E-mail address: xinhu@hebust.edu.cn (Hu Xin)

Abstract: This study investigated the effect of the angle between the solidification direction of columnar grains and the loading direction (ABSL, selected as 0°, 30°, 60°, and 90°) on the high-temperature deformation behavior and microstructural evolution of a 7Mo-0.38N-RE super austenitic stainless steel. The tensile deformation behaviours of samples with different ABSLs were examined in the temperature range of 1150-1250 ℃. The results indicated that as the angles between the direction of columnar grains and the loading direction increased, the hot ductility of the test steel first rose and then declined, reaching a peak at an angle of 60°. An increase in ABSL impeded the inward extension of strain along the columnar grain boundaries, leading to enhanced strain accumulation at these boundaries. This occurred because the columnar grain boundaries parallel to the tensile direction were the main location for sharing accumulated strain. This provided energy for the formation of DRX, resulting in a decrease in the critical stress value for recrystallization. Although the 90° sample recrystallized earliest, its failed the fastest due to the limited ability of strain to be transmitted effectively into the grain interiors. Accordingly, the columnar grain boundaries of 0° sample could more effectively share strain accumulation at the fracture surface than those of 90° sample, thereby improving the hot ductility of the test steel. On this basis, the columnar grain boundaries in more directions of the 30° and 60° samples could disperse the strain accumulation, leading to the formation of more high-dislocation density regions along the boundaries, which promoted DRX initiation. In spite of the similar degrees of DRX observed in 30° and 60° samples, the initiation of DRX in the former occurred at a comparatively later stage, which contributed to the generation of micropores within the matrix during the work-hardening stage. The aggregation of the micropores formed a multitude of cracks in the subsequent deformation process, diminishing the hot ductility. This phenomenon was attributed to the inconsistency in the hot processing maps among samples with different ABSLs. The 30° sample was in the instability region, while the 60° sample was in the stability region. Thereby, the accumulated dislocations led to the formation of instability phenomena, such as micropores during the work hardening stage for 30° sample, reducing the hot ductility. According to the microstructural evolution of DRX, twins, and microcracks under different ABSLs, the grain boundaries of 60° samples could better share responsibility for accumulated dislocations and excessive plastic deformation than others. In addition, higher deformation temperature could promote grain boundary slip, decreasing the critical strain for DRX initiation of 60° sample at 1200 ℃, and promoting the formation of DRX grains. As a result, the 60° sample exhibited the highest DRX degree and hot ductility. However, excessive temperature could lead to abnormal grain growth in the 60° sample at 1250 ℃, which seriously diminished the ability to hinder crack propagation and hot ductility.

Keywords: super austenitic stainless steel; hot ductility; columnar grain; loading direction; microstructural evolution

Distortion Mechanism of Bearing Rings and AI-Driven Techniques for Optimal Flipping Frequency in Mesh Belt furnace Quenching

Li Xusheng[1,2], Deng Xiaohu[3*], Sun Yixiao[4], Xu Weihui[1], Zhao Zhigao[1], Ju Dongying[1*]

[1] Zhejiang XCC Group Co., Ltd., Xinchang, Zhejiang, China

[2] School of Mechanical Engineering, Zhejiang University, Hangzhou, China

[3] School of Mechanical Engineering, Tianjin University of Technology and Education, Tianjin, China

[4] School of Electronic and Information Engineering, University of Science and Technology Liaoning, Anshan, Liaoning, China

Abstract: The mesh belt furnace is an efficient piece of heat treatment equipment widely used for quenching bearing rings. To improve workpiece cooling uniformity, a secondary mesh belt structure is typically added within the oil tank. This enables the bearing ring to alternately contact the secondary mesh belt and the lifting mesh belt at both ends. Since excessive or insufficient contact time prevents achieving the desired cooling uniformity, the ring's residence time on the smaller mesh belt is critical for the intended design effect. Therefore, this study employs a combined approach of experimental investigation, simulation analysis, and AI-driven techniques to systematically explore how the flipping frequency of bearing rings influences post-quenching distortion. First, quenching experiments were conducted under nine different flipping frequencies for three thin-walled bearing products of varying sizes (A, B, and C). Thirty workpieces per batch ensured experimental reliability. The results indicate that distortion characteristics for all three bearing types fluctuate noticeably across different flipping frequencies. Furthermore, flipping frequency significantly affects both the average diameter difference between upper and lower surfaces and the ellipticity of the bearing rings. To further explore the influence mechanism of the small mesh belt during quenching, this study analyzed heat transfer behavior during ring flipping on the mesh belt. It quantitatively derived heat conduction variations associated with different flipping frequencies. The calculated heat transfer coefficients were applied as boundary conditions in the multi-physics heat treatment simulation software COSMAP. Simulations of the quenching process across frequencies revealed how the secondary mesh belt structure mitigates distortion in thin-walled bearing rings. In practical manufacturing, the geometric shapes and sizes of bearing rings vary significantly, leading to different optimal flipping frequencies for different ring types. Additionally, workpiece position within the same batch substantially influences the cooling process. Consequently, determining the optimal flipping frequency solely through simulation analysis is challenging. To address this, experimental results served as training data for a large language model (LLM) to intelligently filter data and identify the flipping frequency minimizing quenching deformation. For bearing ring A, the optimal frequency from simulation closely agreed with the LLM's prediction, confirming the accuracy and practical utility of the AI-driven approach. This method rapidly recommends optimal flipping frequencies for diverse thin-walled bearing rings during quenching, significantly reducing simulation time and improving production efficiency.

Keywords: quenching process; flipping frequency; distortion; heat transfer coefficient; large language models

Excellent Strengthening Effect and Microstructure Control Mechanism of IN718 Alloy Treated by Surface Laser Shock Peening

Liu Mengya, Liu Yang*, Wang Lei*

School of Materials Science and Engineering, Northeastern University, Shenyang, China

E-mail address: liuyang@mail.neu.edu.cn (Liu Yang)

E-mail address: wanglei@mail.neu.edu.cn (Wang Lei)

Abstract: Nickel-based wrought superalloy IN718 exhibits excellent mechanical properties, good oxidation resistance and corrosion resistance in extreme environments. It has been widely used in the manufacture of key component of the hot end of aeroengines. Laser shock peening (LSP) and warm laser shock peening (WLSP), as the promising surface strengthening technologies, can induce severe plastic deformation (SPD) with a strain rate higher than 10^6 s^{-1}, which is particularly effective for improving the mechanical properties of materials. In this study, the uniaxial tensile and fatigue properties of LSP and WLSP-treated IN718 alloy at different temperatures were comparatively investigated. It is found that comparing with the traditional LSP, WLSP shows more significant advantages in maintaining the high-temperature fatigue resistance of the surface hardened layer during long-term service and prolonging the fatigue life of the alloy. After WLSP, a complex structure with gradient distribution of γ'' phase containing stacking faults (SFs) and micro-twins (MTs) surrounded by high-density dislocation is formed in the surface hardened layer. Therefore, comparing with LSP, the remarkable high temperature strengthening effect of WLSP hardened layer is depended on the strengthening mechanisms such as back stress strengthening, dislocation strengthening, and thermal mismatch strengthening. It is worth noting that the complex microstructure formed in the hardened layer of WLSP can not only delay the coarsening of γ'' phase, but also exhibit remarkable strengthening effect stability at high temperature.

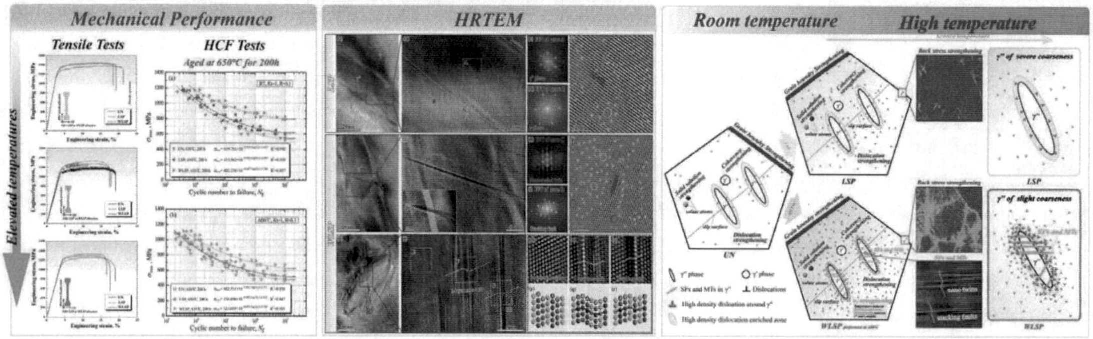

Keywords: superalloy; WLSP; γ'' phase; stacking faults; high-temperature strengthening

Effects of Electropulsing Treatment on Microstructure Reconstruction and Synergistic Control of Strengthening and Toughening in Ni-based Superalloys

Li Jinguo[1*], Liu Jide[1*], Guo Jingdong[2]

[1] Shi-changxu Innovation Center for Advanced Materials, Institute of Metal Research, Chinese Academy of Sciences, China

[2] Shenyang National Laboratory for Materials Science, Institute of Metal Research, Chinese Academy of Sciences, China

E-mail address: jgli@imr.ac.cn (Li Jinguo)

E-mail address: jdliu@imr.ac.cn (Liu Jide)

Abstract: Electropulsing Treatment (EPT), a novel material processing technology integrating Joule heating and non-thermal effects, demonstrates significant advantages in more efficiently regulating the microstructure and optimizing service performance of materials compared to traditional heat treatment. This study focuses on the application of EPT in Ni-based superalloy GH4068 and laser additive manufactured ATI 718Plus. The results show that the targeted dissolution effect of EPT not only significantly shortens the dissolution time of large-sized γ' phases in GH4068 alloy but also effectively coordinates the contradiction between grain boundary strengthening and precipitation strengthening, thus remarkably improving the strength and toughness of the alloy at room temperature and service temperature. For laser additive manufactured ATI 718Plus, EPT treatment rapidly eliminates Laves phases and residual stresses at relatively low temperatures. By increasing grain boundary mobility and reducing matrix stacking fault energy, EPT induces the formation of high-density annealing twins. Furthermore, this study proposes a novel post-treatment strategy combining EPT and traditional heat treatment, which enables the collaborative regulation of precipitation strengthening phases and grain size, effectively balances the inverse relationship between Laves phase content and grain size, and comprehensively optimizes the strength-toughness matching of laser additive manufactured ATI 718Plus.

Keywords: Electropulsing, Ni-based superalloys, microstructure, residual stress, mechanical properties

Microstructural Evolution and Mechanical Properties of Ferrous Medium-Entropy Alloys during Aging

Liao Yu, Li Chuanwei*, Gu Jianfeng*

School of Materials Science and Engineering, Shanghai Jiao Tong University, China

E-mail address: li-chuanwei@sjtu.edu.cn (Li Chuanwei)

E-mail address: gujf@sjtu.edu.cn (Gu Jianfeng)

Abstract: In recent years, high- and medium-entropy alloys (H/MEAs) have attracted widespread attention due to their unique structures and outstanding properties. Among them, ferrous MEAs (Fe-MEAs) have garnered increasing interest due to their cost-effectiveness. This study systematically investigates the microstructural evolution and mechanical properties of martensitic $Fe_{71}(CoCrNi)_{29}$ (at. %) Fe-MEAs during the aging process. During aging, a high density of coherent nanoscale Cr-rich precipitates uniformly forms within the martensitic matrix and gradually coarsens with increasing aging time. Simultaneously, nanoscale reverted face-centered cubic (FCC) phases precipitate along the lath boundaries, maintaining a Kurdjumov-Sachs (K-S) orientation relationship with the matrix. The formation of Cr-rich precipitates significantly enhances the hardness and strength of the alloy. Owing to their coherent interface with the matrix, these precipitates alleviate stress concentrations, improving the work hardening rate and preserving ductility. This work provides an in-depth analysis of the early-stage formation mechanism of Cr-rich precipitates, their contribution to strengthening, and their influence on work hardening behavior. The findings offer valuable insights into microstructure-property optimization for low-cost, high-performance Fe-MEAs and provide effective guidance for the design of other advanced structural alloys.

Keywords: ferrous medium-entropy alloys; microstructure; mechanical properties; aging

AchievingHigh Damping Capacity in Oxygen-Enhanced BCC Zr-Hf-Ti-Nb Multi-Principal-Element Alloys with Low Young's Modulus

Wang Rui*, Wang Qing

Dalian University of Technology, Dalian, China

E-mail address: wangrui24@ mail. dlut. edu. cn (Wang Rui)

E-mail address: wangq@ dlut. edu. cn (Wang Qing)

Abstract: Multi-principal-element alloys (MPEAs) have gained widespread popularity due to their capabilities to allow for efficient synergetic regulation of mechanical and functional properties in huge compositional space. This work developed novel O-enhanced BCC Zr-Hf-Ti-Nb MPEAs with prominent mechanical and damping properties by the composition formula of $(Zr, Hf, Ti)_{15}Nb_3$. The designed $Zr_{14}TiNb_3$ and $Zr_8Hf_6TiNb_3$ alloys possess low BCC-β structural stability because a minor amount (0. 2 ~ 0. 5 at. %) of oxygen additions can induce the precipitation of α″ and ω. While the $Zr_8Hf_4Ti_3Nb_3$ alloy has a much higher BCC-β stability, as evidenced by the fact that only few α″ and ω precipitates appear in 1. 8 at. % oxygen-added alloy. This alloy exhibits an optimal mechanical property with a higher yield strength (σ_{YS} = 1000 MPa) and larger ductility (ε = 15. 1%), compared to the oxygen-free alloy (σ_{YS} = 704 MPa and e = 9. 2%), which is ascribed to the formation of O-rich clusters in BCC matrix. Moreover, the $Zr_8Hf_4Ti_3Nb_3$ alloys with and without oxygen additions exhibit an excellent damping capacity due to their low Young's modulus (E < 70 GPa), as exemplified with a peak value of $(\tan\delta)_{max}$ = 0. 02 for 1. 8 at. % oxygen-added alloy. Notably, the damping characteristics are prominent over a wide temperature range (550 ~ 800 K), which derives from the occurrence of multiple separated oxygen-rich clusters. Strengthening mechanism for high strength and damping mechanism related to various Snoek-type relaxations were further discussed. The present findings provide an avenue to enhance mechanical and functional performances of high-temperature damping alloys.

Keywords: multi-principal-element alloys; BCC structural stability; mechanical property; damping capacity; Snoek-type relaxation

Heat Treatment Process and Friction-Wear Properties of The Bainite Structure in GCr15 Bearing Steel

Ding Zhen, Fu Lihua*

School of Materials Science and Engineering, Henan University of Science and Technology, Luoyang, 471000, China

E-mail address: d18739573319@163.com (Ding Zhen)

E-mail address: flhustb@haust.edu.cn (Fu Lihua)

Abstract: Aim to the urgent demand for improved hardness, impact toughness, and wear resistance of GCr15 high-carbon bearing steel which is the common steel used in rolling mills and mining machinery, this study achieve get a more refined bainite microstructure in GCr15 steel with an enhancing overall performance. This study investigates the effects of different isothermal quenching durations on GCr15 bearing steel by optimizing the conventional isothermal quenching process parameters. Based on these optimized parameters, pre-quenching treatments at various temperatures are further applied to examine how different pre-quenching temperatures influence the refinement of bainite microstructure, mechanical properties, and friction and wear behavior. The results showedthat: (a) As the isothermal quenching time increased, the toughness bainite content in the microstructure increased, while the martensite content decreased. Then, this microstructure led to a gradual reduction in hardness, an improvement in impact toughness, and a first decrease followed by an increase in wear resistance. When the isothermal quenching time of 30 min, the materials have the beast wear resistance due to the combination of strength and toughness. (b) The formation of pre-martensite could divide the undercooled austenite grains and introduce a large number of dislocations, which can increase bainite nucleation sites. With the increasing of pre-quenching temperature, the pre-martensite content is decreased, while the bainite content is increased, and the residual austenite content initially decreased then increased. This caused a gradual reduction in hardness, an initial rise followed by a decline in impact toughness, and a first decrease followed by an increase in wear resistance. When the pre-quenching temperature is 164 ℃, the wear resistance is optimal due to the best combination of strength and toughness. (c) Comparing the pre-quenched specimens with the conventional isothermal quenching samples, the bainite of the pre-quenched specimens under the same isothermal quenching time was hindered by pre-martensite and austenite during the growth process, which led to the refinement of bainite microstructure, which played the role of fine-grain strengthening, and increased the impact toughness by 39.2% and the wear resistance by 25.0%.

Keywords: pre-quenching; bainite; mechanical properties; friction and wear

Strengthening Andtoughening Mechanisms of Titanium Alloys via the Combined Process of Deformation and Heat Treatment

Song Xiu*, Wang Lei, Liu Yang

The Key Laboratory for Anisotropy and Texture of Materials (Ministry of Education), School of Materials Science and Engineering, Northeastern University, Shenyang110819, China

E-mail address: songxiu@mail. neu. edu. cn (Song Xiu)

E-mail address: wanglei@mail. neu. edu. cn (Wang Lei)

Abstract: Titanium and titanium alloy with relative low Yong's modulus, good corrosion resistance and excellent biocompatibility, has been widely used as one kind of promising biomaterials, such as artificial endovascular stents, bone plates, and dental implants. But the strength of pure titanium and β titanium alloy is relative lower than that of α + β titanium alloy. The improvement in strength and ductility of pure titanium and new developed β titanium alloy has been successfully obtained through the combined process of deformation and heat treatment, and the strengthening-toughening mechanisms has been investigated by TEM and EBSD. The results show that the good match of tensile strength and elongation of pure titanium has been obtained through an optimized process including constrained groove pressing and intermediate heat treatment between 673 K and 873K among different pressing passes. The improvement in strength by above 60% compared with that of pure titanium is ascribed to the more dislocation tangles near the sub-grains which from during the annealing at lower temperature, and the twin variants activated during the process combined the deformation and heat treatment. The tensile strength of b Ti-29Ta-13Nb-4. 6Zr (TNTZ) is highly enhanced by above 80% without losing plasticity so much which is still longer than 10%, via cold rolling and cryogenic rolling followed with short-time aging between 573 K and 773 K. On one hand, more dislocations, twins and strain-induced w phases are activated during the rolling at cryogenic temperatures; on the other hand, although the recovery occur during the aging, the enhanced strength is ascribed to more isothermal ω phases which precipitate during the short-time aging at lower than 673K.

Keywords: strengthening; toughening; heat treatment; titanium alloys; deformation

New Role of Martensitic Transformation of L605 Alloy

Yin Tingyu, Wang Lei*, Song Xiu, Liu Yang

TheKey Laboratory for Anisotropy and Texture of Materials (Ministry of Education), School of Materials Science and Engineering, Northeastern University, Shenyang 110819, China

E-mail address: yinty163@163.com (Yin Tingyu)

E-mail address: wanglei@mail.neu.edu.cn (Wang Lei)

Abstract: Cobalt-based superalloy is one kind of key materials in aerospace applications, and strain-induced martensitic transformation (SIMT, γ-Co: FCC → ε-Co: HCP) plays a crucial role in its application. The mechanisms of the martensite nucleation and growth of L605 alloy during in situ tensile deformation at room temperature were investigated. The results show that martensitic transformation occurs under a critical strain range between 2.5% and 3.5%. The martensite volume fraction increases slowly if the strain is lower than 15% ($\Delta f_{HCP}/\Delta\varepsilon = 0.17448$), then accelerates if the strain is higher than 15% ($\Delta f_{HCP}/\Delta\varepsilon = 1.8695$), and finally approaches a plateau which is about 35% when strain reaches 45%. Martensite initially nucleates at grain boundaries ($\varepsilon = 4\%$). With further deformation of 8%, twin boundaries also serve as nucleation sites. When the strain increases to 15%, martensite can nucleate within the grain interiors, independent of any boundaries. The martensite nucleation mechanisms vary with increasing strain, since the nucleation changes from grain boundaries to twin boundaries and finally to intragranular. Earlier nucleation of martensite at boundaries is attributed to the presence of local stress concentration at boundaries. The interfacial energy combined with elevated distortion energy provides additional driving force for martensitic transformation which can reduce the critical shear stress required for activating phase transformation. Meanwhile, martensite preferentially nucleates at grain boundaries due to higher atomic misarrangement, which promotes stacking faults formation and provides more interfacial energy to overcome the phase transformation barrier. In contrast, grain interiors lack of such interfacial misarrangement structures. Thus, intragranular nucleation relies on the interaction among dislocations within grains under higher strain, leading to the formation of localized high-strain regions. Martensite of L605 alloy forms as laths with the thickness of from about 10 nm under a strain of 4% to 70 nm when strain exceeds 25%. On one hand, as the strain increases, multiple martensitic laths stack together; on the other hand, martensitic lath grows in thickness direction due to the sufficiently high shear stress which is contributed to activating multiple atomic layers to shear.

Keywords: Cobalt-based superalloy, strain-induced martensitic transformation, martensite nucleation, in situ tensile test, microstructural evolution

Nanoparticle Enhanced Microstructuresand Heat Treatment Strategies for High-Performance IN718/Y_2O_3 Superalloy Fabricated by Laser Powder Bed Fusion

Yang Jialiu[a], Li Chuanwei[a,b*], Yang Xudong[c], Gu Jianfeng[a,b]

[a] Institute of Materials Modification and Modelling, School of Materials Science and Engineering, Shanghai Jiao Tong University, Shanghai 200240, China

[b] Shanghai Key Laboratory of Materials Laser Processing and Modification, Shanghai Jiao Tong University, Shanghai 200240, China

[c] Xinchang TianQi Institute of Astronautics and Aeronautics, Zhejiang Shaoxing 312500, China

E-mail addresses: li-chuanwei@ sjtu. edu. cn (Li Chuanwei)

E-mail addresses: gujf@ sjtu. edu. cn (Gu Jianfeng)

Abstract: To overcome the processing bottleneck in laser powder bed fusion (LPBF) of IN718 superalloy caused by complex precipitate management and multi-stage heat treatments, this study systematically investigated the regulatory mechanisms of 1. 5 wt. % Y_2O_3 nanoparticles on microstructural evolution and mechanical properties in as-deposited state and under different heat treatment routes (high-strength aging: HSA; standard heat treatment: STA; direct aging: DA). The as-deposited IN718/1. 5Y_2O_3 alloy exhibited remarkable strength-ductility synergy: yield strength reached 1023. 7 ± 16. 5 MPa (28. 4% improvement over unreinforced alloy), ultimate tensile strength 1294. 6 ±22. 1 MPa (27. 3% improvement), and elongation 25. 8 ±0. 9% (9. 1% improvement). Post-heat treatment performance showed significant path dependence: HSA treatment (1130 ℃/5 min solution + dual-stage aging) enhanced the yield strength to 1425. 2 ±21. 4 MPa (+2. 2%), ultimate strength to 1597. 5 ±12. 7 MPa (+6. 8%), while elongation decreased to 12. 1 ± 1. 1% (−2. 7%). This originated from: (i) Complete dissolution of grain-boundary acicular δ-phase and massive Laves phases during high-temperature solution treatment, eliminating brittle fracture risks; (ii) Precipitation of ultrahigh-density coherent γ″ strengthening phases in γ-matrix during subsequent aging, generating lattice distortion fields. Direct aging (DA) demonstrated unique advantages: yield strength 1518. 4 ±18. 0 MPa (+2. 9%), peak ultimate strength 1613. 7 ±12. 5 MPa (+2. 5%), with only 0. 7% reduction in elongation (9. 39 ±0. 7%). This path utilized LPBF's inherent fine-grained substructure, where Y_2O_3 particles and high dislocation density jointly served as precipitation nuclei, promoting preferential γ′/γ″ phase nucleation at dislocation lines to form "dislocation-precipitate synergistic strengthening networks". Standard treatment (STA) (980 ℃/1 h solution + aging) suffered from 20. 1% elongation loss (13. 5 ± 0. 4%) due to residual δ-phase and undissolved Laves phases acting as microcrack sources, highlighting the necessity of complete harmful phase elimination. Mechanical analysis further revealed significantly enhanced strain hardening rate (SHR) during plastic deformation after Y_2O_3 addition. The underlying mechanism involves: nanoparticles and γ′/γ″ phases forming multi-scale obstacle systems that force dislocations to bypass via Orowan looping or shearing, increasing dislocation line tension and promoting new dislocation multiplication. Concurrently, strong strain gradient fields at Y_2O_3/matrix interfaces induced geometrically necessary dislocations (GNDs), whose interactions with statistically stored dislocations (SSDs) provided additional strengthening. Nanoparticles also delayed dynamic recovery, suppressing dislocation cell formation and sustaining work hardening to larger strains. This mechanism effectively retarded necking initiation, enabling the rare strength-ductility synergy in DA-treated specimens. This study establishes a theoretical foundation for developing high-performance additive-manufactured superalloys through nanoparticle enhancement and innovative heat treatment design.

Keywords: laser powder bed fusion (LPBF); IN718 superalloy; Y_2O_3 nanoparticle; heat treatment Strategies; strength-ductility synergy

Unraveling Microstructure Evolution During Isothermal Annealing Treatment in Dual-Phase $FeNiCr_{0.8}Al_{0.8}$ High Entropy Alloy and Its Effect on Corrosion Behavior

Cui Puchang*

School of Materials Science and Engineering, Liaoning University of Technology, China

E-mail address: msecpch@163.com (Cui Puchang)

Abstract: In this work, a systematic investigation of the evolution of dual-phase microstructure, corrosion behavior of $FeNiCr_{0.8}Al_{0.8}$ high entropy alloy after isothermal annealing was conducted and corrosion mechanisms related to microstructure evolution and Cl diffusion behavior in as-annealed dual-phase alloys were revealed. The dual-phase microstructure characteristics in the as-annealed alloy were studied and the relationship between microstructure and corrosion behavior was established. The corrosion mechanism of dual-phase as-annealed alloys served in Cl-containing environments is revealed by combining experimental characterizations with first-principle calculation approaches. The as-cast $FeNiCr_{0.8}Al_{0.8}$ alloy possesses a two-phase body-centered cubic (BCC) microstructure. The BCC1 phase increases from the fraction of 48.78% for as-cast specimen to 70.91% for the as-annealed specimen after annealing treatment at 1200 ℃ for 100 h, which is in agreement with the predicted phase results. The coarsening behavior of the hierarchical phases after annealing and the appearance of the secondary nano-precipitates are observed. Meanwhile, the enhanced corrosion resistance of the as-annealed alloy was examined after prolonged time annealing, which is related to increased fraction of BCC1 phase and the intrinsic nature of the two-phase. The KPFM results and the first-principle calculations proved that the presence of BCC2 phase with a lower work function of 4.53 eV accelerates the occurrence of localized selective corrosion. The diffusion barrier of Cl in the BCC2 phase with lower value of 1.98 eV further interprets the mechanism of improved anti-corrosion behavior in the current as-annealed alloy.

Keywords: high entropy alloys; annealing; corrosion mechanism; diffusion

Formation and Role of TiN in K411 Superalloy Under a Simulated Blade Service Condition

Yu Ren[1], Wang Yao[2], Dong Jiasheng[2], Wang Lei[1]*

Key Lab for Anisotropy and Texture of Materials, Northeastern University, China

Institute of Metal Research, Chinese Academy of Sciences, Shenyang, China

E-mail address: 2010201@stu.neu.edu.cn (Yu Ren)

E-mail address: wanglei@mail.neu.edu.cn (Wang Lei)

Abstract: Heavy-duty gas turbine blades, critical components in electric power generation systems, endure extreme conditions including high temperatures, mechanical stress, and corrosive environments. The persisting knowledge gap in corrosion-creep synergy mechanisms under a simulated blade service condition prevents rational design of nickel superalloys for applications. The precipitation behaviour of TiN and its effects on creep failure of K411 nickel-based superalloy under a simulated turbine blade service conditions (900 ℃/215 MPa with 2 vol. % SO_2) was investigated in present study. The creep properties of K411 alloy exposed to SO_2-containing atmosphere decreased with a 73% reduction in creep life from 2904 h in air to 783 h, and a 46% decrease in total creep strain from 4.5% to 2.4%. Based on the microstructural analysis, it is found that corrosion pits formed in the oxide scale under SO_2 exposure served as preferential channels for nitrogen inward diffusion. Furthermore, applied stress enhanced nitrogen transport through dislocation pipe diffusion mechanisms. This synergistic effect promoted extensive formation of preferentially oriented needle-like and cross-shaped TiN precipitates aligned along maximum shear stress planes. The precipitation process consumed critical γ'-forming elements, generating γ'-free zones measuring approximately 31.6 μm in width. At the TiN/matrix interfaces, strain incompatibility initiated microvoid nucleation, and into microcracking. These microcracks then interconnected with surface corrosion pits, resulting in the formation of surface cracks and premature failure. It has been demonstrated that the combined action of SO_2-induced corrosion and stress-accelerated nitrogen permeation is the fundamental mechanism of the significant creep life reduction. These findings provide essential theoretical and experimental support for developing durable blade alloys and predicting service life, while establishing a framework to optimize blade compositions against failure mechanisms.

Keywords: nickel-based superalloy; corrosion-creep interaction; TiN precipitation; SO_2 environment; turbine blade degradation

Tensile Properties of Al-5.5Zn-2Mg-2Cu Alloy Subjected to Aging Treatment for Different Time

Liu Ruochong*, Chen Lijia

School of Materials Science and Engineering, Shenyang University of Technology, China

E-mail address: lrclyric96@163.com (Liu Ruochong)

E-mail address: chenlj-sut@163.com (Chen Lijia)

Abstract: The influence of aging time on the tensile properties and strengthening effect for the Al-5.5Zn-2Mg-2Cu alloy were investigated by means of scanning electron microscope (SEM), transmission electron microscope (TEM) and tensile tests. The alloy was subjected to the solution treatment at 480 ℃ for 1 h and subsequently aged at 120 ℃ for 10h to 34 h, where the specimen at five typical aging time was selected for the tensile tests and microstructural characterization. The results demonstrate that the tensile properties exhibit the typical age-hardening characteristics. With prolonging the aging time, the ultimate tensile strength and yield strength of the alloy show a tendency of firstly increasing and then decreasing, while the elongation shows a tendency of firstly decreasing and then increasing. Both ultimate tensile strength and yield strength reach the peak values of 488.02 MPa and 428.52 MPa at the aging time of 22 h, followed by a gradual decrease with prolonging the aging time. The microstructural analysis reveals that the strengthening effects are closely related to the precipitation sequence. At the under-aged stage, the strengthening primarily depends on the formation of Guinier-Preston (GP) zones, where solute clusters impede the dislocation motion through the lattice distortion-induced stress fields. At the peak-aged stage, TEM observations reveal that a high density of nanoscale η' phase (metastable $MgZn_2$) uniformly distributes in the α-Al matrix. Concurrently, the grain boundary precipitates undergo significant coarsening and the precipitate free zones (PFZs) widen to approximately 22 ~ 25 nm, leading to maximum strength but minimum elongation of 8.04%. At the over-aged stage, the η' precipitates undergo significant coarsening with the increased interparticle spacing and gradually transform into the equilibrium η phase. This phase transformation leads to the loss of coherency at the precipitate-matrix interfaces, resulting in the diminished strengthening effects due to reduced strain fields and increased obstacle spacing. However, the concomitant reduction in the stress concentration contributes to the improved ductility and fracture toughness. The fractographic analysis results indicate that all specimens exhibit the ductile fracture characteristics, where the under-aged and over-aged specimens show smaller dimples (30 to 40 μm), while the peak-aged specimens display larger and more irregular dimples (40 to 60 μm).

Keywords: Al-Zn-Mg-Cu alloy; aging treatment; aging time; tensile property; tensile fracture

Study on Post Weld Heat Treatment of Residual Stress Reduction for Girth Weld of Long-distance Pipelines

Liu Congyue, Han Bin*, Niu Shengyuan, Li Xueda

Shandong Key Laboratory of Intelligent Energy Materials, School of Materials Science and Engineering, China University of Petroleum (East China), Qingdao, China

E-mail address: 2393752969@qq.com (Liu Congyue)

E-mail address: hbzhjh@upc.edu.cn (Han Bin)

Abstract: The residual stress has a significant impact on the service and performance of the girth weld. However, the residual stress reduction method for girth welds is not mature. Thus, a stress reduction method for low-temperature narrow gaps is introduced in this paper. A model of PWHT is established to simulate stress distribution and analyze the reduction of stress levels of girth weld by various factors, including peak temperature, insulation time, cooling rate, and width of heating band. Through the analysis of Von Mises stress, the optimal stress reduction process is a peak temperature of 250 ℃, an insulation duration of 1500 s, a cooling rate of 3 s/℃, and a heating band width of 233 mm. Under this process, the stress is reduced by 40%. This article promotes the development of stress reduction technology.

Keywords: girth weld; post weld heat treatment; stress reduction; numerical simulation

Research on the Prediction Technology of Hardness for Quenching of Metal Materials

Abstract: This paper focuses on studying the hardness distribution of 45 steel during the heat treatment process, as well as establishing and validating prediction models for this purpose. To achieve high precision in predictions, the research has set two core objectives: constructing a computational model based on the coupling relationship between temperature fields and hardness fields, and developing a software program for simulating the hardness distribution of 45 steels during quenching. In terms of model construction, performance data at different locations were obtained through 45 steel end quenching experiments, and finite element numerical simulation methods were employed to analyze and predict the temperature profile during the end quenching process. Based on the aforementioned data, a neural network algorithm was used for training, and a model capable of accurately predicting heat treatment hardness was established. This model can predict the performance of actual workpieces at different locations, and its reliability and precision have been verified through finite element numerical simulation. In the field of software development, this study custom-developed a simulation software for the hardness distribution of quenching in 45 steel, aimed at quickly and accurately predicting the hardness distribution from the surface to the core. The software integrates temperature field finite element solution algorithms and neural network algorithms, making operation more convenient while ensuring a prediction accuracy of at least 95%. Additionally, the software is highly customizable, allowing heat treatment technicians to adjust and optimize the models according to their specific needs. In the end, a calculation model coupling temperature field and hardness field was successfully established, and it was determined through comparative analysis that the random forest algorithm is the core of the high-precision prediction model. Furthermore, the simulation software developed can not only simulate the temperature evolution of cylindrical workpieces of different sizes and specifications during heat treatment but also intuitively display the hardness distribution. In practical experiments, the model and software were validated by predicting the hardness distribution of cylindrical specimens made of 45 steel with different specifications (e. g., Φ200x200mm), and the average precision reached 95.64%, which verified the effectiveness of the model and software. Through this research, not only has the accuracy of predicting the heat treatment performance of metal materials been improved, but it has also provided reliable technical support for process optimization in actual production, which is of great significance for improving product quality, reducing production costs, and realizing intelligent manufacturing.

The Improvement of the Mechanical Properties of High-Entropy Alloys Induced by Both Precipitation of Cr-rich Phases and Microstructural Regulation

Liu Zhiyu[1], Liu Jingshun[1*]

[1] School of Materials Science and Engineering, Inner Mongolia University of Technology, Hohhot, 010051, China

E-mail address: jingshun_liu@163.com (Liu Jingshun)

Abstract: Thiswork systematically investigates the influence of annealing treatment on the microstructure and mechanical properties of cold-rolled $CoCrFeNiMn_{0.5}$ high-entropy alloys. The results show that the as-cast alloys exhibits a typical dendritic structure, which undergoes grain refinement and transforms into a deformation-banded structure after large-deformation cold-rolling. Annealing treatment can effectively suppress intragranular segregation, release residual internal stress, and significantly improve the plasticity of the alloys. At an annealing temperature of 800 ℃, complete recrystallization occurs in the alloy. After annealing at 1000 ℃ for 1 h, a large number of annealing twinning and twinning boundaries are observed in the alloy. At this time, the plastic deformation mechanism of the alloys is dominated by dislocation slip and twinning, which significantly improves its plasticity (elongation after fracture *A* is 56.1%). The fracture type of the alloy changes from quasi-cleavage fracture in the cold-rolled state to ductile fracture in the annealing state, which is related to the precipitation of annealing twinning in the alloy. In addition, abnormal hardening occurs in the alloy after annealing at 500 ℃ for 1 h, and nanoscale Cr-rich phases precipitate within the alloys, leading to dispersion strengthening and the pinning of dislocations, which significantly enhances the alloy's hardness value to 472.8 HV. In conclusion, compared with other traditional alloys, the $CoCrFeNiMn_{0.5}$ high-entropy alloys obtained by cold-rolling and annealing treatment has important application potential in practical engineering application fields such as special components for equipment and key material devices due to its excellent comprehensive mechanical properties.

Keywords: $CoCrFeNiMn_{0.5}$ high-entropy alloys; cold-rolling; annealing; microstructure; mechanical properties

Preparation and Wave Absorption Performance Study of FeCoNiAl High-Entropy Alloy Powder

Yin Ying[1], Liu Jingshun[1*]

[1] School of Materials Science and Engineering, Inner Mongolia University of Technology, Hohhot, 010051, China

E-mail address: jingshun_liu@163.com (Liu Jingshun)

Abstract: High-entropy alloy powders is a new electromagnetic wave absorbing material with great potential advantages and broad application prospects due to its corrosion resistance, high temperature resistance and excellent soft magnetic properties. Herein, flake $FeCoNiMn_{0.6}Al_{0.4}$ high-entropy alloy powders was prepared by mechanical alloying method, and it was de-alloyed to explore the influence of structural changes and electromagnetic properties on its absorbing properties. The experimental results show that the prepared high-entropy alloy powders has high saturation magnetization and attenuation constant. After dealloying with 3 mol/L NaOH for 5 h, the minimum reflection loss (R_L) of the high-entropy alloy powders with a thickness of 2.5 mm is −50.07 dB, the saturation magnetization is 106.94 Oe, and the attenuation constant can reach 587.73. After the dealloying treatment, the $NiAl_2O_4$ phase is formed in the high-entropy alloy powders, and the surface area ratio of the lamellar particles increases, resulting in more interfaces, enhanced interfacial polarization, improved dielectric constant and permeability, and enhanced microwave absorption performance. In brief, the above mentioned working can provide theoretical basis and technical support for the application of FeCoNiAl high-entropy alloy powders in the field of electromagnetic wave absorption.

Keywords: FeCoNiAl high-entropy alloy powders; de-alloyed; electromagnetic wave; electromagnetic property; reflection loss

Effects of Multiple Hardening Cycles of SCM440 on Microstructures and Mechanical Properties

Rattana Borrisutthekul*, Chutimon Makee, Pusit Mitsomwang, Usanee Kitkamthorn
School of Metallurgical Engineering, Suranaree University of Technology, Thailand
E-mail address: rattana@g.sut.ac.th (Rattana Borrisutthekul)
E-mail address: chutimon.m@sut.ac.th (Chutimon Makee)

Abstract: Under industrial conditions, multiple hardening processes are generally not permitted. However, the specific reasons behind this restriction are not clearly defined. One possible concern is the potential deterioration of mechanical properties. To better understand the effects of multiple hardening treatments on the properties of heat-treated components, this study was conducted using SCM440 steel. In this experiment, SCM440 specimens were subjected to a hardening process consisting of austenitizing at 850 ℃ for 30 minutes, followed by tempering at either 600 ℃ or 650 ℃ for 90 minutes. The samples were divided into three groups based on the number of hardening cycles: 1) Single hardening process. 2) Double hardening process and 3) Triple hardening process. After heat treatment, the specimens were evaluated through microstructural observation, hardness testing, and impact testing. The results indicated that increasing the number of hardening cycles led to a refinement of the austenite grain size, which contributed to improved impact toughness, while maintaining similar hardness values across all tempering conditions.

Keywords: Multiple Hardening; SCM 440 Steel; Hardness Test; Impact Test; Microstructures

TheEffects of Austempering Process on Low-Alloy Steel SCM 440

Chutimon Makee*, Rattana Borrisutthekul, Pusit Mitsomwang, Usanee Kitkamthorn
School of Metallurgical Engineering, Suranaree University of Technology, Thailand
E-mail address: chutimon. m@ sut. ac. th (Chutimon Makee)
E-mail address: rattana@ g. sut. ac. th (Rattana Borrisutthekul)

Abstract: SCM 440 steel is a popular alloy steel used in the automotive and machinery industries due to its high strength and toughness, as well as its high hardenability. One of the processes commonly used to improve the mechanical properties of SCM 440 steel is quenching and tempering, which results in high strength steel. According to TTT diagram of SCM440, the austempering process is feasible to process and may improve its strength, toughness, or impact resistance of SCM 440 steel. To investigate the properties of SCM440 after austempering, this research is carried out. The experiments were started by austenitizing the samples at 850 ℃ for an hour, later samples were cooled by dipped into the salt bath at temperatures of 290 ℃, 310 ℃, and 330 ℃ for 90 and 120 minutes of soaking time. After austenitizing process, the sample were investigated the microstructures and were tested the tensile and hardness test. From the microstructure observations, it was found that the sample dipped in a salt bath at 290 ℃ mainly consisted of tempered martensite with a small amount of bainite. At higher austempering temperatures, up to 330 ℃, bainite formation was promoted, while the martensite content decreased. Moreover, increasing the austempering temperature resulted in lower hardness, whereas the elongation of the samples did not change significantly. Furthermore, under the experimental conditions, austempering time had no significant effect on the mechanical properties or microstructure of SCM440 steel.

Keywords: austempering; quenching; SCM 440 steel; hardness test; tensile test; mechanical properties

Rapid Cooling of Fe-based Alloy Melts and their Nanocrystallization Heat Treatment

Yao Kefu

School of Materials Science andEnginering, Tsinghua University Society, Beijing, China

E-mail address: kfyao@ tsinghua. edu. cn (Yao Kefu)

Abstract: Amorphous-nanocrystalline soft magnetic alloys have attracted lots of attentions due to their high efficiency when used as high-frequency devices, especially in the application area of third generation semiconductors. However, the preparation of this two-phases alloys is complicated. First, the Fe-based alloys should be melted and rapidly cooled down to obtain amorphous ribbons. Then, suitable heat treatment technology will be employed to get large amount precipitates of nanocrystalline bcc α-Fe phase within the amorphous matrix. In order to enhance the soft magnetic property, that is low coercivity and low core loss of the two phase alloy, the grain sized needs to be kept smaller than 20 nanometres. Therefore, increasing the number of crystalline nucleus and retarding the growing rate are necessary. It is obvious that the heat treatment technology is quite important. Here, recent progress in this area will be introduced and summarized. The potential technologies and developing tendency will be discussed.

Keywords: soft magnetic property, Fe-based alloys, amorphous-nanocrystalline alloys, heat treatment

An Overview of the Solution to Heat Treatment Stains by Traceless Water-Based Cleaning Technology

Abstract: The problem of blotching caused by incomplete cleaning or residue of heat treatment workpiece has always been one of the problems in the industry of heat treatment product quality, especially for the nitriding process and carburizing process with strict surface quality requirements of the workpiece. Generally, the contaminants on the surface of workpieces mainly include anti-rust oil, mechanical oil, cutting fluid, anti-rust agents, inorganic salts, metal shavings, and non-metallic particles. , at present, hydrocarbons, kerosene, alcohol and other solvents or alkaline cleaning agents are commonly used to clean them. Solv-based cleaning products have strong cleaning ability on oil substances, but no obvious effect on cutting fluid and other substances. And the traditional alkaline cleaning agent, itself is easy to produce white spot residue, and eventually produce bright spots, white spots, black spots after heat treatment, In addition, this kind of material on carburizing, nitride process also has a certain impermeability effect, easy to form a local "soft spot". Solution -limit source and clean up. The Forest © developed new cutting fluid and gear grinding oil without impermeable substance, which can fundamentally solve the problem of introducing impermeable elements and residual problems. During the cleaning stage, The Forest © has developed a water-based traceless cleaning process technology. This technology has an excellent ability to dissolve and remove dirt, especially for hard-to-clean areas such as narrow crevices and deep holes. It can quickly dissolve oil stains and make them float on the water surface. After washing, there are no white spots or water marks remaining on the surface of the workpiece. After rinsing and rust prevention with a special anti-rust agent for heat treatment, the problem of surface mottling (bright spots, black spots, and white spots) on the workpiece after carburizing and nitriding can be completely solved. Some products can achieve the surface free shot blasting, the appearance of the workpiece is consistent, there is no visible residue on the surface, and the hardness test of the workpiece meets the quality requirements. It has been widely applied in the nitriding, carburizing and other cleaning processes in industries such as gears, bearings and military industry.

Microstructural and Textural Evolution of High-Purity Niobium under Deformation and Thermal Fields

Wang Jiale[1], Wang Aijuan[1*], Li Zhaobo[2], Zhang Jiuhai[2]

[1] School of Materials Science and Engineering, Xi'an University of Technology, Xi'an, Shaanxi, China

[2] Ningxia Orient Tantalum Industry Co, Ltd, Shizuishan, Ningxia, China

E-mail address: 710427420@qq.com (Wang Jiale)

E-mail address: ajwmxl@163.com (Wang Aijuan)

Abstract: Niobium (Nb) is a body-centered cubic (bcc) elemental superconductor with a critical temperature of 9.3 K, the highest among all pure metals. Due to its excellent superconducting properties, chemical stability, good plastic formability, and suitability for deep drawing, Nb has long been widely used in the fabrication of superconducting radio-frequency (SRF) cavities for particle accelerators. In recent years, advances in electron beam melting (EBM) and high-efficiency purification techniques have significantly improved the purity of Nb, providing a solid foundation for in-depth studies on its microstructural evolution mechanisms. In this study, high-purity Nb (> 99.95 wt%) was selected as the material. Before plastic deformation, the sample had a thickness of 50 mm. It was subjected to unidirectional rolling at room temperature, resulting in a final thickness of approximately 3.5 mm (a reduction of about 93 %). Subsequent annealing treatments were carried out at four different temperatures: 500 ℃, 600 ℃, 800 ℃, and 1000 ℃. The evolution of microstructure and texture during deformation and heat treatment was systematically investigated. Through optical microstructural analysis and electron backscatter diffraction (EBSD) techniques, the recrystallization behavior at different temperatures was revealed, with a focus on grain orientation, misorientation distribution, texture intensity, and related defect structures. The results indicate that significant texture evolution occurs during the rolling process. Grains with different orientations exhibit varied deformation behaviors, leading to spatial heterogeneity in stored energy and consequently resulting in non-uniform recrystallized microstructures during annealing. The {001} and {111} components in the deformation texture exhibit high stability, while recrystallized textures after annealing tend to be dominated by the {111} orientation. Moreover, the recrystallization process is governed by the combined effects of the initial texture, stored energy distribution, and deformation band structure, which together determine nucleation and grain growth behavior. In summary, this study systematically elucidates the microstructural and textural evolution of high-purity Nb during deformation and annealing at various temperatures, providing theoretical support and technical guidance for optimizing the processing and performance of Nb, especially for high-performance SRF applications in the superconducting field.

Keywords: Niobium; Microstructure; Texture; Recrystallization; EBSD

Mechanisms of Microstructure Evolution in High-Purity Tantalum During Deformation and Annealing

Zhao Pei[1], Wang AiJuan[1]*, Li Zhaobo[2]

[1] School of Materials Science and Engineering, Xi'an University of Technology, Xi'an, Shaanxi, China

[2] Ningxia Orient Tantalum Industry Co, Ltd, Shizuishan, Ningxia, China

E-mail address: 1033811743@qq.com (Zhao Pei)

E-mail address: ajwmxl@163.com (Wang AiJuan)

Abstract: Tantalum (Ta), with its high melting point, excellent electrical conductivity, outstanding corrosion resistance, and biocompatibility, is widely used in advanced manufacturing fields such as the electronics industry, aerospace, military engineering, and semiconductor integrated circuits. Particularly in the semiconductor field, as a barrier layer material between Cu metal interconnects and Si substrates, it can effectively inhibit Cu-Si interfacial interdiffusion. The uniformity of the tantalum barrier layer is closely related to the orientation dependence of stored energy and the uniformity of texture distribution in the target, directly affecting the resistivity of integrated circuits and the consistency and stability of chip performance. To improve the sputtering performance of high-purity tantalum targets and optimize the quality of deposited films, this study comprehensively adopted Electron Backscattered Diffraction (EBSD) to characterize the microstructure and texture along the thickness direction of high-purity tantalum targets subjected to 90% deformation via 135° circumferential rolling and annealed at different temperatures, and conducted an in-depth analysis of their deformation and recrystallization behaviors. The results show that: During the rolling process, the stress state of the sample varies along the thickness direction, leading to an increase in the intensity of {111} fiber texture from the surface to the core with aggregation in the core region. The {100} texture is stable, while the {111} texture is unstable and tends to rotate toward stable orientations. Through SFDR (Schmidt Factor Difference Ratio) calculation of grains with different orientations and TEM analysis of dislocation morphologies, it is revealed that {111} oriented grains form high-density dislocation tangles due to single slip system activation, resulting in higher deformation stored energy, whereas {100} oriented grains exhibit lower splitting degree and lower stored energy due to multiple slip systems activation forming regions without dislocation tangles, thus leading to the orientation dependence of deformation stored energy. During the heating process from 700 ℃ to 1050 ℃, the contents of {111} and {100} textures show wave-like changes with opposite trends. After high-temperature annealing, the {111} texture has higher intensity, resulting in uneven microstructure distribution. In the recrystallization stage, {100} oriented recrystallized grains mainly nucleate at the grain boundaries of the {111} oriented deformed matrix via the strain-induced grain boundary migration mechanism and grow directionally due to the orientation dependence of the driving force.

Keywords: tantalum target; texture; schmid factor; stored energy; recrystallization

Computer Simulation of Gear distortion of Vacuum Carburizing and Quenching

Zhao Huihao, Wang Hao, Yan Mufu, Zhang Yanxiang*

School of Materials Science and Engineering, Harbin Institute of Technology, Harbin, China

E-mail address: zhhhit2021@163.com (Zhao Huihao)

E-mail address: hitzhang@hit.edu.cn (Zhang Yanxiang)

Abstract: To enhance the load-bearing capacity, fatigue resistance and service life of gears, carburizing and quenching has become the dominant heat treatment process for high-performance hard-toothed gears. However, due to the long high-temperature process cycle and the different temperature changes and microstructure transformations experienced during this period, complex thermal and microstructure stresses are generated, which are prone to causing gear distortion and seriously affecting the gear's performance and manufacturing cost. In this paper, a numerical simulation method is utilized. According to the technical requirements of the carburizing and quenching process, while comprehensively considering the multi-field effects of temperature, phase transformation and stress, the hardness field is introduced into the model. The main factors causing distortion, such as the process and performance, are associated with the distortion, and an analysis and calculation model of the carburizing and quenching process and its distortion is established. The numerical simulation results show that when the cooling rate increases from 1800 W/m · K to 3600 $W/m^2 \cdot K$, the deformation increases sharply from 10^{-1} mm/m to 1 mm/m. Continuing to increase the cooling rate from 3600 $W/m^2 \cdot K$ to 7200 $W/m^2 \cdot K$, the deformation does not further increase and remains at 1 mm/m. The influence of constraint conditions on the deformation of typical parts during vacuum carbon nitriding was studied. It was found that reasonable setting of constraints is beneficial for reducing deformation. The influence of flow field uniformity on the deformation of typical parts during vacuum carbon nitriding was studied. It was found that, according to the shape of the gear, reasonable utilization of the non-uniform flow field is beneficial for reducing deformation. In addition, the deformation after vacuum carbon nitriding of shaft gear parts was compared when using the traditional quenching process and the equal Biot synchronous quenching process respectively. The study found that, compared with the traditional quenching process, the micro-deformation control technology of equal Biot synchronous quenching reduced the radial displacement of 20Cr2Ni4 gears from 0.20-0.24 mm to 0.07-0.09 mm.

Keywords: carburizing; quenching; heat treatment; finite element method; near-zero quenching distortion

The Improvement of Bending Fatigue Properties of Carburized Gears Based on Multi-Dimensional Synergistic Strengthening Mechanism

Long Bangzhuan[1], Li Lingze[1], Zhang Yanxiang, Shi Lubing[2], Yan Mufu[1*]

[1] School of Materials Science and Engineering, Harbin Institute of Technology, Harbin, Heilongjiang 150001, China

[2] China Academy of Machinery ZhengzhouRescarch Institute of Mechanical Engineering Co., Ltd, Zhengzhou, Henan 450052, China

E-mail address: 13206523735@163.com (Long Bangzhuan)

E-mail address: yanmufu@hit.edu.cn (Yan Mufu)

Abstract: As the core component of the transmission system, the gears continue to withstand periodic impact loads and alternating bending stress during service. After long-term service, bending fatigue microcracks are easily generated on the surface of the tooth root and subsurface. Statistics show that more than 80% of mechanical failures are caused by fatigue failure, so the bending fatigue resistance of gears is always a key evaluation indicator of structural design. This study conducted a comparative experiment of rare earth equivalent carburizing (Re^eC) and traditional gas carburizing (GC) on carburized gears. Combined with the carburized layer gradient microscopy, organization characterization hardness gradient and residual stress field distribution, to systematically explain the multi-dimensional synergistic strengthening mechanism of RE equivalent carburizing treatment on bending fatigue performance, and the following conclusions are as follows: The Re^e carburizing induces dual-gradient refinement of both grain microstructure and carbides within the effective carburized layer of the gear. The ultrafine martensitic matrix with an average grain size of 0.82 μm forms at the carburized layer surface, accompanied by the dispersion of submicron spherical carbides, and the sizes of the two change continuously in a gradient along the depth direction of the effective carburized layer. This microstructure consisting of ultrafine grains and submicron carbides increases the microhardness to 842 $HV_{0.1}$ and forms a gradient compressive residual stress field in the carburized layer with a peak value of −667 MPa and a width of 1000 μm. Compared with gas carburized gears, the ultimate bending fatigue strength of equivalent Re^e carburized gears is increased by 33% to 616 MPa, and the fatigue life is increased by an order of magnitude. The enhanced bending fatigue resistance originates from the multi-dimensionally synergistic effect of crack initiation retardation by high gradient residual compressive stress, crack energy dissipation via ultrafine-grained microstructure, and fatigue striation segmentation through submicron spherical carbide pinning effects. This synergistic strengthening mechanism— stress regulation, microstructure refinement and second-phase strengthening— establishes an optimal surface condition for subsequent shot peening.

Keywords: gradient residual compressive stress field; submicron spheroidal carbide; ultrafine grain microstructure; synergistic strengthening; bending fatigue performance

Numerical Simulation of Carbide Precipitation in Carburized Surface Layer and Dual-Gradient Quenching Distortion for 18CrNiMo7-6 Steel Gears

Wang Hao, Zhao Huihao, Zhang Yanxiang, Yan Mufu*

School of Materials Science and Engineering, Harbin Institute of Technology, Harbin, China

E-mail address: 1075453920@qq.com (Wang Hao)

E-mail address: yanmufu@hit.edu.cn (Yan Mufu)

Abstract: Carburizing-quenching is a critical technology for gear strengthening, where dispersed carbides precipitation in the carburized layer enhance matrix strength and fatigue life, while quenching distortion control determines machining precision and fatigue performance. This study employs phase-field method and finite element method to simulate the carburizing combined with dual-gradient quenching process for 18CrNiMo7-6 steel, elucidating precipitation evolution of fine, dispersed carbides during carburizing, as well as the microstructural evolution and deformation patterns during quenching. Ultimately, synergistic control over the carburizing-quenching microstructure and quenching distortion is achieved by optimizing process parameters. The results demonstrate that during carburizing, the diffusion of Cr in the matrix governs the carbide coarsening rate, while the atmosphere carbon potential and matrix carbon content jointly determine the carbide precipitation depth. The phase-field simulation results incorporating interfacial mass transfer demonstrate excellent agreement with experimental observations. Compared to quenching at 880 ℃, the dual-gradient quenching process reduces the maximum deformation by 67.35% and increases the surface hardness from 795.82 HV to 812.47 HV (+2.1%). Compared to quenching at 790 ℃, the maximum deformation after dual-gradient quenching is only slightly higher, while the core hardness increases from 415.68 HV to 429.94 HV (+3.4%). This process successfully realizes the synergistic matching of high-hardness surface (>812.47 HV) and tough core, ultimately accomplishing near-zero quenching distortion.

Keywords: carburizing; quenching; heat treatment; phase-field method; finite element method; near-zero quenching distortion

A Generalized Marcus Theory for Electron Transfer Reactions

Zhang Yanxiang*, Yan Mufu*

Harbin Institute of Technology, Harbin, Heilongjiang, China

E-mail address: hitzhang@ hit. edu. cn (Zhang Yanxiang)

Abstract: The current-voltage relationship is a fundamental topic in the corrosion science of steels and beyond, but is usually not understood clearly by the present theoretical models. In practice, the Butler-Volmer model and the Marcus theory for electron transfer usually can not reproduce the experimental current-voltage relationship, especially under large overpotentials. Here we report a new approach on the basis of the Marcus theory, and show that the approach is quite universal that can reproduce preciously the current-voltage relationships in the whole range of overpotentials in many systems such as corrosions, electrocatalysis, fuel cells, and bimolecular reactions. By this approach, the information for the fundamental processes can be obtained from the current-voltage relationship.

Keywords: corrosion; metals; surface modification

Dual Martensite Layers Prepared byalternate Carburizing and Nitriding Treatment

Li Lingze, Yan Fuyao, Zhang Yanxiang, Yan Mufu*
School of Materials Science and Engineering, Harbin Institute of Technology, Harbin, Heilongjiang 150001, China
E-mail address: 23b909051@stu.hit.edu.cn (Li Lingze)
E-mail address: yanmufu@hit.edu.cn (Yan Mufu)

Abstract: This paper addresses the problems of low case hardness and difficulty in synergistically improving comprehensive properties in traditional carburizing and carbonitriding processes. It proposes an innovative variable-frequency cyclic carbon-nitrogen dual-infiltration treatment technology. This technology successfully prepared high-performance alternate Carburizing and Nitriding layers on 18Cr2Ni4W steel by dynamically and periodically regulating temperature, atmosphere, and atmospheric pressure. Using techniques such as scanning electron microscopy (SEM), transmission electron microscopy (TEM), X-ray diffraction (XRD), and nanoindentation, the evolution of the microstructure of the infiltrated layer and its influence mechanism on mechanical properties were systematically studied. The results show that the microstructure of the alternate Carburizing and Nitriding layers consists of a fine lath martensite matrix (lath width < 100 nm), dispersed carbon/nitrogen compounds, and a small amount of retained austenite. This fine lath martensite is a "dual-phase martensite structure ($\alpha'_C + \alpha'_N$)". Crucially, a gentle carbon concentration gradient distribution was achieved throughout the carburizing and nitriding duplex layer. The microstructure was significantly refined, with prior austenite grain size controlled to within 10 Mμm. High dislocation density was observed within the martensite laths. This alternate Carburizing and Nitriding layer structure significantly increased the surface hardness of the infiltrated layer to over 880 HV_1. The cross-sectional hardness distribution showed a wide "high hardness plateau region", effectively eliminating the "head-down phenomenon" in traditional carburizing treatments. Wear resistance was greatly improved, with the friction coefficient reduced to between 0.3-0.4, and the wear rate decreased by an order of magnitude compared to conventional carburizing.

Keywords: Alternate Carburizing and Nitriding treatment; dual-phase martensite structure; carbon/nitrogen compounds; high hardness

Proceedings of IFHTSE 2025

The 30th World Congress of International Federation for Heat Treatment and Surface Engineering
August, 19-22, 2025, Suzhou, China

Research Progress on Surface Strengthening Technology of Heavy-Duty Gears with High Fatigue Resistance

Wang Xiaonan*, Chen Qiyuan, Shen Xinjun, Zhang Qingyu
School of Iron and Steel, Soochow University, China
E-mail address: wxn@suda.edu.cn (Wang Xiaonan)
E-mail address: qychen1992@suda.edu.cn (Chen Qiyuan)

Abstract: High-performance gears demand exceptional resistance to wear and contact fatigue. This study systematically investigates the influence of microstructure, carbon concentration, and hardness distributions on these critical properties in deep-carburized 18Cr2Ni4W gear steel through experiments in conjunction with finite element simulations. Initial experiments focused on conventional carburizing at temperatures of 930 ℃, 950 ℃, and 970 ℃. Results confirm that surface carbon concentration and hardness increase progressively with higher carburizing temperatures. Tensile tests on the carburized surface layers revealed impressive strengths exceeding 2000 MPa across all temperatures; however, a significant reduction in ductility was observed as temperature increased. Sliding wear tests demonstrated that the sample carburized at 970 ℃ exhibited the lowest wear rate compared to those processed at 930 ℃ and 950 ℃. Analysis identified the primary wear mechanisms as adhesive oxidation wear coupled with abrasive wear. Crucially, energy dispersive spectroscopy mapping revealed the most pronounced aggregation of oxygen on the counterface opposing the 930 ℃ carburized sample, indicating the most severe adhesive oxidation wear in this case. Supporting finite element simulations of transient temperature fields during sliding contact predicted a maximum contact point temperature approximately 50 ℃ lower for the 970 ℃ sample compared to the 930 ℃ sample. This significant temperature reduction effectively mitigates oxidation severity. Consequently, the enhanced wear resistance at 970 ℃ is attributed to the synergistic effect of an increased volume fraction of hard carbides within the case and the lower frictional heating suppressing oxidation. Building upon these findings, a novel two-stage heat treatment combining deep carburizing followed by carbonitriding was specifically designed to further elevate the performance limits of 18Cr2Ni4W steel. Microhardness profiling showed that conventional carburizing yielded a surface hardness of 671 HV1 and a core hardness of 465 HV1. In striking contrast, the novel carburizing-carbonitriding process significantly increased the surface peak microhardness to 748 HV1 (representing an ~11% improvement) while slightly reducing the core hardness to 440 HV1 (a ~6% decrease). The wear rate with novel carburizing-carbonitriding heat treatment is the lowest, with the decrease in wear rate from 8.73×10^{-7} $mm^3 \cdot N^{-1} \cdot mm^{-1}$ to 5.89×10^{-7} $mm^3 \cdot N^{-1} \cdot mm^{-1}$, corresponding to a 33% improvement in wear resistance. The cycle time of contact fatigue stress increases from 1.60×10^7 to 3.91×10^7, an increase of 144%. Compared to conventional carburizing process, the carburizing-carbonitriding heat treatment significantly increases carbon concentration within the surface layer. The higher carbon concentration promotes the formation of carbides, and the proportion of large-sized carbides increases. The interstitial nitrogen atoms induce potent solid solution strengthening within the martensitic matrix, significantly increasing its intrinsic microhardness and thereby contributing substantially to the improved resistance against both wear and fatigue crack initiation/propagation. The synergistic combination of refined carbide distribution and nitrogen-strengthened martensite underpins the exceptional gains in both wear resistance and contact fatigue life. This work demonstrates that optimizing carburizing temperature effectively enhances wear resistance by controlling carbide formation and minimizing oxidation through reduced frictional heating. More significantly, the newly developed carburizing-carbonitriding heat treatment offers a highly effective strategy for dramatically improving the surface integrity and service longevity of critical gear components made from 18Cr2Ni4W steel, achieving substantial simultaneous improvements in both wear resistance and contact fatigue performance.

Keywords: Heavy-duty gears; carburizing process; carbonitriding process; microstructure; wear resistance; fatigue resistance

Two-Layer Nanostructuring of AISI316/316L Stainless Steels for Medical and Special Tooling via Low temperature Plasma Nitriding

Tatsuhiko Aizawa

Surface Engineering Design Laboratory, Shibaura Institute of Technology Ota, 144-0056, Tokyo, Japan

E-mail address: taizawa@ sic. shibaura-it. ac. jp (Tatsuhiko Aizawa)

Abstract: High density plasma immersion nitriding was developed for low temperature surface treatment of metals and alloys. The austenitic stainless steels in series of AISI316 and AISI316L were plasma nitrided at 673 K and 623 K to have a nano-grained, two-phase surface layer. In these low temperatures, no chromium and ion nitrides were synthesized as a precipitate in the surface layer. It was massively nitrogen supersaturated (MNSed) to induce the expansion of g-phase supercells via nearly full occupation of vacancy sites by nitrogen solutes. This expansion in the MNSed layer also induces the plastic straining just below the MNS front in the matrix to preserve the strain compatibility through the MNS front. After precise EBSD analyses, this plastic straining was represented as dense slip-line field with high KAM profile. Two-phase structure analyzed by EBSD was in correspondence to two cluster system, where nitrogen rich phase with high chromium content is neighboring within 5 nm range to nitrogen poor phase with high iron and nickel contents. Two applications were explained using this two-layer nanostructured AISI316 and AISI316L by MNS. The MNSed AISI316L substrate was machined and finished by using the PCD-chipped cutting tool to yield a mold for hot-stamping the oxide glass preform to optical lenses and elements. The MNSed AISI316 rod had a two-layered microstructure in similar manner to the Japan sward where the nanostructured surface layer via MNS is continuously connected to the ductile matrix of AISI316. A proto type of micro-surgery knife was produced by this treatment.

Keywords: low temperature plasma nitriding; massive nitrogen supersaturation; multi-dimensional microstructure; two-phase layered structure; hot-stamping mold for oxide glasses; micro-surgery knife

Innovative Materials Solutions with Thermochemical Surface Engineering of Steels

Marcel A. J. Somers
Technical University of Denmark
E-mail address: somers@ ipl. dtu. dk (Marcel A. J. Somers)

Abstract: Thermochemical surface engineering of steels provides a sheer endless source for innovative materials solutions. In this lecture a cornucopia of materials solutions relying on low temperature surface hardening (LTSH) and high temperature solution nitriding (HTSN) is presented, in some cases in combination with additive manufacturing. The following examples are included: —LTSH of conventional and additively manufactured stainless steels by carburizing; Simultaneous LTSH and bulk hardening of maraging steel and precipitation hardening stainless steel; HTSN for improving austenite stability and preventing strain-induced martensite formation during severe deformation; HTSN of duplex stainless steels for improved resistance against crevice corrosion; Combatting white-etch cracking in bearings with HTSN and eutectoid decomposition to obtain a fine-grained martensitic case. For the examples given it is demonstrated that a deep understanding of materials microstructure development and its relation to materials performance is a prerequisite for successful innovation.

Keywords: surface engineering; high temperature solution nitriding; carburizing; bulk hardening

Research on Machine Learning Design of Heat Treatment Process for High-Performance Aluminum-Lithium Alloys

Liu Yan, Chen Junfeng*, Deng Chuangseng, Zou Linchi

School of Materials Science and Engineering, Fuzhou University, Qishan Campus, Fuzhou, China

E-mail address: 221810012@fzu.edu.cn (Liu Yan)

E-mail address: chenjunfeng@fzu.edu.cn (Chen Junfeng)

Abstract: This study focuses on the machine learning design of heat treatment processes for high-performance aluminum-lithium alloys. Aiming at the high cost and long cycle of traditional trial-and-error methods in alloy process design, a design approach combining data-assisted modeling and genetic algorithm (GA) optimization of BP neural networks is proposed. The study constructs a dataset using literature data and accumulated data from the research group. Through GA optimization of BP neural networks, relationship models between process parameters-elongation (EL) and process parameters-ultimate tensile strength (UTS) are established. The results show that both the EL model and UTS model exhibit satisfactory performance prediction accuracy, with data predominantly concentrated within the 95% confidence interval. Based on the models, target properties are designed to optimize heat treatment process parameters. After experimental verification, new performance data are iteratively added to the dataset for new-round screening, achieving the enhancement of the alloy's mechanical properties. This research validates the feasibility of data-assisted modeling combined with genetic algorithm optimization for the rapid design of aluminum-lithium alloy process parameters, providing a new approach for heat treatment process optimization of similar materials and holding significant significance for improving the comprehensive properties of alloys.

Keywords: machine learning; heat treatment; aluminum-litium alloys; mechanical properties

Ion Bombardment Heat Treatment of High-Speed Heavy-Duty Gears

Lu Jinsheng, Xu Hongxiang

Zhengzhou Transmission Technology Co., Ltd. No. 56, Lumei Street, Zhengzhou

E-mail address: lujinsheng2008@126.com (Lu Jinsheng)

E-mail address: 13838005005@163.com (Xu Hongxiang)

Abstract: To explore ways to enhance the load-carrying capacity of nitrided gears, common nitriding gear steels 42CrMo, 40CrNiMo, and 25Cr2MoV. Ion bombardment nitriding was applied to achieve different nitrided layer depths and white bright layer microstructures based on varying core hardness levels. Roller contact fatigue tests and analyses were conducted. The results indicated that increasing the core hardness from 240-260 HBW to 310-330 HBW could enhance the fatigue limit by 30%. the nitriding layer depth increased from 0.5 mm to 0.8 mm, resulting in a 25% increase in contact fatigue limit, and the surface γ' phase compared to the ε phase showed a 40% increase in contact fatigue limit. Gear bench tests were conducted using an optimized deep-penetration controlled ion nitriding process. Compared to carburized gears, the contact fatigue strength increased by 10-15%, the bending fatigue strength was comparable, and the resistance to adhesion was significantly higher than that of carburized gears. Under certain module values and application conditions, nitrided gears exhibit advantages in terms of mechanical properties, heat treatment distortion, and manufacturing costs. They have achieved widespread engineering applications in high-speed, heavy-duty gears.

Keywords: high-speed heavy-duty gears; plasma heat treatment; nitrided gears; Gear strength

Research on the Residual Stress Distribution and Fatigue Performance of Carburizing, Nitriding and Carbo-nitriding

Gao Yukui

Tongji University, 201804, School of Materials Science and Engineering 4800, Caoan Road, Jiading, SH, China

E-mail address: yukuigao@ tongji. edu. cn (Gao Yukui)

Abstract: Chemical heat treatment is a widely used surface modification technique that can significantly improve the wear, corrosion and fatigue properties of components. However, due to non-uniform temperature distribution during the chemical heat treatment process, residual stress is generated on the material surface, which may adversely affect the performance and reliability of the components. In order to elucidate the correlation between key chemical heat treatment parameters, such as carburizing, nitriding, and carbo-nitriding, the resulting residual stress distribution and fatigue performance on the surfaces, this report investigates the variation patterns of carbon, nitrogen, and other elemental concentrations as well as microstructural changes along the depth direction following treatments under various process conditions. Several residual stress measurement techniques are introduced and applied to characterize the residual stress distribution on the treated surfaces. Furthermore, the fatigue properties of the treated materials are evaluated, and the underlying mechanisms by which chemical heat treatment parameters influence fatigue performance are clarified. Based on this analysis, an intrinsic relationship framework of "chemical heat treatment-microstructure distribution-residual stress-fatigue properties" is established.

Keywords: chemical heat treatment; carburizing; nitriding; carbo-nitriding; residual stress; fatigue

Effects of Theprior Forming Process on Nitrocarburized Low-Carbon Steels

Natnaree Senaweenin[1*], Nipon Taweejun[1], Teerapat Mungwattana[1], Torranin Chairuangsri[2], John TH. Pearce[2], Sankum Nusen[2]

[1] ThaiTohken Thermo Co. Ltd, Chonburi 20160 Thailand

[2] Department of Industrial Chemistry, Faculty of Science, Chiang Mai University, Chiang Mai 50200 Thailand

E-mail address: natnaree@ thaitohken. co. th (Natnaree Senaweenin)

E-mail address: nipon@ thaitohken. co. th (Nipon Taweejun)

Abstract: It is well-known that preparation of surfaces before surface treatment of steels is necessary. If there is dirt, rust or oxide contamination on the surface of steels, it may affect nitrogen or carbon penetration, and the properties of the treated surface. Some prior processes before surface treatment, such as hot forming or hot rolling, have been applied to particular steel grades leading to a possible formation of oxide layers and some research has reported that pre-oxidation before nitrocarburizing of steels may form a thicker compound layer with more uniform and higher hardness. Cold-rolled, low-carbon steel (SPCC) and hot-rolled, low-carbon steel (SPHC) have similar compositions, but were produced by different forming processes prior to nitrocarburizing. The hot-rolling process can form oxide layers on steel surfaces, while the cold-rolling process may not. Therefore, in the present study, it is of interest to study effects of the prior-forming process on nitrocarburized low-carbon steels. Nitrocarburizing at 570 ℃ for 2. 5 hours in a controlled atmosphere of ammonia and endothermic gas in the ratio 8:7 was performed on cold-rolled low-carbon steel (SPCC) and hot-rolled low-carbon steel (SPHC) workpieces, both contained around 0. 040-0. 043 wt. % C. These steels have similar compositions, but were produced by different forming processes prior to the nitrocarburizing process. The aim was to study the hot-and cold-rolled external surfaces, the thickness and nature of iron-oxide surface layers, as well as their effects on the nitrocarburized compound layers. Vickers microhardness testing at 100 g load was used to measure the surface hardness of all the test samples. Nano-hardness testing at 1 mN load was used to measure the depth-hardness profile. Optical microscopy (NIKON DS-Ri1, NIS-Elements software) was used to measure the thickness of the compound layers. Scanning electron microscopy (SEM) and wavelength dispersive X-ray spectroscopy (JEOL, JXA-ISP100) were used to investigate the microstructure and porosity in each layer, as well as the elemental distribution. Optical microscopy revealed that the microstructure at the surface of the as-received SPHC has a top oxide layer formed by hot rolling, 6. 59 μm thick, while that of the as-received SPCC has not. After nitrocarburizing, the surface hardness of the nitrocarburized SPCC is HV(0. 1) 696, higher than that of the nitrocarburized SPHC (HV(0. 1) 651). However, the compound-layer thickness of SPCC is 27. 1 μm, lower than that of SPHC (27. 8 μm). The compound-layer in the case of SPHC is underneath the prior oxide layer and it is believed that the presence of the prior oxide layer led to the lower surface hardness of the nitrocarburized SPHC. From SEM and WDS results, carbon aggregation was found in the prior oxide layer of the nitrocarburized SPHC. From the Ellingham diagram, direct reduction of iron oxides by carbon atoms is possible at temperatures higher than about 250 ℃. In the present experiment, the nitrocarburizing temperature was 570 ℃, therefore carbon atoms could directly reduce iron oxides to iron metal and react with oxygen forming carbon monoxide (CO) or carbon dioxide (CO_2), which caused high porosity in the iron-oxide layer and less carbon atoms diffusing into the iron matrix of the nitrocarburized SPHC. These possible phenomena led to a lower surface hardness of the nitrocarburized SPHC. On the other hand, nitrogen atoms did not react with iron-oxides and hence nitrogen atoms can penetrate through the iron-oxide layer and diffuse into the iron matrix forming a compound layer underneath the prior oxide layer. In the SPCC, there is no such iron-oxide layer that would cause carbon aggregation, so carbon could diffuse into the iron matrix, causing higher carbon content in the epsilon phase of the nitrocarburized SPCC. From the hardness profile, the hardness of the epsilon phase containing a higher amount of carbon in the nitrocarburized SPCC was harder than that of the epsilon phase containing lower carbon in the nitrocarburized SPHC, the latter having a surface hardness near to the typical level of the gamma prime phase. The prior oxide layer may also retard penetration of nitrogen atoms so that the compound layers in SPHC contained more gamma prime phase (less nitrogen content and hardness than epsilon phase). In conclusion, the present study showed that the oxide layer formed by prior hot rolling of the low-carbon steels can lead to a lower surface hardness after nitrocarburizing due to high porosity, low carbon content in the epsilon phase of the compound layers and high amount of the gamma prime phase in the compound layers.

Keywords: nitrocarburizing; low carbon steels; prior oxide layer; surface hardness; compound layers

Research on Ion Permeation for Thin-Walled Parts

Chen Shengyi[1], Zhu Wei[1], Long Liwei[1], Luo Defu[2*]

[1] Sichuan Initial Metal technology Co., Ltd., 610100, Department of Research and Development, Longquanyi District, Chengdu, Sichuan Province, China

[2] XiHua University, 611700, School of Materials Science and Engineering, Pixian District, Chengdu, Sichuan Province, China

E-mail address: 23498192@qq.com (Chen Shengyi)

E-mail address: cd_df@126.com (Luo Defu)

Abstract: The ion permeation was applied to the thin-walled 17Cr2Ni10Co14Mo steel, and the impact of varying compound layer thicknesses on the mechanical properties of this steel was analyzed. The results demonstrate that after ion permeation treatment, the surface hardness of the test steel can exceed 700 HV0.1; with the increase in compound layer thickness, the tensile strength, yield strength, elongation, and impact energy of the test steel exhibit a declining trend, with a more pronounced decrease observed as the proportion of compound layer thickness to the sample thickness increases. In addition, the fatigue resistance of nitride-treated samples is significantly enhanced, peaking when the compound layer thickness reaches 0.5% of the specimen thickness. The fracture surfaces of both tensile and impact tests reveal that the nitrided layer area exhibits brittle fracture characteristics, while the base material maintains ductile fracture properties.

Keywords: ion permeation; thin-walled parts; mechanical property

Underlying Mechanisms of La-mediated Carbon Redistribution at Bcc Iron Grain Boundaries

Chen Baofeng[a,b], Wang Hao[b], Li Xinlei[b], Zhang Wenliang[a], Li Xianjun[a], Jiang Chao[a*], Yan M. F.[b*]

[a] China Academy of Machinery Beijing Research Institute of Mechanical & Electrical Technology Co., Ltd., Beijing, China

[b] National Key Laboratory for Precision Hot Processing of Metals, School of Materials Science and Engineering, Harbin Institute of Technology, Harbin Heilongjiang, China

E-mail address: jiangchao@139.com (Jiang Chao)

E-mail address: yanmufu@hit.edu.cn (Yan M. F.)

Abstract: This study employs first-principles calculations to elucidate the atomic-scale mechanisms by which lanthanum (La) doping regulates carbon diffusion kinetics and redistribution at Fe Σ3(111)[110] grain boundaries (GBs). We systematically investigate the preferential substitution behaviour of La atoms at GB sites, their impact on carbon migration energetics, and the resulting interfacial electronic interactions. Results reveal that La atoms thermodynamically stabilize at substitutional GB positions, inducing lattice expansion and generating a repulsive chemical environment for carbon. This environment asymmetrically modulates carbon diffusion barriers: the energy barrier for carbon migration from grain interiors to GBs increases by 20%, while the reverse process (GB-to-interior diffusion) exhibits a 10% reduction. Such dual modulation establishes a thermodynamic driving force that expels carbon from GB segregation zones and promotes its redistribution into grain interiors. Electronic structure analysis demonstrates that the repulsive La-C interaction arises from orbital hybridization and charge redistribution, creating a chemical potential gradient that drives carbon dispersion. This mechanism directly explains experimental observations of suppressed intergranular carbide coarsening and the formation of refined, dispersed carbides in La-doped carburized steels. Furthermore, La doping alleviates interfacial stress concentrations by mitigating carbon segregation while preserving metallic bonding continuity at GBs, potentially enhancing both strength and ductility. The work establishes a paradigm for solute interaction-driven GB engineering, highlighting the critical role of rare earth (RE) elements in reconstructing interfacial diffusion kinetics. By correlating atomic-scale electronic interactions with macroscopic solute redistribution, this study bridges the knowledge gap regarding the catalytic effects of RE elements at GBs. The findings provide fundamental insights into designing advanced carburized steels/layer with optimized carbide distributions, where RE-mediated suppression of GB segregation enables synergistic improvements in mechanical properties. Future work should extend this framework to other RE systems and explore synergistic doping strategies combining La with carbide-forming elements to further enhance precipitate dispersion and interfacial stability. Integrating computational predictions with in situ experimental validation will advance the development of high-performance materials for extreme service conditions in aerospace and energy applications.

Keywords: grain boundary; rare earth doping; carbon diffusion; first-principles calculations; carbide refinement

Application of Low Temperature Gas Nitrocarburising of 316L Austenitic Stainless Steel in Hydrogen Energy Equipment

Li Dan

Qingdao Fengdong Heat Treatment Co. , Ltd. , Qingdao, China

Abstract: The cylinder sleeve material of ultra-high pressure hydrogen compressor requires high hardness, wear resistance, hydrogen embrittlement resistance, hydrogen corrosion resistance, and low surface roughness. This paper uses the low-temperature gas nitrocarburising process to modify the surface performance of 316L stainless steel cylinder sleeves, which effectively solves the problem of low service life of untreated 316L cylinder sleeve. This process has been applied in the production of ultra-high pressure hydrogen compressors.

Keywords: 316L austenitic stainless steel; low temperature gas nitrocarburising; cylinder sleeve; ultra-high pressure hydrogen compressor

Influence of Temperature on Surface Properties of DLC/Carbonitride Composite Prepared on Pre-nitrided M50 Steel by Plasma Carburizing

Yao Jiawei[1], Cong Peiwu[1], Lu Wenlin[1], Du Chunhui[1], Yan Mufu[2*], Ma Jingbo[1*]

[1] China Academy of Machinery Beijing Research Institute of Mechanical & Electrical Technology Co, Ltd., Beijing, China

[2] School of Materials Science and Engineering, Harbin Institute of Technology, Harbin, China

E-mail address: yanmufu@hit.edu.cn (Yan Mufu)

E-mail address: majingbo1991@163.com (Ma Jingbo)

Abstract: M50 steel is widely used in the manufacture of high-end bearing steel for aero-engine shafts, which needs excellent surface performance to meet harsh service conditions. In this study, a self-lubricating co mposite layer was in-situ fabricated on pre-nitrided M50 steel through plasma carburizing at different temperatures varying from 410 to 570 ℃ to investigate the response of the microstructure and mechanical properties to carburizing temperature. The composite layer was fully characterized by optical microscopy, scanning electron microscopy, transmission electron microscopy, grazing incidence X-ray, electron microprobe analysis, Nano-indentation tester, Vickers hardness tester, and Pin-on-disc tribometer. The results revealed that the microstructure and mechanical properties of the self-lubricating composite layer significantly respond to carburizing temperatures. The composite layers were composed of carbon-rich film, dense transitional layer, and diffusion layer in-depth direction. The carbon-rich film with nano-domains contains a large amount of sp^3 bonds at low carburizing temperatures due to the generation of Fe_5C_2 and exhibits DLC character. The composite layer carburized at 490 ℃ owns remarkable hardness and wear resistance due to the lubricating effect of the carbon-rich film with DLC features.

Keywords: Diamond-like carbon; gradient structure; plasma carburizing; plasma nitriding, tribology

Numerical Simulation and Experimental Study on Austempering of 18CrNiMo7-6 Case Hardening Steel

Qin Shengwei, Shao Jinhui

School ofMechanical and Power Engineering, Zhengzhou University, China

E-mail address: qinsw@ zzu. edu. cn (Qin Shengwei)

E-mail address: shaojh2001@ 163. com (Shao Jinhui)

Abstract: Aiming at the key scientific issues in the austempering process of carburized gears, this study designed three test steels with carbon contents of 0. 21wt. %, 0. 39wt. %, and 0. 8wt. % to obtain simulation parameters through experiments. A multi-field coupling model of carbon diffusion-phase transformation-stress was constructed based on Deform-HT software, elucidating the regulatory mechanism of carbon concentration gradient in the carburized layer on bainite nucleation/growth kinetics. The influence of different austempering temperatures on hardness gradient and residual stress was systematically studied. The results show that with the increase of austempering temperature: (1) The hardness significantly decreases from 683HV to 437HV, attributed not only to the increase of soft-phase austenite but also to grain coarsening; however, the effective hardened layer depth of the sample increases by 21% (from 0. 95 to 1. 15 mm). (2) When the quenching temperature rises from 200 ℃ to 360 ℃, the maximum compressive stress on the surface first increases and then decreases, reaching the maximum value of-428. 64 MPa at 240 ℃. The multi-field coupling simulation by DEFORM-HT shows that the average deviation between simulation and test results is controlled within 8%, verifying the reliability of the material parameter system.

Keywords: 18CrNiMo7-6 alloy steel; austempering; DEFORM-HT; numerical simulation

Hollow Cathodic Plasma Source Modification of Titanium Alloy

Li Yang[1], He Yongyong[2]

[1] School of Nuclear Equipment and Nuclear Engineering, Yantai University, Yantai, China

[2] State Key Laboratory of Tribology, Tsinghua University, Beijing, 100084, China

E-mail address: metalytu@163.com (Li Yang)

E-mail address: heyy@mail.tsinghua.edu.cn (He Yongyong)

Abstract: With a high strength-to-weight ratio, good biocompatibility, and corrosion resistance, titanium and titanium alloys are essential materials for aerospace manufacturing, biomedical implants, and automotive products. However, their poor tribological properties often limit titanium and titanium alloys. In this work, a new structure consisting of hollow cathode metal plate can produce a hollow cathode discharge and heat the workpiece by radiation. This principle is used to perform hollow cathode source plasma nitriding (HCSPN). The key findings of this study are as follows: (1) Electron beam melting (EBM) titanium alloy were strengthened by carrying out a hollow cathodicplasma source nitriding treatment. The nitrided layers formed on the EBM-TC4 alloy were identified as TiN, Ti2N, and nitrogen-stabilized α(N)-Ti phases. Potentiodynamic polarization and electrochemical impedance spectroscopy measurements were carried out to evaluate the electrochemical behavior of the samples in Hank's balanced salt solution. The relationship between the corrosion resistance of the nitrided layer and its micro-structure was explored. Thus, the improved properties of the nitrided alloys can be attributed to the formation of a Ti-N layer composed of an outermost nanocrystalline/amorphous TiN layer and a crystalline TiN sub-layer on the top of the Ti2N crystal layer. The novel Ti-N layer structure with improved corrosion resistance provides broad opportunities for the application of commercial Ti alloys and improving their durability. (2) Ti-N-O layers were fabricated on the surface of TC4 alloy using a two-step hollow cathode plasma source oxynitriding technique. This resulted in the formation of TiN, Ti2N, TiO2, and nitrogen-stabilized α(N)-Ti phases on the TC4 alloy, forming a Ti-N-O modified layer. The microhardness of the samples treated with plasma oxynitriding (PNO) was found to be 300-400% higher than that of untreated (UN) samples. The experimental conditions were set at 520 ℃, and the corrosion current density of the PNO sample was measured to be 7.65×10^{8} A/cm^2, which is two orders of magnitude lower than that of the UN sample. This indicates that the PNO-treated TC4 alloy exhibited significantly improved corrosion resistance in the artificial saliva solutions. (3) It prepared the compound layers and coatings with the same thickness on TA2 pure titanium by plasma nitriding and multi-arc ion plating. The results reveal that nitrided and coating samples showed the friction coefficients (0.11-0.12) were significantly decreased compared to that (~0.33) of the TA2 pure titanium matrix. It was also found that the wear mechanism is different after nitriding or coating, which is caused by the surface phases, chemical composition, and surface roughness under other processes.

Keywords: Surface modification; hollow cathode plasma nitriding; titanium alloy; corrosion; wear

Particle-in-Cell/Monte Carlo Collision Simulation of Carbon Species Dynamics in Magnetron Sputtering for DLC Film

Ma Yiming, Cong Peiwu, Lu Wenlin, Ma Jingbo, Yao Jiawei, Lu Wenlin*

China Academy of Machinery Beijing Research Institute of Mechanical & Electrical Technonlogy CO., Ltd, Beijing, China

E-mail address: mymbs0@163.com (Ma Yiming)

E-mail address: 1361054310@qq.com (Lu Wenlin)

Abstract: This study presents a two-dimensional Particle-in-Cell/Monte Carlo Collisional (PIC/MCC) simulation to explore the behavior of carbon species during the direct current magnetron sputtering (DCMS) process for depositing diamond-like carbon (DLC) films. Due to the remarkable properties, such as a low friction coefficient, high hardness, and excellent anti-adhesion characteristics, DLC films have been extensively used in various industrial sectors, especially in improving the surface performance of cutting tools. Among the numerous factors influencing the quality of DLC films, the energy and flux of carbon species, particularly carbon ions, generated during sputtering significantly influence determining film density, hardness, and deposition rate. PIC/MCC provides an observational means for tracking the movement and distribution of charged and neutral particles in complex electromagnetic fields It is especially suitable for modeling the transport of sputtered carbon species from the graphite target to the substrate. In this research, we employ the PIC/MCC approach to delve into the dynamics of carbon species and to explore the influence of Different process parameters and magnetic field distribution on their behavior. The primary aim is to regulate and optimize the transport of carbon species to enhance film quality and deposition efficiency. In the collision process, the electron correlation collision is mainly considered. To improve the physical fidelity of the simulation, the collision model encompasses not only elastic and ionizing collisions but also excitation and de-excitation processes involving metastable carbon states: $C(2s^2 2p^{2\,1}D)$, $C(2s_2 2p^{2\,1}S)$, and $C(2s^2 p^{3\,5}S^{\circ})$. The de-excitation cross sections are ascertained based on the principle of detailed balance, ensuring thermodynamic consistency in the reverse processes. At the same time, it also includes the ionization of carbon atoms and carbon metastable particles. The sputtering process is caused by argon ions and carbon ions, and carbon atoms and secondary electrons are generated. The motion of charged particles is actuated by electromagnetic forces. The three-dimensional magnetic field model was obtained by simulating the actual magnetron sputtering target structure. It is then integrated into the two-dimensional PIC/MCC model to determine the spatial distribution of the magnetic field. The electric field is calculated by solving the Poisson equation, which self-consistently couples the externally applied circuit voltage with the space charge distribution within the plasma. Due to the complex distribution of the electromagnetic field, the Boris method is employed to compute particle motion and enhance computational efficiency. Based on this model, this work systematically investigates the influence of varying magnetic field distributions, electric field conditions, and working gas pressure on the motion state, spatial distribution, and flux of carbon particles. The research results shed light on key physical mechanisms regulating particle transport and identify potentially improved discharge conditions for achieving enhanced film performance. Moreover, the simulation provides insights for the structural optimization of the magnetron sputtering target to enhance process uniformity and deposition efficiency. This research shows that PIC/MCC-based modeling, when combined with a detailed analysis of metastable states and realistic field configurations, can be a powerful means for predicting and optimizing the behavior of carbon particles in magnetron sputtering systems used for DLC film deposition.

Keywords: PIC/MCC; DLC; magnetron sputtering

Research on Local Nitriding Technology of High Alloy Steel

Zhang Chengsong, Yu Wenjie*

SouthwestJiaotong University ,Chengdu, China

E-mail address: cszhang@ swjtu. edu. cn (Zhang Chengsong)

E-mail address: wjyu@ my. swjtu. edu. cn (Yu Wenjie)

Abstract: The key components (such as shaft necks and bushings) in aviation turbofan engines require high wear resistance and fatigue performance at the contact surfaces during service. Traditional techniques use copper plating protection and anti-nitriding coatings for local nitriding of the components, but this involves the application and removal of anti-nitriding coatings, making the process complex and costly. In response to the technical requirements for local nitriding of typical components in aviation engines, an integrated design is needed to prepare an applicable anti-nitriding composite functional layer and develop a corresponding nitriding process, providing a new technical approach for the rapid and efficient processing and manufacturing of typical components. Therefore, the main research content is how to design a rare earth perovskite-passivation film composite functional layer to achieve the purpose of anti-nitriding. Taking 1Cr11Ni2W2MoV as the research object, an innovative improvement strategy using chemical methods was adopted, and the "gel layer-passivation film" dual-effect anti-nitriding mechanism of $LaCrO_3$ was clearly defined: both passive and active anti-nitriding exist simultaneously. The anti-nitriding agent was applied to the surface of the nitrided parts to promote the formation of the surface passivation film, physically isolating the diffusion of nitrogen atoms, passively hindering nitriding, and simultaneously forming a functional layer that selectively adsorbs N atoms to actively hinder nitriding, reducing the penetration of the base material. The advantage of this method is that nitriding can be carried out directly after coating, without the need for a removal process, making the process simple and the cost low. Based on the characteristic that 1Cr11Ni2W2MoV stainless steel contains Cr elements, after comparing $La(NO_3)_3$ and $Cr(NO_3)_3$, $LaCrO_3$ was selected as the main component of the anti-nitriding agent due to its superior performance in component compatibility, catalytic activity, and stability. Based on the analysis of the microstructure infiltration layer thickness results, it was found that the$Cr(NO_3)_3$ coating has an infiltration-promoting effect, while $LaCrO_3$ and $La(NO_3)_3$ have obvious anti-infiltration effects. This confirms the role of rare earth elements in anti-infiltration. Through XRD phase analysis, $LaCrO_3$ and $La(NO_3)_3$ can effectively inhibit the formation of nitrides. Based on the XPS elemental state analysis, the $Cr2p^{3/2}$ peak in $Cr(NO_3)_3$ shifted to the left relative to the peak of $LaCrO_3$, indicating a decrease in bond energy and a more unstable passivation film structure, thereby promoting nitriding. The O1s peak in $La(NO_3)_3$ shifted to the right relative to the peak of $LaCrO_3$, suggesting that it is conducive to the combination of oxygen vacancies and active N, thereby hindering nitriding. The results show: Firstly, the $LaCrO_3$ and $La(NO_3)_3$ coatings can significantly inhibit the nitriding process of high alloy steel and reduce the thickness of the infiltration layer; Secondly, the $Cr(NO_3)_3$ coating has an accelerating effect on infiltration, increasing the depth of the infiltration layer, which is contrary to the effect of rare earth elements; Thirdly, the coating time has an impact on the anti-seepage effect. The $LaCrO_3$ coating has a stable anti-seepage effect at different times (10 minutes/30 minutes), while the $La(NO_3)_3$ and $Cr(NO_3)_3$ coatings lose their effect with the extension of time. The successful development of a rare earth perovskite-passivated film composite functional layer (with $LaCrO_3$ as the core) enables a "no removal" integrated process for local nitriding of high-alloy steel, replacing traditional copper plating and anti-nitriding coating technologies. This new method can eliminate the coating and removal processes of the anti-nitriding layer in traditional processes, reducing energy consumption and shortening the cycle. It solves the local nitriding problem of key components such as shaft necks and bushings of aircraft engines, and extends the service life by more than twice.

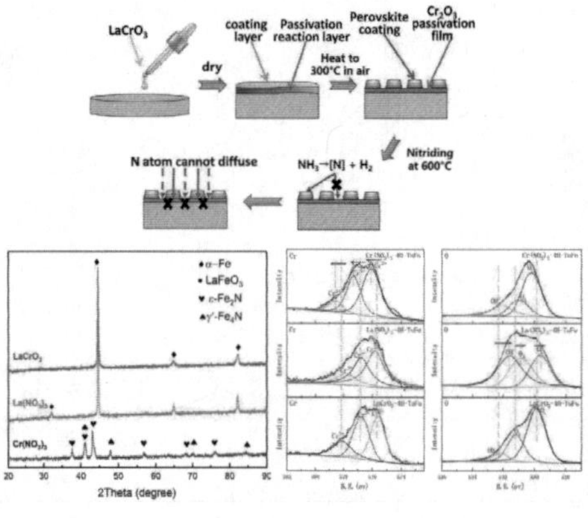

Fig. 1 Mechanism of anti-nitriding of perovskite-passive oxide composite coating

Keywords: 1Cr11Ni2W2MoV; local gas nitriding; ABO_3 rare earth perovskite; barrier formation mechanism; barrier formation conditions

The Effect of the Second Generation of QPQ Salt Bath Composite Treatment Technology on the Organization and Properties of 5CrMnMo Hot Forging Dies

Xie Mingqiang, Wei Jue*, Wang Hui, Wen Can*, Wang Qiang, Gan Qianrong, Li Kongjun, Dai Yanqin, Chen Bo
Chengdu Tool Research Institute Co., Ltd., Chengdu, Sichuan
E-mail address: 1939379800@qq.com (Wei Jue)
E-mail address: wencan1991@126.com (Wen Can)

Abstract: The utilization of hot forging dies manufactured from 5CrMnMo is a pervasive phenomenon within the automotive and machinery manufacturing sectors, where they function as pivotal constituents of the fundamental process equipment. However, failure caused by thermal fatigue significantly reduces the service life of the components. Consequently, it is imperative that effective measures are implemented to prevent such occurrences. It has been demonstrated through experimentation that utilizing the second-generation QPQ salt bath composite treatment technology to modify the surface of 5CrMnMo, a special material for hot forging dies, is an effective solution to the problems of unstable service life and reduced service life that are caused by gas nitriding and ion nitriding. Furthermore, this treatment technology maximizes the thermal fatigue performance, wear resistance, and corrosion resistance of hot forging dies with a nitriding time of 2.5-3 h.

Keywords: 5CrMnMo hot forging die; QPQ technology; Thermal fatigue; Wear resistance; Corrosion resistance

Analytical Model of Low-Pressure Carburizing Process

Ma Jingbo

China Academy of Machinery Beijing Research Institute of Mechanical & Electrical Technology Co., Ltd., Beijing, Beijing, China

E-mail address: ma_jing_bo@163.com (Ma Jingbo)

Abstract: Carburizing is a vital case hardening process by injection of carbon into the surface of the workpiece. C_2H_2 is a unique carburizing medium due to the catalyze of Fe. The pulsed carburing process proves its effectiveness in practice. However, the pulsed process results in the lack of an analytical equation of the pressure carburizing process. In this paper, we derived the analytical solver of Fick's law under boost-diffusion pulsed process. In the boost stage, carbide emerges on the surface of the workpiece, and the carbon potential is mainly 6.67 wt.%. The high carbon potential results in high carburizing efficiency in each boost stage compared to gas carburizing. The surface transfer coefficient and diffusion coefficient are crucial parameters for carburized material, which could be derived from the carburized result, such as surface carbon content and depth. The minimum carbon content is a crucial parameter for carburizing preparation. Each min carbon content is responding to a pulsed number. With the increase of min carbon content, the pulsed number increases, and the carburized time decreases.

Preparation of Carbon-Based Coatings with the High Hardness and Toughness As Well As High Corrosion Resistance

Liu Quanli*, Li Zhen, Liu Ruiliang

College of Materials Science and Chemical Engineering, Harbin Engineering University, China

E-mail address: liuquanli@hrbeu.edu.cn (Liu Quanli)

E-mail address: lizhen199907@hrbeu.edu.cn (Li Zhen)

Abstract: In this study, the liquid-phase plasma electrolytic deposition (LPED) technology was utilized to rapidly prepare carbon-based coatings on the surface of AISI 304 stainless steel with acetonitrile as the carbon source. This method has the advantages of low equipment cost, high deposition efficiency, low environmental pollution, preparation at normal temperature and pressure, a wide variety of films that can be prepared, and convenient doping. However, this method has relatively strict requirements for process parameters. Based on this, in this paper, the effects of voltage and conductive salt content on the microstructure and properties of carbon-based coatings during the deposition process were investigated by scanning electron microscopy, X-ray photoelectron spectroscopy, X-ray diffraction, Raman spectroscopy, Vickers hardness tester and electrochemical workstation systems. The results show that the carbon-based coating on the surface of stainless steel is mainly composed of nanoparticles, with particle sizes ranging from 400 to 1500 nm. The Raman spectra have obvious characteristic peaks near 1350 cm^{-1} and 1600 cm^{-1}, which are typical DLC structures. With the increase of voltage and the content of conductive salts, the surface particles and coating thickness of the coating gradually increase, while the proportion of sp^3 hybrid carbon bonds in the coating, the hardness and corrosion resistance of the coating all show a changing trend of first increasing and then decreasing. When the deposition voltage is 260 V and the content of conductive salt is 3 g/L, the coating thickness is approximately 2.4 μm, the hardness can reach 972 $HV_{0.05}$, the corrosion potential can reach −0.42 V, and the corrosion current density is about three times lower than that of the untreated material. Doping Cr, Ni and Ti can all change the microstructure of carbon-based coatings to a certain extent. After doping chromium, the particle size and gap on the coating surface decrease, and the arrangement becomes more dense, thereby enhancing the hardness (1015.17 $HV_{0.05}$) and corrosion resistance of the carbon-based film. The toughness of the coating and its adhesion to the substrate are relatively good. When the voltage is 280 V or the conductive salt is 5 g/L, the coating thickness can reach 10.3 μm and 15.2 μm respectively. However, due to the intense oxidation reaction occurring on the material surface, this leads to an increase in the content and size of oxides within the carbon-based coating, making the coating loose and porous, and thereby reducing its hardness and corrosion resistance. Finally, based on the results of electrochemical tests and the analysis of corrosion morphology, the corrosion resistance mechanism of carbon-based coatings deposited by liquid-phase plasma electrolytic deposition was expounded.

Keywords: plasma electrolytic deposition; AISI 304 stainless steel; carbon-based coating; microstructure; corrosion resistance

Effects of the Oxidizing Layer on Microstructure and Properties of Plasma Nitriding-Ion Plating Composite Layers

Li Fulin*, Liu Ruiliang

College of Materials Science and Chemical Engineering, Harbin Engineering University, China

E-mail address: lifulin@hrbeu.edu.cn (Li Fulin)

E-mail address: liuruiliang@hrbeu.edu.cn (Liu Ruiliang)

Abstract: This paper explores the influence of the oxide layer as a transition layer on the microstructure and properties of the nitrided composite TiCN coating of AISI 9310 steel through four different processes: plasma nitriding, oxidizing after nitriding, plasma nitriding-ion plating composite, and composite TiCN ion plating after nitriding oxidation. The microstructure under different processes was characterized by methods such as X-ray diffractometer, scanning electron microscope and surface analysis system (XPS). The hardness, wear resistance and corrosion resistance of the nitrided layer were tested and evaluated by using a Vickers hardness tester, a ball and disk friction and wear testing machine, an electrochemical workstation, a nano-indentation instrument, etc. The experimental results show that: After plasma nitriding at 420 ℃ for 8 hours, uniform-sized nitride particles were formed on the surface of the sample. A 1.24 μm compound layer containing Fe4N was formed on the surface. The surface hardness reached 971 $HV_{0.1}$, and the corrosion potential increased from −0.764 V to −0.249 V. And the corrosion current density has decreased by more than three times. After oxidation, a uniform oxide film of several nanometers in size was formed on the surface, making the surface more normal. The surface hardness reached 950 $HV_{0.1}$, the corrosion potential reached −0.267 V, and the corrosion current density further decreased. A composite coating with a multi-layer structure of TiN/TiCN was deposited on the surface through multi-arc ion plating. The oxide layer, as a transition layer, is conducive to the deposition of the coating layer. It can make the deposited coating layer (0.97 μm) thicker than the coating layer directly deposited after nitriding (0.75 μm), and the surface microhardness increases from 1378 $HV_{0.1}$ to 1390 $HV_{0.1}$. According to the nanoindentation structure, it is found that After oxidation, due to the thicker obtained coating, the nano-hardness and elastic modulus of the surface increased from 29.8 GPa and 347 GPa to 36.5 GPa and 404 GPa, and the anti-plastic deformation ability of the surface could be further improved. According to the electrochemical test, the corrosion potential of the surface after the coating further increased. The corrosion potential and corrosion current density of the surface of the nitriding composite TiCN coating were −0.160 V and 1.199 uA/cm^2 respectively, and the corrosion potential of the composite TiCN coating after nitriding oxidation was −0.226 V and 1.390 uA/cm^2 respectively. And when judging the bonding force by the indentation method, it was found that the oxide layer, as the transition layer of the nitrided composite TiCN coating, could improve the bonding force between the interfaces. In conclusion, the oxide layer, as a transition layer in plasma nitriding-ion plating composite layer, can improve the comprehensive performance of the surface to a certain extent.

Keywords: AISI 9310 steel; plasma nitriding-ion plating composite layer; oxidizing layer; TiCN coating; property

Anisotropic Response of Residual Stress in Low-Temperature Gaseous Carburized LPBF-316L Steel: Experimental and Numerical Analysis

Peng Yawei[a,b*], Feng Yajian[a,b], Zhao Zhenxu[a,b], Gong Jianming[a,b]

[a] School of Mechanical and Power Engineering, Nanjing Tech University, China

[b] Jiangsu Province Key Laboratory of Reliability-centered Manufacturing of Energy Power Equipment, China

Abstract: Low-temperature gaseous carburization (LTGC) is a surface modification technique that produces expanded austenite with enhanced hardness and compressive residual stress under a controlled atmosphere. It is particularly effective for improving the surface performance of complex geometries produced by additive manufacturing. However, 316L stainless steel fabricated via laser powder bed fusion (L-PBF) exhibits crystallographic texture and microstructural anisotropy due to rapid solidification, which may influence the LTGC response. This study investigates the effect of inherent microstructural anisotropy on residual stress development in L-PBF 316L stainless steel subjected to LTGC. Experimental results revealed that anisotropic compressive residual stresses across different planes—top (perpendicular) and side (parallel) to the build direction—within the expanded austenite case. This anisotropy was further examined via nanoindentation and in-situ EBSD. Building upon these experimental findings, a finite element analysis (FEA) was performed. Through the synergy of experimental observations and numerical simulations, the mechanisms responsible for the anisotropic response of residual stress in expanded austenite are revealed.

Keywords: additive manufacturing; low-temperature gaseous carburization; anisotropic response; microstructure; residual stress

Deformation Response of Meta-stable and Nitrogen-Stabilized Austenitic Stainless Steel Subjected to Surface Rolling at Elevated Temperature

He Yun[1], Fu Chenyue[1], Wang Bo[1]*, Zhou Chen[2], Shi Yousen[2]*

[1] School of Materials Science and Engineering, Shanghai University, Shanghai, China

[2] Jiangsu Fengdong Heat Treatment& Surface Engineering Research Co., Ltd., Jiangsu, China

E-mail address: bwg@shu.edu.cn (Wang Bo)

E-mail address: shi.yousen@fengdong.com (Shi Yousen)

Abstract: AISI 304 stainless steel plates were austenitized and solution nitrided in Ar and in N_2 atmosphere, respectively, followed by severe mechanical deformation via surface rolling at room and elevated temperatures. By combining composition and straining gradients, an attempt to improve the understanding of the role of nitrogen in deformation under low and elevated temperature conditions is made. Irrespective of deformation temperature, the strain transfer is deeper in the austenitized material than for the solution nitrided condition. While surface rolling at room temperature leads to a deformation-induced transformation of austenite into martensite in the austenitized specimen, the solution nitrided specimen remains austenitic at the surface and martensite appears close to the case/core interface where the nitrogen content is relatively low. For both austenitized and solution nitrided conditions, no martensitic transformation occurs throughout the materials upon surface rolling at an elevated temperature. Eventually, the role of nitrogen in thermal degradation resistance of the deformed materials is interpreted in relation to the microstructural evolution including grain coarsening and defect recovery.

Keywords: AISI 304 stainless steel; high-temperature solution nitriding; deep rolling; deformation response; thermal resistance

Numerical Simulation and Experimental Analysis of Temperature Field Evolution in Multi-Pass Laser Quenching

Liu Xin, Wang Yixue, Wang Xiao, Li Changcong, Yu Shu, Li Jintao

Shanghai Universityof Engineering Science

E-mail address: liuxin980730@ 163. com (Liu Xin)

E-mail address: sunnywang2013@ sina. com (Wang Yixue)

Abstract: Laser quenching is an efficient surface strengthening technique for enhancing material wear resistance and service life. However, the influence mechanisms of the thermal accumulation effect during multi-pass laser quenching on the temperature field, microstructural transformation, and hardness gradient of 1Cr17Ni2 martensitic stainless steel remain unclear, and the process optimization lacks systematic support. This study employs a combined approach of numerical simulation and experimentation, investigating the temperature fields and hardened layer properties under different quenching passes. A thermal conduction model was established based on ABAQUS, utilizing a Gaussian heat source to simulate the temperature fields for various passes and analyze the impact of thermal accumulation on the modified layer. Experiments involving microstructural observation, hardness testing, and layer depth measurement were conducted to validate the simulations and elucidate the underlying performance influence mechanisms. The results indicate that single-pass quenching achieved a surface hardness >550 HV and a hardened layer depth of 1. 65 mm; due to the thermal accumulation effect, multi-pass quenching increased the surface peak hardness to 620 HV and extended the layer depth to 1. 2 mm, resulting in a more gradual cross-sectional hardness gradient and a refined martensitic microstructure. Simulations and experimental results demonstrated good agreement, confirming that multi-pass quenching, through controlled heat input, promotes sufficient martensitic phase transformation at the surface and delays the hardness decline in deeper regions.

Keywords: laser Quenching; finite element simulation; hardened layer depth; multi-pass

Study on Improvement Measures of Internal Oxidation in Carburizing and Quenching Process

You Yitian[1], Liu Deyun[1], Xu Zhibin[1], Zhang Xiangyu[1], Wu Maoyong[1]

[1] Qingdao Fengdong Thermal Technology Co., LTD, Shandong, China

E-mail address: yitian@fengdong.com (You Yitian)

Abstract: When using methanol, nitrogen-methanol, and RX gas as the carrier gas for carburizing treatment, internal oxidation increases during the process. Adjusting common process parameters to control internal oxidation is ineffective. This paper proposes a method of periodically adjusting the furnace atmosphere with nitrogen, which ensures that the workpiece meets other technical requirements while effectively suppressing internal oxidation growth and reducing its depth.

Keywords: carburizing; RX gas; internal oxidation; nitrogen

Surface Nanocrystallization of Pure Iron Achieved by Anodization-Reduction Treatment and Its Effect on Lower-Temperature Nitriding Behaviors

Fu Hao, Zhu Xiangcun, Sun Jian*

Department of Materials Science and Engineering, Hefei University of Technology, Hefei 230009, China

E-mail address: 3181695073@qq.com (Fu Hao)

E-mail address: sunjian@hfut.edu.cn (Sun Jian)

Abstract: Due to the rapid diffusion effect of atoms in nanostructures, the surface nano pretreatment technology caused by mechanical plastic deformation has a significant effect on reducing the nitriding treatment temperature of steel materials. However, due to its inability to handle complex shapes or large-sized workpieces, this technology is difficult to apply in practical production. In response to this issue, this article adopts a composite treatment method of anodizing and hydrogen reduction to achieve surface nanoscaling of pure iron, and studies its effect on low-temperature nitriding behavior. The experimental results show that a nano porous oxide layer with a thickness of about 1.3 μm and a pore size of about 23.61 nm can be prepared on the surface of pure iron through anodizing process. After being reduced by hydrogen gas at 400 ℃ for 2 hours, the microstructure of the surface layer of the sample changed from nanoporous to stacked nanoparticles with a particle size of 50-100 nm. There were also several nanocrystals with a grain size of 2-10 nm inside the particles, achieving surface nanosizing of pure iron. After low-temperature ion nitriding at 400 ℃, a nitride layer with a thickness of about 4 μm was formed on the surface of the nanoscale pure iron, which is thicker than the nitride layer formed on the original coarse-grained pure iron (about 2.5 μm). Mechanism analysis shows that the oxidation-reduction sample has a good nitriding effect due to the increase in surface area, which can adsorb more nitrogen atoms, and the formation of nanocrystals can provide a fast channel for nitrogen atom diffusion.

Keywords: anodization; reduction; surface nanocrystallization; iower-temperature ntriding; pure iron

The Causes of Low-temperature Impact Energy Fluctuations in the 20CrMnMo Mud Pump Valve Seat and the Improvement Process

Xue Xing[1]

[1] Qingdao Fengdong heat treatment Co. , LTD, Shandong, China

E-mail address: xing. xue@ fengdong. com (Xue Xing)

Abstract: After carburizing, quenching, and tempering, the low-temperature impact energy of the 20CrMnMo mud pump valve seat fluctuates significantly. The metallographic structure, chemical composition, and inclusions were analyzed for the impact test samples. The results indicate that the primary cause of the fluctuation in low-temperature impact energy is the variation in manganese sulfide inclusions. By using a heat treatment process that combines carburizing slow cooling with sub-temperature quenching, fine strip-like tough ferrite was obtained, which improved the low-temperature impact toughness of the valve seat.

Keywords: Mud Pump Valve Seat; impact energy; manganese sulfide inclusions; sub-temperature quenching

A Novel High-Activity Nitrocarburizing Process Utilizing Acetylene

Wang Haocun

[1] Weifang Fengdong heat treatment Co., LTD, Shandong, China

E-mail address: wang.haocun@fengdong.com (Wang Haocun)

Abstract: Workpieces treated with the acetylene-cracked high-activity nitrocarburizing process exhibit significantly enhanced surface hardness compared to traditional CO_2-based nitrocarburizing. This method generates a dense and uniform compound layer ("white layer") with markedly superior corrosion resistance. This study elucidates how acetylene cracking promotes diffusion layer growth, refines the microstructure of the compact compound layer, and synergistically improves both hardness and corrosion resistance. The process offers a novel pathway for high-performance surface engineering with substantial industrial applicability.

Keywords: Acetylene cracking; Nitrocarburizing

Proceedings of IFHTSE 2025

The 30th World Congress of International Federation for Heat Treatment and Surface Engineering
August, 19-22, 2025, Suzhou, China

An Accuracy Calculation Method for Low-Pressure Carburizing Dynamic Parameter

Cong Peiwu[1,2], Ma Jingbo[2]

[1] Jilin University, Changchun, Jilin Province, China

[2] China Academy of Machinery Beijing Research Institute of Mechanical & Electrical Technology Co., Ltd., Beijing, Beijing, China

E-mail address: congpeiwu@126.com (Cong Peiwu)

E-mail address: ma_jing_bo@163.com (Ma Jingbo)

Abstract: Vacuum low-pressure carburizing technology has extensive applications in transmission parts such as bearings and gears. The precise adjustment of the carburized layer depends on dynamic parameters such as surface transfer systems and diffusion coefficients. However, the pulse carburizing process is complex and the calculation of carbon concentration is slow, making it difficult for traditional algorithms to achieve rapid calculation of dynamic parameters. This paper proposes a method for calculating dynamic parameters based on genetic algorithms. Firstly, a functional relationship between the pulse carburizing results and dynamic parameters is established. Then, the surface transfer coefficient and diffusion coefficient are calculated based on genetic algorithms. The accuracy and effectiveness of the calculation method are verified through parameter calculations and experiments on various carburized gear steels. The research shows that the diffusion coefficient of carbon atoms increases with the increase of carburizing temperature, and the surface transfer coefficient increases with the increase of carburizing pressure.

Simulation Study on the Uniformity of Atmosphere During the Working Process of Large-Scale Carburizing Equipment

Li Fei[1], Cong Peiwu[1*], Lu Wenlin[2*]

[1] Jilin University, Jilin, China[1], China Academy of Machinery Beijing Research Institute of Mechanical & Electrical Technology Co., Ltd., China

[2] China Academy of Machinery Beijing Research Institute of Mechanical & Electrical Technology Co., Ltd., China

E-mail address: congpeiwu@126.com (Cong Peiwu)

E-mail address: 776628887@qq.com (Lu Wenlin)

Abstract: In response to the uneven distribution of acetylene gas during the carburizing process in traditional large-sized carburizing furnaces, this work investigates the effects of nozzle form, nozzle arrangement, and gas supply flow rate on the uniformity of the atmosphere inside the carburizing furnace through numerical simulation. Among them, research on nozzle forms has adopted two forms: straight nozzle and branched nozzle. The design of branched nozzle can significantly optimize the distribution of acetylene gas in the horizontal direction, which is more conducive to the diffusion of acetylene gas; Regarding the nozzle arrangement, a side arrangement scheme was ultimately selected by designing different quantities, spacing, and unknown surfaces, which can effectively reduce the acetylene concentration difference between the core and external workpiece surfaces; Regarding the study of gas supply flow rate, five inflation flow rates of 100 L/min, 200 L/min, 300 L/min, 400 L/min, and 500 L/min were set to compare the distribution of acetylene under different flow rates. The research results showed that as the inflation flow rate decreased, the uniformity of the atmosphere in the furnace could be effectively improved.

Keywords: numerical simulation; nozzle form; nozzle arrangement; gas supply flow rate

Research on the Application of Digital Twin System for Vacuum Carburizing Equipment

Zhang Jiarui, Lu Wenlin

China Academy of Machinery Beijing Research Institute of Mechanical & Electrical Technology, Beijing, China

E-mail address: zhangjiaxiaorui@126.com (Zhang Jiarui)

E-mail address: 76628887@qq.com (Lu Wenlin)

Abstract: In order to promote the informatization and intelligence development of vacuum heat treatment equipment, this article proposes a method to build a digital twin system for vacuum heat treatment equipment and presents several typical application scenarios. To make the method have more reference value, this article selects the WZST type vacuum carburizing and oil quenching furnace as the research object, which has the common characteristics of multiple types of vacuum heat treatment equipment and can perform various heat treatment such as carburizing, carbonitriding, gas cooling, oil quench, annealing, etc. Based on the five-dimensional digital twin model which is the most mature and widely used in different industries, this article builds a digital twin system for vacuum heat treatment equipment from five dimensions: Physical Entity (PE), Virtual Entity (VE), Digital Twin Data (DD), Services (Ss), Connection (CN).

Keywords: vacuum heat treatment equipment; digital twin; quenching; carburizing

Microstructure and Performance of Low Temperature Plasmanitrided AISI 304 Stainless Steel for Bipolar Plate of PEMFC

Liu Ruiliang*

College of Materials Science and Chemical Engineering, Harbin Engineering University, China

Abstract: In the present investigation, effects temperature on low temperature plasma nitrided (LTPNed) AISI 304 stainless steel for bipolar plates (BPPs) of proton exchange membrane fuel cell (PEMFC) were studied systematically. After low-temperature plasma nitriding, the modified layer on the surface of the sample consists mainly of nitrogen-expanded austenite phase (γ_N) with a small amount of nitrides, with the increase of temperature, the content of nitrogen element in the γ_N phase is increased. Compared with the untreated stainless steel, the corrosion resistance of the sample nitrided at 360 ℃ for 4h are improved significantly. The surface morphologies of the LTPNed samples after potentiostatic polarization test shows that for the untreated sample, pitting corrosion is mainly happened, while for the LTPNed samples, the corroded mechanism changes from slight pitting corrosion to mainly grain boundary corrosion with the increase of nitriding temperature. Moreover, interfacial contact resistance of the LTPNed stainless steel is also lower than that of the untreated one. In addition, the LTPNed stainless steel also shows the better mechanical properties than those of the untreated one. Therefore, LTPN technology can be used to improve the performance of AISI 304 stainless steel for BBP of PEMFC with the high comprehensive property and the low cost.

Keywords: Stainless steel; low temperature plasma nitriding; bipolar plate; proton exchange membrane fuel cell; property

Influence Mechanism of Complete Austenitization on Mechanical Properties of M50NiL Carburized Steel

Xie Bo*, Gai Dengyu, Qiao Yingjie

College Of Materials Science and Chemical Engineering, Harbin Engineering University, Harbin, China.

E-mail address: shepherd@hrbeu.edu.cn (Xie Bo)

E-mail address: gaidengyu@hrbeu.edu.cn (Gai Dengyu)

Abstract: This paper investigates the mechanism by which complete austenitization influences the mechanical properties of M50NiL carburized steel, a material widely used in high-stress mechanical components due to its superior wear resistance. Through controlling carbide content and grain size,, we systematically analyzed phase transformations and microstructural evolutions during sequential heat treatments, including austenitization, solution treatment, and tempering. Particular emphasis was placed on in-situ characterization of carbide precipitation dynamics, enabling quantitative analysis of carbide morphology and distribution. Results demonstrate that fully normalized samples exhibit prolonged transition to bulk diffusion states, where initial diffusion kinetics are profoundly governed by grain size. During bulk diffusion, growth rates of nascent carbides are suppressed by undissolved pre-existing carbides, which simultaneously impede carbon migration and reduce carburized layer depth. At 920 ℃ carburization, complete austenitization specimens showed a 12.5% increase in surface hardness, 17.9% enhancement in effective case depth. Concurrently, the average friction coefficient and wear rate were reduced by 8.9% and 27.8%, respectively. The comprehensive mechanical property improvement is primarily ascribed to the formation of nanoscale, homogeneously dispersed carbides in surface and subsurface regions. which enhance dislocation pinning and hinder crack propagation.

Keywords: M50NiL steel; complete austenitization; mechanical properties

Low Temperature Nitriding of Specialty Austenitic Alloys

Xiao Tao

School ofMaterials Science and Engineering, Southeast University, Nanjing, 211189, China

E-mail address: x. tao@ seu. edu. cn (Xiao Tao)

Abstract: Low-temperature nitriding (LTN) enhances the surface hardness and tribological properties of austenitic stainless steels (SSs) without compromising (and, sometimes, improving) their corrosion resistance, owing to the formation of an extreme nitrogen supersaturated and expanded austenite (γ_N, or known as S phase) case on the alloy surface. However, current understanding of γ_N is often limited to the LTN studies on AISI grade 304 and 316 SSs. Apart from 304/316 SSs, there is a wide range of austenitic alloys used in marine, chemical processing, and energy industries. The influences of the different composition of these specialty austenitic alloys to the nitrogen absorption and microstructure of the γ_N cases formed following LTN remain largely unknown. Austenitic medium/high entropy alloys (MEAs/HEAs) have been rapidly developed, which show superior mechanical and corrosion properties than 304/316 SS and great potentials to be adopted for demanding environments in the future. In comparison with the 316 SS, two austenitic MEAs-i. e., a ternary FeCrNi and a quandary FeCoNiCr-were plasma nitrided at the typical low-to-intermediate temperatures at 430-480 ℃. Compared with the nitrogen level at 16-32 at. % in the 430 ℃ nitrided 316 SS, the nitrided case in FeCoNiCr HEA shows N concentrations at 20-28 at. %. Analogous to that in γ_N-316, cellular Cr-nitride precipitates were observed starting at the grain boundaries in the nitrided case in FeCoNiCr HEA after nitriding at 450 ℃ for 10h. In contrast, the 430 ℃ nitrided case in the ternary FeCrNi MEA exhibits significantly higher N concentration at 30-36 at. % and microcracking. Intriguingly, finely dispersed Si-rich precipitates were observed in all the nitrided cases in FeCrNi MEA at 430-480 ℃. These works pave the way to the development of nitriding process to enhance the wear/corrosion performance of MEAs/HEAs and composition complex alloys.

Keywords: Plasma nitriding; expanded austenite; surface hardening; medium entropy alloys; high entropy alloys

Research on the Influence of Different Heat Treatment Processes on Gear Residual Stress

An Kang[1], Hu Zhaonan[1], Yang Baoqing[1], Wang Li[1], Li Xiaoxue[1], Deng Chunyan[1], Zhao Long[1]

[1] Chongqing Gearbox Co., Ltd., Chongqing, China

Abstract: Carburizing quenching stands as the core process for enhancing the fatigue resistance of gears, with its residual stress exerting a pivotal influence on fatigue life. For 18CNiMo7-6 material gears, the surface residual stress under conventional carburizing quenching is at the −50 MPa level. Vacuum carburizing quenching can achieve −180 MPa, while cold treatment after carburizing quenching reaches −380 MPa. Vacuum carburizing quenching offers an optimal stress gradient distribution, and cold treatment enables the subsurface to attain higher residual compressive stress. Following different processes, cleaning shot peening significantly elevates the surface residual stress to approximately −400 MPa.

Keywords: Gear; Residual stress; Heat treatment process; Carburizing and quenching; Fatigue life

Innovative Duplex Treatment by Ti-enhanced Plasma Nitriding and AlCrN Coating with Excellent Performances

Wu Jiqiang, Hu Jing*

School of Materials Science and Engineering, Changzhou University, Changzhou, China

E-mail address: wujiqiang88@126.com (Wu Jiqiang) E-mail address: jinghoo@126.com (Hu Jing)

Abstract: In order to obtain a gradient composite coating with excellent-performance, including high surface hardness and strong adhesion, an innovative titanium-enhanced plasma nitriding process was adopted as a pretreatment prior to the deposition of AlCrN coating for 38CrMoAl steel. Meanwhile, a comparison was made by adopting traditional plasma nitriding as a pretreatment. The microstructure and properties of the gradient composite coating were evaluated by scanning electron microscopy (SEM), X-ray diffraction (XRD), scratch tester, hardness tester, wear tester, and three-dimensional profilemeter. The research results showed that the compound layer obtained by titanium-enhanced plasma nitriding is thinner and denser, and thus can support AlCrN coating more strongly and own better performance compared to the conventional plasma nitrided compound layer, with the adhesion strength increasing from 16.8 N to 29.4 N, the surface hardness increasing from 3650 HV0.05 to 3780 HV0.05, the friction coefficient and wear rate reducing from 0.64 and 5.4849×10^{-6} $mm^3/(N*m)$ to 0.61 and 2.3060×10^{-6} $mm^3/(N*m)$, respectively. Additionally, the mechanism why titanium-enhanced plasma nitriding enhances the performance of the coating has been analyzed and it is considered to be attributed to the fact that TiN phase formed in the titanium-regulated nitrided layer has the same FCC crystal structure as that of CrN in the coating, and both lattice constants are close, which is conducive to the epitaxial diffusion growth of the coating, promoting the mutual diffusion of N element during AlCrN coatings deposition, and thus increase the growth rate and adhesion strength of the coating and improve the toughness as well.

Keywords: plasma nitriding; AlCrN coating; duplex treatment; adhesion strength; compound Layer

Wear Behavior of a Novel $Al_2O_3/La_xP_yO_{13}/MoS_2$ Composite Layer In-situ Prepared by Micro-arc Oxidation

Shang Jian, Li Qian, Liu Tao

Liaoning University of Technology, 121001, School of Materials Science and Engineering, 169, Shiying Street, Jinzhou, Liaoning Province, China

E-mail address: shangjan@lnut.edu.cn (Shang Jian)

E-mail address: liqian19991024@163.com(Li Qian)

Abstract: This work present a novel composite layer aiming at improving wear and corrosion resistance of 6082-T6 alloy used in high-speed railway catenary systems. The $Al_2O_3/La_2P_4O_{13}/MoS_2$ composite layer was successfully in-situ prepared on 6082-T6 alloy surface utilizing micro-arc oxidation (MAO) technology by introducing Na_2MoO_4, Na_2S, and $C_2H_6LaO_3$ into the electrolyte. The microstructures, compositions, morphologies and tribological behavior of the composite layer are characterized by XRD, XPS, SEM, EDS, TEM and UMT. The results show that: with the increase of $C_2H_6LaO_3$ concentration, the surface roughness changed from 1.447 to 2.588 μm, the porosity changed from 3.79 to 2.45 % and the hardness rises from 737.66 to 1177.41-1260.44 HV1; the self-lubricating phases $La_2P_4O_{13}$ and MoS_2 were dispersed between the hard phase Al_2O_3; when the concentration was 5 g/L, the average friction coefficient is reduced by 41.66 % compared with the Al_2O_3/MoS_2 composite layer at 0 g/L. There was no obvious mass or volume changes of coupled body before and after sliding wear. The in-situ formation of MoS_2 and $La_2P_4O_{13}$ in the MAO layer can have a coordinate effects for improving the compactness, hardness and self-lubricating behavior.

Keywords: 6082-T6 alloy; micro-arc oxidation; composite layer; wear behavior

Experimental Study on Local Nitrocarburizing of Maraging Stainless Steel Part

Ma Shengge

Shanghai Advanced Metallurgical Technology Corporation, Shanghai, China

E-mail address: ma. shengge@ samtc. com (Ma Shengge)

Abstract: Maraging stainless steel features high strength, excellent toughness, and corrosion resistance. Nitrocarburizing treatment can further enhance its wear resistance while ensuring good toughness of the case layer. The research object of this paper is a maraging stainless steel part with a cam. The cam needs nitrocarburizing treatment for high wear resistance, while other areas of the part, especially the thin-walled positions with a thickness of only 0. 5 mm, are required not to be nitrocarburized to prevent increased brittleness after nitrocarburizing, which may lead to fractures during use. To achieve local nitrocarburizing of the part, the part is first immersed in an electroless plating solution for electroless nickel plating. The thickness of the nickel layer is about 7. 2 μm. Then a laser cleaning machine is used to clean the cam surface, aiming to locally remove the nickel layer. After laser cleaning, samples are taken to prepare metallographic specimens to observe the cam surface and ensure that the nickel layer has been completely removed. Next, the laser-cleaned part is placed in a nitriding furnace for gas nitrocarburizing. During nitrocarburizing, the gas nitriding furnace is first evacuated, and then a special activating gas is used to activate the part's surface. The purpose of activation is to remove the passivation layer on the part's surface, that is, the dense chromium oxide layer, and eliminate the barrier to active nitrogen and carbon. After the part is activated, ammonia gas and carbon-containing gas are used for nitrocarburizing at a temperature of 550-600 ℃ for 20 hours. After nitrocarburizing, a micro-hardness tester is used to measure the surface hardness of the cam. The result shows that the hardness is 940-992 HV0. 05. Metallographic specimens are prepared to observe the nitrocarburizing effect in the cam area and the shielding effect in the non-nitrocarburizing area. The depth of the case layer on the cam surface is 106-124 μm, and there are no cracks. There is no case layer in the non-nitrocarburizing area, and the nickel plating layer remains intact.

Keywords: maraging stainless steel; nitrocarburizing; local nitrocarburizing; gas nitrocarburizing; electroless nickel plating; laser cleaning

Rare Earth-Enhanced Nitrogen-Expanded Austenite Stability in Plasma-Nitrided High-Entropy Alloys at High Temperature

Wang Yifan[a], You Yuan[a*], Yan Mufu[b], Zhang Yanxiang[b], Sheng Wenping[a], Wang Yan[a],
Shimiao Xu[a], Feng Gu[a], Nan Bian[a], Chaohui Wang[a], Weidong Cheng[a]

[a] School of Materials Science and Engineering, Heilongjiang Provincial Key Laboratory of Polymeric Composition materials, Qiqihar University, Qiqihar 161006, China

[b] National Key Laboratory for Precision Hot Processing of Metals, School of Materials Science and Engineering, Harbin Institute of Technology, Harbin 150001, China

E-mail address: greatyouyuan@163.com (You Yuan)

Abstract: High-temperaturecan improve nitriding efficiency of austenitic alloys, but it also causes the decomposition of expanded austenite phase, reducing mechanical properties and corrosion resistance. Therefore, a novel method is needed to enhance the stability of the austenite phase during nitriding. This paper delves into the influence of lanthanum (La) on the plasma nitriding of CoCrCuFeNi high-entropy alloys (HEAs) at temperatures ranging from 440 ℃ to 500 ℃, with treatment durations spanning from 8 to 24 h. The investigation focuses on elucidating the strengthening mechanisms that occur within the samples during La nitriding. Notably, the presence of La atoms was found to augment the diffusion coefficient of nitrogen (N) atoms during rare earth nitriding. The incorporation of La into the nitriding process effectively alleviates the "trapping effect" of Cr, thereby enhancing the stability of the expanded austenite phase during high-temperature treatment. Moreover, the addition of La modifies the phase composition of the specimen surface, thereby improving its corrosion and wear resistance. The main wear mechanisms observed are abrasive, oxidative, and adhesive wear. This work concludes by presenting a methodology that not only elevates the nitriding temperature for austenitic alloy but also markedly improves their surface properties.

Keywords: rare earth; microstructure; plasma nitriding; high entropy alloy; wear

Research Progress on Surface Strengthening Technology of Heavy-Duty Gears with High Fatigue Resistance

Wang Xiaonan*, Chen Qiyuan, Shen Xinjun, Zhang Qingyu

School of Iron and Steel, Soochow University, China

E-mail address: wxn@suda.edu.cn (Wang Xiaonan)

E-mail address: qychen1992@suda.edu.cn (Chen Qiyuan)

Abstract: High-performance gears demand exceptional resistance to wear and contact fatigue. This study systematically investigates the influence of microstructure, carbon concentration, and hardness distributions on these critical properties in deep-carburized 18Cr2Ni4W gear steel through experiments in conjunction with finite element simulations. Initial experiments focused on conventional carburizing at temperatures of 930 ℃, 950 ℃, and 970 ℃. Results confirm that surface carbon concentration and hardness increase progressively with higher carburizing temperatures. Tensile tests on the carburized surface layers revealed impressive strengths exceeding 2000 MPa across all temperatures; however, a significant reduction in ductility was observed as temperature increased. Sliding wear tests demonstrated that the sample carburized at 970 ℃ exhibited the lowest wear rate compared to those processed at 930 ℃ and 950 ℃. Analysis identified the primary wear mechanisms as adhesive oxidation wear coupled with abrasive wear. Crucially, energy dispersive spectroscopy mapping revealed the most pronounced aggregation of oxygen on the counterface opposing the 930 ℃ carburized sample, indicating the most severe adhesive oxidation wear in this case. Supporting finite element simulations of transient temperature fields during sliding contact predicted a maximum contact point temperature approximately 50 ℃ lower for the 970 ℃ sample compared to the 930 ℃ sample. This significant temperature reduction effectively mitigates oxidation severity. Consequently, the enhanced wear resistance at 970 ℃ is attributed to the synergistic effect of an increased volume fraction of hard carbides within the case and the lower frictional heating suppressing oxidation. Building upon these findings, a novel two-stage heat treatment combining deep carburizing followed by carbonitriding was specifically designed to further elevate the performance limits of 18Cr2Ni4W steel. Microhardness profiling showed that conventional carburizing yielded a surface hardness of 671 HV1 and a core hardness of 465 HV1. In striking contrast, the novel carburizing-carbonitriding process significantly increased the surface peak microhardness to 748 HV1 (representing an ~11% improvement) while slightly reducing the core hardness to 440 HV1 (a ~6% decrease). The wear rate with novel carburizing-carbonitriding heat treatment is the lowest, with the decrease in wear rate from 8.73×10^{-7} $mm^3 \cdot N^{-1} \cdot mm^{-1}$ to 5.89×10^{-7} $mm^3 \cdot N^{-1} \cdot mm^{-1}$, corresponding to a 33% improvement in wear resistance. The cycle time of contact fatigue stress increases from 1.60×10^7 to 3.91×10^7, an increase of 144%. Compared to conventional carburizing process, the carburizing-carbonitriding heat treatment significantly increases carbon concentration within the surface layer. The higher carbon concentration promotes the formation of carbides, and the proportion of large-sized carbides increases. The interstitial nitrogen atoms induce potent solid solution strengthening within the martensitic matrix, significantly increasing its intrinsic microhardness and thereby contributing substantially to the improved resistance against both wear and fatigue crack initiation/propagation. The synergistic combination of refined carbide distribution and nitrogen-strengthened martensite underpins the exceptional gains in both wear resistance and contact fatigue life. This work demonstrates that optimizing carburizing temperature effectively enhances wear resistance by controlling carbide formation and minimizing oxidation through reduced frictional heating. More significantly, the newly developed carburizing-carbonitriding heat treatment offers a highly effective strategy for dramatically improving the surface integrity and service longevity of critical gear components made from 18Cr2Ni4W steel, achieving substantial simultaneous improvements in both wear resistance and contact fatigue performance.

Keywords: heavy-duty gears; carburizing process; carbonitriding process; microstructure; wear resistance; fatigue resistance

Effects of Rare Earth Sol-Gel Films on the Microstructure and Properties of Ion-Nitrided Layers: Mechanisms of Nitriding Enhancement

Chen Hongtao[a*], Zhang Yuxiang[a], Zhu Lei[b], Yan Mufu[b], Feng Yicheng[a]

[a] School of Materials Science and Chemical Engineering, Harbin University of Science and Technology. Harbin 150080, China

[b] School of Materials Science and Engineering, Harbin Institute of Technology. Harbin 150001, China

Abstract: This study explores the effect of Al-La-O sol-gel films on the ion nitriding efficiency, microstructure, and properties of the nitrided layer in 16Cr3NiWMoVNbE steel. Different Al-La-O sol films with varying Al-La ratios were prepared on the surface of 16Cr3NiWMoVNbE steel via dip-coating method. The effective hardened layer thickness is significantly increased in all sol film-coated samples, with improvements of 5.5% (pure Al), 38.9% (4:1), 27.8% (8:1), 16.7% (12:1), and 11% (16:1) compared to uncoated samples, confirming the accelerated nitriding effect of Al-La-O films. Both Al and Al-La-O sol films suppress the formation of a white layer in nitrided 16Cr3NiWMoVNbE steel, resulting in a diffusion zone composed mainly of γ'-Fe_4N and α'_N phases, with γ'-Fe_4N dominant in pure Al sol-gel films and α'_N predominant in Al-La-O sol-gel films. Although the coated samples exhibit slightly higher friction coefficients, their wear track width is significantly reduced, indicating improved wear resistance. Additionally, the sol-gel films and surface nitride layers enhance corrosion resistance by increasing the corrosion potential and reducing passive current density. The pure Al sol film forms a nano-lamellar structure that increases the surface area of 16Cr3NiWMoVNbE steel, providing more adsorption sites for active nitrogen atoms. The resulting high nitrogen potential at the surface facilitates nitrogen diffusion. For Al-La-O sol-gel films, the microcracked and porous structures serve as diffusion channels for active nitrogen atoms into the substrate. During nitriding, Fe atoms from the substrate migrate into the sol film, forming a porous composite structure that further optimizes nitrogen diffusion pathways. Simultaneously, La atoms incorporated into the substrate surface introduce lattice defects. These synergistic effects collectively enhance nitrogen permeation.

Keywords: Al-La-O sol-gel film; 16Cr3NiWMoVNbE steel; plasma nitriding; wear; corrosion resistance

Arc Plasma Assistednitriding Process and Its Application

Zhao Yanhui*, Yu Baohai, Xu Yanliang, Wang Xianghong

Sino-German Institute of Engineering, Shanghai Technical Institute of Electronics & Information, Shanghai, 201411, China

Abstract: This work presents an arc plasma assisted nitriding process, and it was used to nitriding 40Cr, 304 stainless steel, Cr12MoV steel, and W18Cr4V high-speed steel at different temperatures. It was found that the thickness of the nitriding layer increased with the increase of nitriding temperature, and there was basically no compound layer at lower temperatures. The thickness of the compound layer increased with the increase of nitriding temperature. The microstructure of the nitriding layer treated at different temperatures is closely related to the nitriding temperature and the substrate material; The maximum thickness of the nitride layer reaches 50mm, indicating a high nitriding rate. And this technology can be combined with vacuum arc deposition to deposit hard coatings to improve the properties of these steels.

High Temperature Ablative and Wear Resistance of Titanium Alloy by Shot Peening and Subsequent Nitriding

Yang Yang

Key Laboratory of Green Fabrication and Surface Technology of Advanced Metal Materials, Ministry of Education,

Anhui University of Technology, Maanshan 243002, PR China

Abstract: The ablation behavior of TC11 titanium alloy was investigated in this study using an oxypropane flame at 1100 ℃, 300 s. A combination of plasma nitriding and shot peening was used for surface modification of TC11 titanium alloy. Plasma nitriding increased the surface hardness of TC11 titanium alloy to 1000 HV, and different intensity shot peening pretreatments produced many crystal defects, including twins, dislocation arrays, dislocation cells, and high-density dislocation tangles. These defects can add elemental diffusion paths to the surface, forming aluminum diffusion zones after nitriding. In addition, the refinement of the high free energy grain size on the top surface is promoted after different intensity shot peening, which accelerates nitrogen diffusion and thus increases the thickness of the nitrided layer. After ablation, TiO_2 and Al_2O_3 could be detected on the ablated surface, and the 0.45 mmA shot peening intensity had the minimum mass ablation rate among the different shot peening intensities. The ablation of the surface layer was limited to a very limited thickness due to the enhancement layer introduced by the combination of shot peening and plasma nitriding which significantly hindered the penetration of reactive oxygen species. The grain refinement and the presence of dense grain boundaries contribute to the ablation resistance.

Study on Microstructure and Wear Resistance of α + β Titanium Alloy by Deformation Plasma Nitriding

Fu Yudong

School of Materials Science and Chemical Engineering, Harbin Engineering University, China

E-mail address: fuyudong@hrbeu.edu.cn (Fu Yudong)

Abstract: Titanium alloys have important application value in aerospace, marine vessels and other fields due to their excellent specific strength and corrosion resistance. However, its defects such as low surface hardness and poor wear resistance seriously restrict its service life and performance. Nitriding technology can form a high-hardness TiN ceramic layer on the surface of titanium alloys and is one of the most used chemical heat treatment methods. Although the traditional high-temperature nitriding process can effectively improve the surface hardness, it is easy to cause problems such as coarsening of the matrix structure and hydrogen embrittlement, destroying the overall mechanical properties of the material (such as strength and fatigue life), and the large hardness difference between the surface and the matrix is also easy to cause brittle cracking of the coating. Low-temperature nitriding processes generally have problems such as easy formation of barrier layers, limited diffusion of N atoms, and insufficient quality of the nitrided layer, which makes it difficult to meet the performance requirements under extreme service environments. Therefore, it is urgent to develop a new low-temperature nitriding technology that can take into account the improvement of both surface and matrix performance. In this context, this paper proposes and implements the deformation plasma nitriding process. Taking the dual-phase α + β titanium alloy TC4 as the research object, the element migration, microstructure evolution and friction and wear behavior under deformation are systematically studied by combining experimental and computational methods, aiming to reveal the deformation plasma nitriding mechanism in single-phase and dual-phase titanium alloys. During the nitriding process, the content of precipitated β phase in the surface area of the solid solution sample decreases, and a discontinuous compound layer is formed on the surface, in which the nitride grain size is large. With the increase of deformation, the surface nitriding structure changes significantly: the content of precipitated β phase in the surface area of the 10% deformed sample does not decrease compared with the matrix, and the grain size of the surface compound layer is refined; when the deformation is further increased, the density of β phase in the surface area increases, and the crystallinity of nitride decreases significantly. After N enters the α′ phase, it is easier to approach V to form V-N bonds to anchor V, inhibiting the aging precipitation process dominated by V diffusion. The deformed sample contains many dislocations, which enhances the diffusion ability of N, thereby weakening the barrier to V diffusion and promoting the aging precipitation process dominated by V diffusion. Deformation affects the nucleation and growth behavior of surface nitrides by regulating the aging precipitation process. Among them, the 10% deformed sample has the best wear resistance, which is mainly attributed to the formation of high-hardness nanocrystalline nitride on its surface. This structure significantly improves the surface wear resistance.

Keywords: titanium alloy; plastic deformation; plasma nitriding; aging precipitation

Metal Surface Fe_3O_4/Fe_2O_3 Oxide Layer Phase Study on Controlled Preparation and Corrosion Resistance

Sun Rui, Zhao Yibo
College of Material Science and Technology, Shandong University of Science and Technology, Qingdao 266510, China
E-mail address: 603562234@ qq. com (Sun Rui)
E-mail address: zybgigi@ 163. com (Zhao Yibo)

Abstract: With the rapid development of smart manufacturing, the requirements for corrosion resistance and wear resistance of metal surfaces are increasing significantly. Green and low-carbon surface treatment technologies are a crucial component in enhancing the core competitiveness of intelligent manufacturing equipment. The "14th Five-Year Plan" also explicitly proposes that developing green and low-carbon heat treatment industries is a key direction for the sector's advancement and the corrosion resistance cannot meet the technical requirements of higher-grade corrosion-resistant working conditions. The study on the preparation process and corrosion resistance mechanism of the oxide layer formed by post-nitriding oxidation is of significant theoretical and engineering importance. This study employed DT4C pure iron, 45# steel, and 40Cr steel as experimental materials. By systematically optimizing post-oxidation parameters including oxidation medium, temperature, and duration, the preparation process and corrosion resistance mechanism of nitrided oxide layers were investigated. The microstructure of the nitrided layer was characterized using optical microscopy and field emission scanning electron microscopy (SEM), fine phase analysis was conducted using X-ray diffraction (XRD), Raman spectroscopy, and electron backscatter diffraction analysis (EBSD), mechanical properties and corrosion resistance of the oxide layer were evaluated using a Vickers hardness tester, scratch tester, potentiodynamic polarization testing, and neutral salt spray test, respectively. Through process optimization of gas nitriding followed by oxidation, a green surface treatment technology named Nitriding-Aqueous-Oxidation (NAO) was developed. This technology enables precise regulation of Fe_3O_4/Fe_2O_3 phase composition, resulting in a dual-phase oxide layer structure consisting of a 1 μm-thick Fe_2O_3 phase surface layer and an Fe_3O_4 phase sub-surface layer. The treated specimens exhibited no rust formation after 1080 h of neutral salt spray testing. The research systematically investigated the influence of C and Cr elements in the steel on the microstructure and phase composition of the oxide layer formed during the post-vapor oxidation treatment. The growth mechanism of the oxide layer was elucidated, along with detailed exploration of the corrosion mechanism underlying the oxide layer's protective properties.

Keywords: gas nitriding; post-oxidation; steam; adhesion; corrosion resistance

The Fabrication and Heat Dissipation Performance of PEO High Emissivity Ceramic Coatings for thermal management application

Wang Yaming[a,b*], Wang Shuqi[a,b], Zou Yongchun[a,b,c], Chen Guoliang[a,b,d], Zhou Yu[a,b]

[a] Institute for Advanced Ceramics, School of Materials Science and Engineering, Harbin Institute of Technology, Harbin, China

[b] Key Laboratory of Advanced Structure-Function Integrated Materials and Green Manufacturing Technology, Harbin Institute of Technology, Harbin, China

[c] Center of Analysis and Measurement, Harbin Institute of Technology, Harbin, China

[d] School of Energy Science and Engineering, Harbin Institute of Technology, Harbin, China

E-mail address: wangyaming@ hit. edu. cn (Wang Yaming)

Abstract: Metals (Al, Ti and Nb alloys et al) are widely used as heat exchanger and radiator for aerocrafts, internal combustion engines and electronic device (3C products). To improve the thermal radiative and heat dissipation properties of light metals (Al, Ti and Nb alloys), the high emissivity ceramic coatings were designed on the metals surface by Plasma Electrolytic Oxidation (PEO) method. The effects of uncoated and PEO coated metals on infrared emissivity and heat dissipation properties were comparatively investigated. The results show that PEO coated Al, Ti and Nb alloys exhibit a high emissivity up to 0. 85, while the values are only 0. 1-0. 3 for the metals substrate. The SiC or HfC particles doped PEO coating enhance the spectral emissivity value, especially in the wavelength of 3-8 μm. The heat dissipation results indicate that the PEO and SiC doped PEO coated alloy enabled a significant temperature drop. PEO method is expected to be a promising candidate for thermal management applications of elevated temperature metal components.

Keywords: metals; PEO; high emissivity coating; thermal management

PEO Technology for Surface Multi-Functions of Light Alloys

Liu Baodan

Foshan Gradate School of Innovation, Northeastern University FoshanDongfo Surface Technology Ltd Shunde District, Foshan, China

E-mail address: baodanliu@hotmail.com (Liu Baodan)

Abstract: This work presents the recent process of PEO technology developed in our group with the aim of surface multi-functions of light alloys. By designing specific electrolytes and optimizing the plasma discharge processed, we have realized the surface functions of light alloys towards diverse applications. Especially, extensive efforts have focused on the electrical isolating coatings of Al and Ti-based alloys, and the maximum breakdown voltage of Al alloys has been improved from 500 V to 5000 V through the integration of PEO methods with other technology. This enables much more broad applications of Al-based electronics in extremely critical conditions in comparison with conventional organic isolating coatings, which can not work at the temperature exceeding 300 ℃. In addition, we also developed series of PEO technologies for gas catalysis, anti-bacterial coating, metal surface colouring etc and these technologies have been widely used in industries.

Keywords: plasma electrolytic oxidation; light alloys; surface treatment; electrical isolation; anti-bacterial coating

Non-Electrode Plasma Electrolysis for Fibers and Powders for Aero-Industrial Applications

Chen Weiwei

Department of Materials Science and Engineering, Beijing Institute of Technology, 100083, China

E-mail address: wwchen@ bit. edu. cn (Chen Weiwei)

Abstract: In response to the urgent need for domestic substitution and performance improvement of carbon fiber composites, focusing onthe key technical bottlenecks such as fiber damage, agglomeration and coating uniformity during the surface modification process of carbon fibers, a new non-electrode surface modification method was proposed, aiming to break through the limitations of traditional electrode processes and achieve the independent and controllable preparation of high-performance carbon fiber composites. A new cathodic polarization process based on the synergy of plasma field, temperature field, electric field, and fluid field was proposed innovatively. A low-temperature liquid environment effectively modified the carbon fiber surface to avoid the damage to the fiber structure caused by high-temperature treatment. Meanwhile, the Coulombic agglomeration of ultrafine short fibers in the electrolyte was inhibited by multi-physics field coupling, and the dispersion of fibers was improved by more than 30%. A non-electrode grate-tooth double-motion continuous deposition device was designed to achieve highly uniform nickel plating on the surface of ultrafine short fibers (carbon-nickel mass ratio 0. 1-0. 15), and deformable dynamic sealing materials with a density of 0. 5-1 g/cm^3 were successfully prepared. A jet-type liquid-phase plasma device was developed, which broke through the limitation of uniform deposition of ceramic coatings on the complex structures of wide two-dimensional fabrics and three-dimensional preforms. When applied to carbon phenolic heat-resistant sleeve components, the temperature resistance performance was greatly improved. This study provides a brand-new solution for preparing high-performance chopped carbon fiber composites, significantly enhancing the independent support capacity for aero materials.

Keywords: carbon fiber; non-electrode modification; cathodic polarization; liquid phase plasma; dynamic sealing material

Effect of ZrO_2 Reinforcement Concentration on the Growth of Plasma Electrolytic Oxide Layer on Nanocomposite Mg-Zn-Y/ZrO_2

Anawati Anawati

Physics Department, Faculty of Mathematics and Natural Sciences, Universitas Indonesia

E-mail address: anawati@ sci. ui. ac. id (Anawati Anawati)

Abstract: Magnesium alloys Mg-Zn-Y are potentially applied as biodegradable and bioabsorbable implant materials. The mechanical properties of the alloys need to be improved. The addition of reinforcement particles is proposed to enhance the mechanical properties of the alloy. Nanoparticle ZrO_2 with concentrations of 0. 6, 0. 8, and 1 wt% was used as reinforcement for preparing nanocomposite Mg-1Zn-2. 9Y/ZrO_2 by spark plasma sintering (SPS). The fabrication temperature was at 500 ℃ followed by solution treatment at T6 condition. The compressive test revealed significant improvement in the ultimate strength and yield strength of the alloy by the presence of reinforcement particles. However, no significant improvement was obtained in the degradation resistance of the alloys. Plasma electrolytic oxidation (PEO) is applied to the composites in 0. 25 M Na_3PO_4 + 10 g/l KOH for 5 min to improve the corrosion resistance. The variation of reinforcement concentration on the development of PEO coating was investigated using SEM and XRD instruments. The corrosion resistance was evaluated by the polarization test and electrochemical impedance spectroscopy measurements (EIS) in 0. 9 wt% NaCl solution at 37 ℃. The resulting PEO coatings were composed of amorphous magnesium phosphate and oxide phases. Voltage drop was observed during the coating formation on the Mg-Zn-Y/0. 6ZrO_2 and 0. 8ZrO_2 specimens while a relatively stable voltage was demonstrated for the Mg-Zn-Y and Mg-Zn-Y/1ZrO_2. The PEO coating consistently ennobled the corrosion potential of all the specimens.

Keywords: magnesium, composite, Ti foil; plasma electrolytic oxidation; electrolyte concentration; discharge voltage; lithium ion battery anode; corrosion

Fabrication and Corrosion Behavior of Plasma Electrolytic Oxidation Coatings on Zr_3Al Intermetallic Compound

Xue Wenbin

School of Physics and Astronomy, Beijing Normal University, Beijing, China

E-mail address: xuewb@bnu.edu.cn (Xue Wenbin)

Abstract: . The Zr_3Al intermetallic compound has potential application in the nuclear energy and aerospace industries. It is expected to improve the wear and corrosion resistance of Zr_3Al based alloy through appropriate surface modification treatment. In this work, a thin and compact film was fabricated by plasma electrolytic oxidation (PEO) on Zr_3Al based intermetallic alloy. The morphology, microstructure and composition of PEO film on Zr_3Al based alloy were analyzed. The surface charge of Zr_3Al based alloy and its PEO film were measured in different solutions by streaming current method. In-situ 768 h electrochemical impedance spectroscopy (EIS) tests of bare and PEO-coated Zr_3Al based alloys were performed in 300℃/14MPa lithium borate buffer solution. It was found that the PEO film on Zr_3Al intermetallic compound mainly consisted of t-ZrO_2 phase, m-ZrO_2 phase and a small amount of Al_2O_3 phase. The Al_2O_3 phase could stabilize t-ZrO_2 phase in film and reduce the phase transformation from t-ZrO_2 phase to m-ZrO_2 phase. The LiCl solution and NaCl solution behaved as inert electrolyte for PEO film. In addition, the low zeta potential reduced the adsorption of chloride ions and made the PEO film less susceptible to chloride attack. The PEO film with proper surface charge can also protect Zr_3Al alloy from the attack of hydrogen ions and lithium ions. The high temperature EIS tests indicated that the impedance of PEO film was higher than that of bare Zr_3Al based alloy in high temperature and high pressure lithium borate solution. Due to the partial chemical dissolution of outer layer, the impedance of outer layer of PEO film decreased while the impedance of inner layer increased until the end of 768 h exposure time. The uniform compact inner layer played a key role in improving the corrosion resistance of Zr_3Al based alloy.

Keywords: Zr_3Al intermetallic compound; plasma electrolytic oxidation; surface charge; high temperature in-situ electrochemical test; electrochemical impedance spectroscopy

Microstructure and Properties of a Tri-layered Micro-arc Oxidation Coating on Hf-Nb-Ta-Zr Refractory High-entropy Alloy

Xu Chi*, Xue Wenbin

Key Laboratory of Beam Technology of Ministry of Education, School of Physics and Astronomy,

Beijing Normal University, Beijing, China

E-mail address: xuchi@bnu.edu.cn (Xu Chi)

E-mail address: xuewb@bnu.edu.cn (Xue Wenbin)

Abstract: . This work presents the microstructure and properties of miro-arc oxidation (MAO) coatings on Hf-Nb-Ta-Zr high-entropy alloy (HEA). MAO has traditionally been applied to valve metals such as Al, Mg, Ti, and Zr, attracting significant attention due to its ability to fabricate uniform ceramic coatings on intricate metal parts. With increasing research focus drawn on HEAs, applying the MAO coating technique to HEAs is of great interest to further improve their hardness, wear or corrosion resistance, and other properties. In this work, MAO coatings were prepared on the Hf-Nb-Ta-Zr HEA at different treatment durations (20, 40 and 60 min) in a phosphate electrolyte. Their microstructure and phase compositions were characterized using scanning electron microscopy / transmission electron microscopy (SEM/TEM), X-ray diffraction (XRD), and glow discharge optical emission spectroscopy (GDOES) techniques. The MAO coating thicknesses range from ~4 to ~7.5 μm with increasing treatment durations. The coatings consisted of tetragonal and monoclinic ZrO_2/ HfO_2 phases, together with Nb_2O_5 and Ta_2O_5 phases. TEM analysis revealed that the MAO coating consists a tri-layered structure: an inner amorphous layer, a polycrystalline middle layer and an outer amorphous layer. The amorphous layers contained uniform distributions of Hf, Nb, Ta, Zr elements, forming compact, continuous layers. Whereas the middle layer comprised inner equi-axed grain zones and outer columnar grain zones, both containing mixed oxides of ZrO_2/ HfO_2 phases and Nb_2O_5/Ta_2O_5 phases. Surface roughness and hardness were significantly improved for the MAO coated samples compared to the HEA substrate. The coatings exhibited strong adhesion strengths ranging from ~8.40 to ~12.85 N, measured using the electrical contact scratch test method. The corrosion behavior of the MAO coatings was assessed using electrochemical impedance spectroscopy (EIS). The MAO treated Hf-Nb-Ta-Zr HEA exhibited a lower corrosion current density and higher electrochemical impedance. The MAO surface treatment markedly enhanced corrosion resistance, with the fully amorphous outer layer and amorphous barrier layer playing crucial roles in corrosion protection. The coating prepared at +480 V/ −100 V for 60 min, with a high corrosion potential of −0.28 V and low corrosion current density of 2.1×10^{-9} A · cm^{-2}, showed the best corrosion resistance.

Keywords: Hf-Nb-Ta-Zr; high entropy alloy; micro-arc oxidation; microstructures; properties; electrochemical corrosion

Evolution of B_4C Reinforcements During the PEO of B_4C/Al Matrix Composite and Its Effect on Properties

Song Yingwei*, Zhu Mingyu, Dong Kaihui, Han Enhou

Key Laboratory of Nuclear Materials and Safety Assessment, Institute of Metal Research, Chinese Academy of Sciences, Shenyang 110016, China

Abstract: B_4C/Al matrix composite (AMC) with excellent neutron absorption property can be used as grid materials for storage and transportation of spent fuel in nuclear power plants. However, the grid materials need to have good thermal control and long-term durability due to the long service life more than 60 years at a high temperature of nearly 400 ℃, so surface protection is required for AMC. PEO is an effective method commonly used to improve the thermal control and corrosion resistance of Al alloys and their composites. However, the reinforcements in the AMC have a great influence on the microstructure and properties of the PEO film. In this paper, the quality of PEO film on AMC with different B_4C content was compared, and the evolution of B_4C reinforcements during the PEO process was studied. The results show that the B_4C reinforcements, as a semiconductor structure, can inhibit the spark discharge at the initial stage of oxidation and hinder the formation of oxide film. With the increase of oxidation voltage, B_4C reinforcements are oxidized to B_2O_3, and the compactness of PEO film is improved. Thermal control and long-term durability properties can satisfy the requirement for actual application.

Keywords: B_4C reinforcements; B_4C/Al matrix composites; PEO; thermal control; corrosion resistance

Sol-Gel Derived Diopside Nano Powder Coatings with Multifunctional Elements on Plasma Electrolytic Oxidized Surface

So Yun Joo*, Han Cheol Choe

Advanced Functional Surface and Biomaterials Research Lab, Department of Dental Materials and Convergence Research Center for Treatment of Oral Soft Tissue Disease (MRC), College of Dentistry, Chosun University, Korea

E-mail address: wnthdbs1933@naver.com (So Yun Joo)

E-mail address: hcchoe@chosun.ac.kr (Han Cheol Choe)

Abstract: Known for its excellent mechanical strength, biocompatibility, and ability to induce hydroxyapatite formation, diopside ($CaMgSi_2O_6$) has emerged as a promising bioceramic for biomedical applications such as bone and dental implants. It offers superior mechanical properties compared to hydroxyapatite while maintaining favorable biological interactions. Substituting functional elements like Zn and Sr into the diopside structure has been reported to further improve its biological activity and mechanical performance, making it an attractive material for biomedical surface coatings. In this study, sol-gel synthesized diopside and Zn/Sr-substituted diopside nanoparticles were incorporated onto titanium alloy surfaces via plasma electrolytic oxidation (PEO). This approach aims to enhance bioactivity, corrosion resistance, and osseointegration by leveraging the properties of both pure and modified diopside. Characterization techniques including SEM, AFM, contact angle analysis, XPS, corrosion testing, and *in vitro* experiments were used to evaluate the coated surfaces. The results demonstrated that the addition of nanosized diopside and its substituted variants contributed to improved surface morphology, enhanced biological response, and increased corrosion protection, highlighting the potential of diopside-based PEO coatings for improving the performance of titanium implants. (Supported by National Research Foundation of Korea: 2023R1A2C1005748; RS-2023-00222390; RS-2023-00285353; hcchoe@chosun.ac.kr).

Keywords: diopside; plasma electrolytic oxidation; sol-gel; multifunctional element; titanium implant

Exploration of Laser Hybrid Micro-Arc Oxidation Technology for Light Alloys

Wu Guolong[1,2]

[1] College of Mechanical Engineering, Zhejiang University of Technology, Hangzhou 310014, China

[2] Institute of Laser Advanced Manufacturing, Zhejiang University of Technology, Hangzhou 310014, China

Abstract: Although micro-arc oxidation (MAO) technology shows unique advantages in light alloy surface modification, its inherent defects in difficult processing in the constituent area, high energy consumption and limited performance enhancement have restricted its application in high-end equipment fields such as aerospace, marine engineering and biomedical. To solve these technical bottlenecks, the emerging laser hybrid micro-arc oxidation (LHM) technology has gained rapid development in recent years through the synergistic coupling of laser energy and electrochemical processes. Current hybrid strategies for LHM technology can be divided into two types: process sequential hybrid and energy field synchronized hybrid. This technology not only breaks through the limitations of the single MAO process in terms of high energy consumption, low processing efficiency and limited enhancement of coating performance, but also realizes the directional regulation of electrochemical processes through the laser-induced thermal-force-photoelectric effect modulating the thermodynamic and kinetic properties of the electrode surface reaction. Based on this technological framework, this work systematically explored the fundamental mechanisms and implementation methods of LHM technology in light alloy surface engineering, focusing on three breakthrough application directions: 1) the combination of MAO and laser additive manufacturing to realize the precision modification of complex alloy components, 2) the development of gradient structural coatings by laser surface-strengthened hybrid MAO technology, and 3) laser-MAO synchronous hybrid to realize real-time interaction with multiple energy fields. In addition, through in-depth analysis of the improvement mechanism of functional coatings on light alloy surfaces, this study further proposed the future development trend from the dimensions of processing efficiency, performance enhancement, process strengthening and equipment innovation.

Fabrication and Characterization of Microarc Oxidation Coatings on Al-50Si Alloy with Ultra-high Si Content

Li Xin[1], Wang Lijiao[1], Chai Luqi[1], Jin Xiaoyue[2], Xu Chi[1], Du Jiancheng[1], Zhang Yongzhong[3], Xue Wenbin[1,2]*

[1] Key Laboratory of Beam Technology of Ministry of Education, School of Physics and Astronomy, Beijing Normal University, Beijing 100875, China

[2] Institute of Radiation Technology, Beijing Academy of Science and Technology, Beijing 100875, China

[3] General Research Institute for Nonferrous Metals, Beijing 100088, China

E-mail address: lx492562143@163.com (Li Xin)

E-mail address: xuewb@bnu.edu.cn (Xue Wenbin)

Abstract: The Al-Si alloys with ultra-high silicon content are strongly concerned in the fields of electronic equipment and electronic packaging due to their low thermal expansion coefficient, excellent thermal conductivity and good processing ability. Thus the high-silicon Al alloys become one of the most important materials in the field of modern electronic packaging. However, they also face great limitations in applications, especially in harsh environments. The Si particles promotes the occurrence of local corrosion of the high-silicon Al-Si alloys. In addition, the insulation properties are also critical for electronic package materials. Microarc oxidation (MAO) presents a promising surface modification technique to enhance these properties. It is found that the MAO technology is well suitable for the low-silicon aluminum alloys (<20 wt% Si) to fabricate the MAO coatings with high compactness and excellent wear and corrosion resistances. When the Si content increases, the compactness of MAO coatings becomes worse and their properties are strongly affected by Si content. These Si particles have low oxidation rate due to their low conductive property. In this study, the ceramic coatings on Al-50wt. % Si alloy with ultra-high silicon content were fabricated using microarc oxidation (MAO) technology under different voltages in silicate solution. Their morphology, microstructure, element composition and phase constituents were analyzed, meanwhile, the dielectric property, electrochemical corrosion behavior and tribocorrosion performance of the coatings were evaluated. Moreover, the growth mechanism of MAO coatings on high-silicon aluminum alloy was also discussed. It is found that the Si particles in Al-50Si alloy have been oxidized, but they also hinder the growth of MAO coatings to some extent. Some fine residual Si particles are remained in the MAO coating close to the coating/alloy interface at +480 V/-100 V, but they are hardly observed in the coatings while the positive voltage rises and the spark discharge enhances. The MAO coatings under different voltages consist of γ-Al_2O_3, mullite and amorphous SiO_2 phases, meanwhile, the amorphous phase increases with the increase of positive voltage. The coatings under different voltages reveal good hydrophilicity, and their surface free energy are at about 70 mJ · cm^{-2}. When the voltage is +480 V/ -100 V, the dense coating was prepared, which has the better insulation, corrosion resistance and tribocorrosion performance.

Keywords: microarc oxidation; high-silicon aluminum alloy; ceramic coatings; preparation and characterization; corrosion resistance

Preparation and Characterization of the Low-energy Plasma Electrolysis Oxide Coatings on Mg-Li Alloy

Chen Fei*, Tian Haoyue, Dou Zheng, Wei Yuankun, Zhao Zicong

College of New Materials and Chemical Engineering, Beijing Institute of Petrochemical Technology, Beijing 102617, China

Abstract: Plasma electrolytic oxidation (PEO) is a promising surface treatment method to improve the surface properties of light alloys. However, the high operating voltages led to significant power consumption and a burden on the grid, which limited its application. This work, designed to reduce the energy consumption of PEO on Mg-Li alloy, successfully formed the low energy plasma electrolysis oxidation (LEPEO) coating at a relatively low voltage (~100 V) by using an organic-inorganic mixed electrolyte solution. The energy consumption per unit volume (ECPUV) of the LEPEO process is 8.3 kJ/(dm2 · μm), which is approximate 57.0 % energy consumption savings compared with the PEO process (19.3 kJ/(dm2 · μm), NaOH-Na2SiO3 electrolyte). Results showed no remarkable difference between the two coatings in terms of morphologies, thickness, element type and distribution. The fracture process and corrosion protection performance of the coatings were evaluated by means of in-situ SEM tensile test and electrochemical impedance spectroscopy (EIS). The LEPEO coating only consists of MgO and amorphousSiOx, showing higher tensile strength, deformation displacement and better corrosion resistance compared with the PEO coating. The improved surface properties and lower energy consumption of LEPEO coatings will facilitate the application and development of PEO technology.

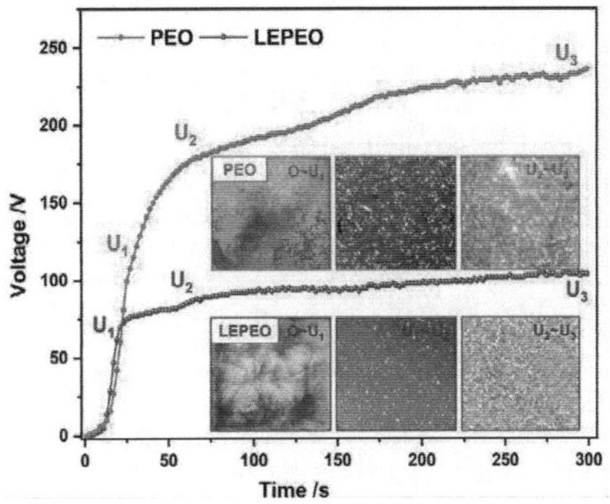

Fig. 1 Correlation of the PEO/LEPEO voltage and the specimen morphology with the PEO/LEPEO time at a constant current 3 A

Superior Tribocorrosion Resistance of Mg Metal Matrix Composite by Self- Sealing PEO Coating

Xie Yufei, Lu Xiaopeng*

School of Materials Science and Engineering, Northeastern University, Shenyang, China

E-mail address: xieyufei@ stumail. neu. edu. cn (Xie Yufei)

E-mail address: luxiaopeng@ mail. neu. edu. cn (Lu Xiaopeng)

Abstract: SiC/Mg metal matrix composites (MMCs) offer notable mechanical benefits, yet the reinforcing particles adversely impact corrosion resistance of Mg matrix. To concurrently enhance corrosion, wear, and tribocorrosion properties, self-densified plasma electrolytic oxidation (PEO) coatings were applied using phosphate-zirconium-fluoride (P-Zr-F) electrolytes. The findings revealed a two-order increase in corrosion resistance and hardness enhancement from 1. 5 to 5-6 GPa after PEO treatment. Compared to the porous PEO coating with traditional silicate electrolyte, the post-deposition behavior of low-melting point MgF2 in the sintering process results in a 39% reduction in the porosity of the P-Zr-F coating. The increased hardness caused by partial reactive incorporation of SiC and MgF_2-induced stable friction led to superior tribocorrosion performance of coatings on SiC/Mg MMC compared to base Mg alloy. This work presents novel perspectives to optimize MMC surface properties for extreme environments via PEO treatment

Keywords: Mg; tribocorrosion; PEO coating; metal matrix composite; corrosion

Tuning Corrosion Performance of Mg Alloy by Inhibitor and PEO Coating

Lu Xiaopeng

School of Materials Science and Engineering, Northeastern University, Shenyang, China

E-mail address: luxiaopeng@mail.neu.edu.cn (Lu Xiaopeng)

Abstract: Poor corrosion resistance is a critical factor that limits the wide range of applications for Mg and its alloys. In the present study, various kinds of corrosion inhibitors and plasma electrolytic oxidation (PEO) coatings were used and fabricated to tune and increase the corrosion performance of Mg surfaces. The influence of inhibitor on the morphology and chemical composition of the corrosion product layer was investigated to disclose the underlying inhibition mechanism. Inhibitor loaded nano/micro-containers were produced and subsequently incorporated into PEO coatings to enhance the corrosion resistance of Mg alloy.

Keywords: Mg; PEO coating; corrosion; inhibitor

Surface Modification via Plasma Electrolytic Oxidation for Dental Implant

Han Cheol Choe*

Advanced Functional Surface and Biomaterials Research Lab, Department of Dental Materials and Convergence Research Center for Treatment of Oral Soft Tissue Disease (MRC), College of Dentistry, Chosun University, Korea

E-mail address: hcchoe@ chosun. ac. kr (Han Cheol Choe)

Abstract: Pure titanium and titanium alloys have been regarded as a replacement for damaged bones and joints due to their excellent strength, fracture toughness, high corrosion resistance, and biocompatibility. Owing to the poor osseointegration capability, titanium implants may fail and are accompanied by bone resorption. As a result, the functional coatings for the implant's surface have been deliberated as an indispensable and inventive approach for improved bone-implant contact by generating micro- and nano-scale surface topographies. Hence, in this lecture, funtional coating method will be introduced for dental implant. In this study, the characteristics of dental implants doped with functional substances through various surface treatments were researched. In the case of femtosecond laser treated and TiO2/hydroxyapatite coated surface, these studies aimed to examine the effect of femtosecond laser (Fs) treatment, functional coatings, and plasma electrolytic oxidation (PEO) coatings on the surface characteristics of Ti-6Al-4V alloy with the evaluation of surface characterization, phase analysis, surface roughness, wettability, hardness, corrosion behavior, and adhesion potential. Also, a cell culture test was performed using MC3T3-E1 cells that indicated that Fs texturing and combined Fs texturing and PEO-coating can promote high cellular proliferation. The PEO oxide layer was doped with an osteoconductive bioinorganicfunctional elements, to enhance its biological performance. The coatings exhibited excellent hydrophilicity, owing to their porous morphology. MC3T3-E1 cell studies indicated that functional elements promoted high cell proliferation. (Supported by National Research Foundation of Korea: 2023R1A2C1005748; RS-2023-00222390; hcchoe@ chosun. ac. kr).

Keywords: Plasma electrolytic oxidation, Functional Surface Modification, Dental Implnat, Biocompatibility, Ti-alloys

Effects of Phytate-Containing Electrolyte on Micro-Arc Oxidation Coating Properties and Its Spent Solution on Pengze Crucian Carp Growth

Yue Dizhu, Zhang Shufang, zhao Rongfang, Zhang Rongfa

[1] School of Materials and Energy, Jiangxi Science and Technology Normal University, Nanchang 330013, China

[2] Jiangxi Province Key Laboratory of Surface Engineering, Nanchang 330013, China

Abstract: We employed sodium phytate hydrate (Na12Phy) as the electrolyte for micro-arc oxidation (MAO) to systematically investigate its effects on the coating characteristics of magnesium and titanium alloys as well as the MAO solution failure process, while simultaneously evaluating the impact of spent electrolyte on the growth of Pengze crucian carp (Carassius auratus var. Pengze). The results demonstrated that Na12Phy effectively increased both phosphorus (P) and calcium (Ca) content in MAO coatings while enhancing coating thickness and bonding strength. The MAO treatment significantly improved the degradation resistance and in vitro cytocompatibility of magnesium alloys. LC-MS/MS analysis revealed that during MAO processing, Na12Phy molecules hydrolyzed from antinutritional compounds into fish-absorbable inorganic phosphates and a series of physiologically active lower inositol phosphates. Although the degradation resistance gradually decreased with increasing MAO treatment cycles, the samples maintained good cytocompatibility even after 50 cycles. Notably, the 96-hour acute toxicity test showed no toxicity of 50% MAO waste solution to Pengze crucian carp, and subsequent aquaculture experiments confirmed that the MAO waste solution actually promoted fish growth. These findings collectively suggest that Na12Phy-containing MAO solutions possess significant potential for future applications.

Keywords: magnesium alloys; titanium alloy; MAO; Na12Phy; degradation resistance; acute toxicity test

Controllable Preparation of High-Performance Thermal Control Coatings by Plasma Electrolytic Oxidation

Liang Jun[1*], Peng Zhenjun[2]

[1] Research Institute of Interdisciplinary Science & School of Materials Science and Engineering, Dongguan University of Technology, China

[2] State Key Laboratory of Solid Lubrication, Lanzhou Institute of Chemical Physics, Chinese Academy of Sciences, China

E-mail address: jlianglicp@ 163. com (Jun Liang)

E-mail address: zhenjunpeng2010@ licp. cas. cn (Zhenjun Peng)

Abstract: Lightweight metals such as aluminum, magnesium, and titanium alloys are widely used in spacecraft structural components, which usually need to be coated with thermal control coatings for the passive thermal management. The commonly used thermal control coatings are organic-containing paints,' but they are prone to degradation in the space irradiation environment, leading to the deterioration of thermal control performance. In recent years, the preparation of inorganic thermal control coatings on aluminum, magnesium, and titanium alloys by plasma electrolytic oxidation to replace organic paints has received increasing attention. In this report, we summarized the research status of thermal control coatings prepared by plasma electrolytic oxidation. And controllable preparation of the black and white coatings with excellent thermal control performance was achieved by regulating the growth process of plasma electrolytic oxidation coatings through electrolyte composition design and process parameter optimization. In addition, the evaluation of the simulated space environment tests show that the coatings have good environmental stability. The high-performance thermal control coatings prepared by this method have been successfully applied to spacecraft components.

Keywords: plasma electrolytic oxidation; thermal control coating; controllable preparation; environmental stability

Plasma Electrolytic Oxidation Coating with Sustainable Corrosion Protection Through Addition of Y_2O_3 Nanoparticles at Ultra-High Frequency pulse current

Dong Shuai

Shanghai Jiao Tong University, China

E-mail address: liuxiaohe@ sjtu. edu. cn (Dong Shuai)

Abstract: This work aims to design and validate the synergistic effects of ultra-high frequency and Y_2O_3 nanoparticles on sustainable corrosion resistance of plasma electrolytic oxidation (PEO) upon Mg alloys. Incorporation of Y_2O_3 nanoparticles into PEO coatings was efficient, with preferential growth on coating surface and within discharge channels. In particular, a high concentration of Y_2O_3 nanoparticles, up to 20. 6 at. %, was observed in PEO coating prepared at the highest frequency (20 kHz) pulse current. Y_2O_3-based compounds such as YOOH and $Y(OH)_3$ present within PEO coating exhibit superior electrochemical stability, thereby enhancing overall stability and corrosion resistance.

Keywords: plasma electrolytic oxidation; ultra-high frequency; Y_2O_3 nanoparticles; magnesium alloys

A Systematic Study on the Temporal Evolution of Soft Sparking Discharge Characteristics and Its Influence on the Growth Behavior of Al_2O_3 Coatings

Lu Xuchen, Liu Cancan
Nanjing Tech University, China
E-mail address: xuchenlu113@ 163. com (Lu Xuchen)
E-mail address: liucan9933@ 163. com (Liu Cancan)

Abstract: To explore the temporal evolution of individual and collective characteristics of soft sparking on the anode, plasma electrolytic oxidation (PEO) process was conducted on 6061 Al alloy in phosphate-silicate composite electrolytes for 9000 s. Effects of the soft sparking regimes on the growth behaviors of the PEO coatings on Al substrate were studied. Results showed that the traditional soft sparking was difficult to sustain. After 3600s, most of soft sparks tended to cluster together and move rapidly as a collective group across the anode surface, entering the cluster soft sparking stage. During the two soft sparking regimes, the thickening of PEO coatings mainly occurred in the dense inner layer, which was beneath the lower boundary of the original MC coating. The two soft sparking regimes facilitated the growth of the inner dense layer of Al_2O_3. About 6300s later, the cluster soft sparking evolved into intense arc, which seriously destroyed the integrity of the dense inner layer. Besides, the mechanisms for forming the thick and dense inner Al_2O_3 layers during soft sparking regimes were proposed.

Keywords: plasma electrolytic oxidation; soft sparking; temporal evolution; growth behavior

A Precise Machining Surface by Electrolyte Plasma Polishing on Titanium Alloy: Microstructure and In-situ Mechanical Behaviors

Zou Yongchun*, Wang Shuqi, Chen Guoliang, Wang Yaming*, Zhou Yu

Institute for Advanced Ceramic, Harbin Institute of Technology, Harbin 150080, Heilongjiang

Abstract: As an innovative finishing technique, electrolytic plasma polishing (EPP) is effective in improving the surface quality of Ti-6Al-4V alloy. In this work, the microstructural evolution of Ti-6Al-4V alloy at different polishing voltages was investigated by means scanning electron microscopy (SEM), confocal laser scanning microscopy (CLSM), transmission electron microscopy (TEM). Furthermore, the *in-situ* tensile test was performed *via* SEM to systematically characterize the mechanical properties and fracture behaviors. According to the research results, EPP significantly reduced scratches and crests on the surface at 300 V, reducing the average surface roughness to 1/24 of its original value. Subsequent to EPP treatment at 300 V, the ultimate tensile strength (UTS) of the Ti-6Al-4V alloy peaked at 1094.5 ± 25.2 MPa, with an elongation increase of approximately 16 % compared to the untreated sample. Finally, fracture modeling was performed through finite element simulation and *in-situ* tensile test. Initiated in surface microcracks, the crack developed rapidly along slip lines. The tensile damage caused by axial stresses extended from surface microcracks to the entire fracture, thus causing the material to fracture. Future research will focus on the effectiveness of EPP in polishing high-roughness and complex 3D-printed surfaces. This would contribute to further expanding the application for EPP technology.

Keywords: electrolyte plasma polishing, titanium alloy, microstructure, surface roughness, mechanical behaviors

Study on the Effect of Anion Type on the Growth and Insulation Property of Plasma Electrolytic Oxidation Ceramic Insulating Film

Zhang Meng

State Key Laboratory of Electrical Insulation and Power Equipment, Xi'anJiaotong University, Xi'an 710049, China

Abstract: In plasma electrolytic oxidation technology, electrolyte anionic types significantly affect the film formation process. However, the specific mechanism by which anions regulate the microarc discharge behavior, crystalline evolution and film properties through interfacial reaction kinetics is not clear. In this study, we systematically analyzed the discharge behavior, dynamic growth of membrane layer and microstructure and material evolution by comparing the differences in the roles of three anions of silicate (SiO_3^{2-}), meta-aluminate (AlO_2^-) and meta-phosphate (PO_3^-) in the PEO process. The constitutive relationship between the chemical characteristics of anions and the film property was elucidated. The results showed that the anions had a significant effect on the arcing voltage, which was 447 V under the passivation of SiO_3^{2-}, while AlO_2^- reduced the arcing voltage to 359 V by providing an additional Al_2O_3 pathway and PO_3^- increased the arcing voltage to 481 V through its strong complexation. The growth rate of ceramic film in silicate electrolyte was the fastest, up to 2 μm/min. But it was easy to form Si-Al-O amorphous composite material on the film surface, which led to the increase of surface roughness to 1.5 μm and easily induced partial discharge and deterioration of the insulation strength. Meta-aluminate (AlO_2^-) accelerated the nucleation and crystallization of Al_2O_3 and the phase transition process. The degree of crystallinity could be up to 75% and the content of α-Al_2O_3 could reach up to 88%, which effectively improved the hardness and insulation property of ceramic film, and the insulation strength could reach 25.87 V/μm. The strong complexation of PO_3^- slowed down the growth of the ceramic film, but the crystallization rate was faster in the early stage of oxidation. Meta-phosphate (PO_3^-) contributed to the formation of ceramic film with uniform microporous structure, avoided the occurrence of boundary effect, effectively improved the surface uniformity of ceramic film, reduced the surface roughness of ceramic film to 0.6 μm. As a result, the surface roughness of the ceramic film was reduced to 0.6 μm, and the dielectric strength was dramatically improved to 41.76 V/μm. This study reveals the differential effects of anions on the film-forming process of PEO and the properties of ceramic films, which provides a basis for the selection of electrolyte for the preparation of functionalized ceramic films for specific applications.

Keywords: plasma electrolytic oxidation; electrolyte anions; ceramic insulating film; crystallinity; alumina phase transition

Fabrication of PEO-Coated Insulating Aluminum Wire and Its Application in Renewable Energy Electrical Equipment

Li Haomin

State Key Laboratory of Electrical Insulation and Power Equipment, Xi'anJiaotong University, Xi'an 710049, China

Abstract: With the development of renewable energy electrical equipment towards high power density, lightweight and miniaturization, conventional organic insulation faces the problems of poor heat resistance and insufficient thermal conductivity. Plasma electrolytic oxidation (PEO) can form highly thermally conductive and heat-resistant ceramic insulating coatings on aluminum wires. However, applying PEO for ceramic insulation on aluminum wires still faces challenges, including non-uniform ceramic coating growth and insulation property optimization. This paper systematically investigates the discharge behavior, growth behavior, organizational structure and insulation property of PEO ceramic film coatings and the correlation between them by regulating the parameters of the pulse power supply and optimizing the composition of the electrolyte. The results showed that during the PEO reaction process, the surface of the aluminum wire undergoes an adaptive voltage-current variation process regulated by the film thickness and electric field differences. In addition, the synergistic effect of high-energy microarc discharge and meta-aluminate ions (AlO_2^-) can significantly promote the nucleation and growth of α-Al_2O_3 phase. By modeling the quantitative relationship between discharge energy-phase composition-insulation performance and using machine learning for performance prediction, the volume resistivity of the ceramic coating was increased to 10^{12} $\Omega \cdot m$ and the dielectric strength was increased to 20 V/μm. Ultimately, the ceramic insulating aluminum wires were applied to the windings of dry-type transformers. Compared with the traditional organic insulated copper wires, it can reduce the volume and weight of the conductors by 30% and reduce the temperature rise by 26%. This technology provides a new type of insulating wire solution for renewable energy electrical equipment with light weight, high insulation and low cost.

Keywords: plasma electrolytic oxidation; renewable energy electrical equipment; ceramic insulating aluminum wire; insulation property

Influence of Plasma Electrolytic Oxidation Duration on Tetragonal Phase Formation and Corrosion Performance of Zr-4 Coatings

MamanKartaman Ajiriyanto[1*], Anawati Anawati[2]

[1] Research Center for Nuclear Material and Radioactive Waste Technology, Tangerang Selatan 15314, Indonesia

[2] Department of Physics, Faculty of Mathematics and Natural Sciences, Universitas Indonesia, Depok 16424, Indonesia

E-mail address: mama014@ brin. go. id (MamanKartaman Ajiriyanto)

E-mail address: anawati@ sci. ui. ac. id (Anawati Anawati)

Abstract: . The corrosion resistance of ceramic coatings produced by plasma electrolytic oxidation (PEO) can be significantly enhanced through particle embedding and process optimization. This study investigates the effect of PEO treatment time on the corrosion resistance of Y_2O_3-embedded coatings on Zr-4 alloy. Using a fixed concentration of Y_2O_3 nanoparticles in the electrolyte, coatings were synthesized with treatment durations of 5, 10, and 15 minutes. The nanoparticles were successfully incorporated into the ZrO_2 - SiO_2 matrix. Characterization revealed that both the amount of embedded yttrium (Y) and the coating thickness increased with longer PEO durations. Notably, the fraction of the zirconia tetragonal (t-ZrO_2) phase increased significantly from 20. 1, 50. 8 and 81. 2% at 5, 10, and 15 minutes, respectively. This phase transformation contributed to improvements in corrosion resistance and mechanical properties. Electrochemical testing showed a significant decrease in corrosion current density from 2.8×10^{-8} A/cm^2 at 10 minutes to 4.5×10^{-9} A/cm^2 at 15 minutes, indicating enhanced corrosion resistance. In parallel, coating hardness increased from 500 to 570 HV as treatment time increased. These results indicate that increasing the PEO treatment time promotes the formation of the tetragonal ZrO_2 phase and improves the coating morphology, both of which contribute significantly to the enhanced corrosion resistance and hardness of the PEO coatings.

Keywords: plasma electrolytic oxidation; zircaloy-4; zirconia; corrosion resistance; micro hardness

Chromium Oxide-Modified Zirconia Coating via Plasma Electrolysis

Juan Carlos Sihotang[1*], MamanKartaman Ajiriyanto[2], Anawati[1*]

[1] Department of Physics, Faculty of Mathematics and Natural Science, Universitas Indonesia, Depok, 16424, Indonesia

[2] Research Center for Nuclear Material and Radioactive Waste Technology, ORTN BRIN, Serpong, 15314, Indonesia

E-mail address: juan004@brin.go.id (Juan Carlos Sihotang)

E-mail address: anawati@sci.ui.ac.id (Anawati)

Abstract: . Plasma Electrolysis (PE) is an advanced surface modification technique employed to form oxide coatings on metallic substrates. To enhance coating performance, various particle additives can be introduced into the electrolyte. In this study, chromium (Cr) was incorporated into the oxide layer on pure zirconium (Zr) by dissolving a Cr-based compound in an electrolyte composed of 30 g/L $NaAlO_2$, 30 g/L KOH, 10 g/L ethanol, and 10 g/L K_2CrO_4. The PE treatments were conducted at a constant electrolyte temperature of 10 ℃, using either fixed current density (50 mA/cm^2) with varying treatment durations (3, 5, and 10 minutes) or varying current densities (50, 75, and 100 mA/cm^2) for 5 minutes. The surface morphology and elemental composition of the coatings were analyzed using scanning electron microscopy (SEM) and energy-dispersive X-ray spectroscopy (EDS). Phase identification was carried out via X-ray diffraction (XRD), Raman spectroscopy, and X-ray photoelectron spectroscopy (XPS). Mechanical performance was evaluated through micro-Vickers hardness testing, and electrochemical behavior was assessed using potentiodynamic polarization (PDP) and electrochemical impedance spectroscopy (EIS). The results showed that coating thickness increased with both current density and treatment time, ranging from 3.10 μm (3 min) to 8.04 μm (100 mA/cm^2). XRD and Raman spectroscopy confirmed tetragonal ZrO_2 as the dominant phase in all the PE coatings. EDS analysis verified successful Cr incorporation, with its content increasing with either current density or duration—reaching 2.73 and 2.77 at% at the highest duration and current density, respectively. XPS analysis further revealed the presence of chromium oxides in the coating. The coating formed at 100 mA/cm^2 or after 10 minutes demonstrated the highest hardness values, up to 763 ± 74 HV, attributed to increased thickness and a more developed oxide structure. Among the coatings produced at 50 mA/cm^2 with varying treatment durations, PE coating after 5 minutes exhibited the best corrosion resistance, with a corrosion current density of 2.33×10^{-10} A/cm^2.

Keywords: plasma electrolysis; zirconium; oxide layer; chromium oxide; corrosion resistance

Nanotube Shapes on the Ti Alloy by Anodization for Drug Doping

YoungEun Choe*, Seok-Jun Kim

Institute of Well-Aging Medicare & CSU G-LAMP Project Group, ChosunUniversity, Korea

E-mail address: yechoe22@gmail.com (YoungEun Choe)

E-mail address: heaven1472@chosun.ac.kr (Seok-Jun Kim)

Abstract: Most recently, to improve the biocompatibility, various processes that aim at coating an implant material withthe bioactive nanoparticles such as synthetic hydroxyapatite have been proposed. For this purpose, barrier and porous/tubular type of anodic oxide films could be formed by electrochemical anodization using a set of specific conditions including optimized potential, electrolyte composition, and temperature. Thus, in this work, the manufacture of the titanium dioxide (TiO_2) multi-layer was accurately performed in a mixed electrolyte containing hydroxyapatite (HAp) nanoparticles. A Ti substrate was first anodized through an electrochemical process with a mixed solution treatment of 1.5M $(NH_4) \cdot H_2PO_4 + 0.5M\ NH_4F$ to create TiO_2 nanotubes. FE-SEM examination revealed that unique geometric morphologies were well-formed on the surface, and the total thickness of the oxidized TiO_2 multilayer film was approximately 800 nm with a HAp layer thickness of approximately 100 nm. The HAp layer that formed on the TiO_2 nanotube film grew vertically, and the ultra-fine HAp nanoparticles were evenly distributed along the whole surface and electrodeposited onto each other. XPS peaks for the Ti-2p, O-1s, and Ca-2p, F-1s, C-1s, and N-2s transitions were continuously detected in the multilayer as the sputtering depth was increased (times of 100, 200, and 300 sec). The dominance of the Ti-2p and O-1s peaks, both before and after the Ar ion sputtering, shows that the surface consists mainly of an oxidized TiO_2 layer. Ca-2p and P-2p peaks were detected from the components of HAp nanoparticles electrodeposited onto the TiO_2 nanotube film during an electrochemical reaction. Thus, the incorporated phosphate species were found mostly in to the forms of $H_2PO_4^-$, HPO_4^{2-}, PO_4^{3-}, and PO_3^-. This interpretation is supported by the result showing that the binding energy of Ca-2p and P-2p in the multilayer surface is agreement with that from a quantitative analysis. (Supported by National Research Foundation of Korea: RS-2023-00285353; heaven1472@chosun.ac.kr).

Keywords: ti alloy; anodization; nanotube; bioactive nanoparticles; multilayer; titanium dioxide

Recent Advances in Mechanistic Understanding of Plasma Electrolytic Oxidation Processes

Aleksey Yerokhin
The University of Manchester, UK
E-mail address: Aleksey. Yerokhin@ manchester. ac. uk (Aleksey Yerokhin)

Abstract: Current trends towards sustainable, resource- and energy efficient manufacturing bring surface engineering of light alloys at the forefront of research interest. Plasma Electrolytic Oxidation (PEO) attracts significant attention as an advanced technology platform for high-performance ceramic coating of Al, Mg and Ti alloys, which offers numerous advantages in lightweighting of structural components, protection form wear and corrosion as well as development of new functional devices and consumer products. Advanced PEO coatings are formed under pulsed reversed polarisation conditions characterised by a large number of experimental variables, which provides significant challenges for process optimisation, diagnostics and control. We attempt to address these challenges by developing a mechanistic understanding of the behaviour of associated metal-oxide-electrolyte systems using original in-operando process diagnostic techniques. Resent results indicate that this behaviour is influenced by dynamic rearrangements in the barrier layer of the anodic oxide film growing under alternating cathodic and anodic polarisation. Although the barrier layer occupies a small portion of PEO coating located at the interface with the metal substrate, its evolution influences both structural and morphological transformations in the whole coating. This presentation will therefore discuss practical significance and implications of these findings for process energy efficiency and real-time control over coating characteristics and properties.

Nanotube Shapes on the Ti Alloy by Anodization for Drug Doping

YoungEun Choe*, Seok-Jun Kim

Institute of Well-Aging Medicare & CSU G-LAMP Project Group, ChosunUniversity, Korea

E-mail address: yechoe22@gmail.com (YoungEun Choe) E-mail address: heaven1472@chosun.ac.kr (Seok-Jun Kim)

Abstract: Most recently, to improve the biocompatibility, various processes that aim at coating an implant material withthe bioactive nanoparticles such as synthetic hydroxyapatite have been proposed. For this purpose, barrier and porous/tubular type of anodic oxide films could be formed by electrochemical anodization using a set of specific conditions including optimized potential, electrolyte composition, and temperature. Thus, in this work, the manufacture of the titanium dioxide (TiO_2) multi-layer was accurately performed in a mixed electrolyte containing hydroxyapatite (HAp) nanoparticles. A Ti substrate was first anodized through an electrochemical process with a mixed solution treatment of 1.5M $(NH_4) \cdot H_2PO_4$ + 0.5M NH_4F to create TiO_2 nanotubes. FE-SEM examination revealed that unique geometric morphologies were well-formed on the surface, and the total thickness of the oxidized TiO_2 multilayer film was approximately 800 nm with a HAp layer thickness of approximately 100 nm. The HAp layer that formed on the TiO_2 nanotube film grew vertically, and the ultra-fine HAp nanoparticles were evenly distributed along the whole surface and electrodeposited onto each other. XPS peaks for the Ti-2p, O-1s, and Ca-2p, F-1s, C-1s, and N-2s transitions were continuously detected in the multilayer as the sputtering depth was increased (times of 100, 200, and 300 sec). The dominance of the Ti-2p and O-1s peaks, both before and after the Ar ion sputtering, shows that the surface consists mainly of an oxidized TiO_2 layer. Ca-2p and P-2p peaks were detected from the components of HAp nanoparticles electrodeposited onto the TiO_2 nanotube film during an electrochemical reaction. Thus, the incorporated phosphate species were found mostly in to the forms of $H_2PO_4^-$, HPO_4^{2-}, PO_4^{3-}, and PO_3^-. This interpretation is supported by the result showing that the binding energy of Ca-2p and P-2p in the multilayer surface is agreement with that from a quantitative analysis. (Supported by National Research Foundation of Korea: RS-2023-00285353; heaven1472@chosun.ac.kr).

Keywords: Ti alloy; Anodization; Nanotube; Bioactive nanoparticles; Multilayer; Titanium dioxide

Revolutionizing Surface Engineering: Synergistic Effects of Ultrasonic Assistance in Plasma Electrolytic Oxidation-Mechanisms, Advancements, and Future Horizons

Zhang Xinxin

Huazhong University of Science and Technology

Abstract: Ultrasound-Auxiliary Micro-arc Oxidation (UMAO) is an emerging composite surface modification technology that significantly optimises the uniformity, compactness, and functional properties of traditional micro-arc oxidation (MAO) coatings through the synergistic interaction of ultrasonic cavitation effects and plasma discharge. In recent years, this technology has demonstrated unique potential in biomedical materials and marine engineering applications. Theultrasonic cavitation effects enhance electrolyte mass transfer, promote discharge uniformity, and suppress coating defects, thereby optimising the growth kinetics of MAO coatings. For instance, in the fabrication of Cu-doped TiO_2 coatings on titanium-based implants, ultrasonic assistance significantly improves the homogeneous incorporation of Cu, forming a dual-layered structure (an outer amorphous Ca-P-Cu composite layer and an inner porous TiO_2 layer). This structure simultaneously enhances corrosion resistance, interfacial adhesion, antibacterial capability, and cytocompatibility. In marine engineering materials, such as manganese-aluminium bronze (MAB) alloys, UMAO technology successfully prepares composite ceramic coatings with an Al_2O_3/SiO_2 matrix embedded with nano-Cu particles by regulating coating thickness (increased by 35%) and porosity (reduced by 32%). These coatings exhibit corrosion resistance improved by two orders of magnitude compared to conventional MAO coatings, along with excellent anti-biofouling capabilities. UMAO technology provides novel insights for designing multifunctional coatings across diverse fields. By exploring thescientific principles, challenges in industrial translation, and the interdisciplinary integration potential of UMAO, this work offers theoretical references for researchers and practitioners in surface engineering.

Environment-friendly and High-performance Post-sealing Technologies for Improving Corrosion Resistance of Anodized Aluminum Alloys

Ma Yanlong*

College of Materials Science and Engineering, Chongqing University of Technology, Chongqing, 400054, China

E-mail address: myl@ cqut. edu. cn (Ma Yanlong)

Abstract: Anodizing surface treatment has been widely used to improve corrosion resistance of aluminum alloys. Apart from the alloy substrate microstructure and anodizing process, post-sealing of the anodized alloy in a specific medium is deemed the key factor affecting the anti-corrosion performance of the coating system. The common anodic film sealing technologies include hot water sealing, chromate sealing, nickel salt sealing, sol-gel sealing, and organic sealing. However, the existing sealing technologies have some problems such as insufficient corrosion resistance or environmental issues. Thus, new post-sealing technologies based on layered double hydroxides (LDHs) and trivalent chromium process (TCP) have been developed. When applied these new technologies to anodized aluminium alloys such as 2024 and 7075, the treated samples can withstand more than 1000 h of neutral salt spray test without showing any visible corrosion pits. Besides, the treated samples also showed satisfying corrosion resistance in acidic environment. The reported technologies will have widespread application in aviation industry, where high corrosion resistance in both neutral and acidic environment is critical.

Keywords: Aluminum alloys; anodizing; post-sealing; corrosion resistance

The Evolution From Conventional Anodizing to Spark Anodizing (Plasma Electrolytic Oxidation) and Their Difference

Cheng Yingliang*
College of Materials Science and Engineering, Hunan University, Changsha, China
E-mail address: chengyingliang@ hun. edu. cn (Cheng Yingliang)

Abstract: Spark anodizing, also called plasma electrolytic oxidation (PEO), is an advanced electrochemically-based surface treatment method that converts the surface of metals into high performance ceramic oxides. PEO is evolved from the conventional anodizing but works under the high voltage end with the occurrence of plasma discharges. This presentation compares the similarity and differences between the two techniques. Although anodizing and PEO are both anodic techniques, they differ fundamentally in processing, mechanism, coating properties, and applications. Conventional anodizing operates at relatively low voltages (typically $\leqslant 100$ V), forming a porous, honeycomb-structured oxide film via electrochemical reactions. In contrast, PEO commonly utilizes pulse high-voltage (up to 600 V or beyond) to induce continuous plasma discharges on the metal surface. This generates localized plasma that momentarily melts the substrate, rending a very complex coating formation mechanism compared to all other coating procedures. Due to employment of pulse electrical regime, PEO is versatile and influenced by a various parameters such as frequency, duty cycle and cathodic polarization. Among these parameters, cathodic polarization is beneficial for improving the growth rate, homogeneity and performance of the resulting coatings. Particularly, soft sparking may occur under specific cathodic-to-anodic currrent/charge ratio, which is of great importance in obtaining high quality coating at lowered energy consumption. Conventional anodizing is usually confined to a few limited valve metals such as Al, Mg, Ti and their alloys. However, the application scope of PEO is broader with reports on some non-valve metals such as carbon steel, brass, copper and zinc alloys. PEO coatings usually exhibits exceptional microhardness, wear resistance, electrical insulation, and corrosion endurance, outperforming the mechanically adhered anodized films in both adhesion strength and long-term protection. Therefore, despite higher energy consumption and processing costs, PEO are driving its adoption in advanced industrial sectors, progressively displacing traditional anodizing where extreme durability is paramount.

Keywords: conventional anodizing; spark anodizing; plasma electrolytic oxidation(PEO)

Photocathodic Protection of LDHs-Coated Stainless Steels

Xu Teng, Zhang Lianjian, Yang Huiqing, Hu Jiming*
Department of Chemistry, Zhejiang University, Hangzhou 310058, China
E-mail address: kejmhu@ zju. edu. cn (Hu Jiming)

Abstract: Current photoanodes in photogenerated cathodic protection systems face persistent technical limitations, including fixed bandgap characteristics, narrow spectral absorption ranges, insufficient operational stability, complex synthesis protocols, dark-state inefficacy, and poor substrate adhesion limitations that restrict practical applications. To address these challenges, this study employed a facile cathodic electrodeposition method to directly fabricate six transition metal-incorporated layered double hydroxides (LDHs) coatings, CoAl, NiFe, NiCo, NiMn, NiCoFe, and NiCoMn LDH, in situ on 304 stainless steel substrates. Mott-Schottky curves and UV-Vis absorption spectra confirmed the semiconducting properties of all synthesized LDHs. Photoelectrochemical tests revealed that these semiconducting LDH films provided photogenerated cathodic protection to the 304 stainless steel substrates. The protection efficacy correlated strongly with their band gap values: narrower band gaps resulted in larger OCP drops, higher photogenerated current densities, and lower corrosion rate of 304SS substrates. Dark-state immersion experiments demonstrated that the semiconducting LDH coatings exhibited comparable corrosion protection performance to non-semiconducting ZnAl LDH and MgAl LDH coatings. The bi-functional advantages of semiconducting LDH films both in dark and light illumination synergistically enhanced the corrosion resistance of the 304 stainless steel substrates.

Keywords: layered double hydroxides; corrosion protection; photogenerated cathodic protection

Discussion on the Growth of Anodic Oxide Films on Light Metals under High Electric Field

Sungmo Moon[1,2*]

[1] Energy and Environment Materials Research Division, Korea Institute of Materials Science, Changwon-si, Gyeongnam, Korea

[2] Advanced Materials Engineering, Korea University of Science and Technology, Daejeon, Korea

E-mail address: sungmo@ kims. re. kr (Sungmo Moon)

Abstract: This presentation discusses growth mechanisms of anodic oxide films on light metals, such as Al, Mg and Ti under high electric field. It focuses on the electrochemical and physical processes that govern the formation and dissolution of the anodic oxide films. The growth behavior is strongly influenced by various factors, including the nature of anions in the electrolyte, alloy composition, temperature, electrolyte convection and solution pH. Under a high electric field, the migration of cations and anions through the anodic oxide film drives not only the growth of oxide layer but also the generation of heat. If the ionic resistance of the oxide film is sufficiently low, thick films can be formed even at relatively low applied voltages. The ionic resistance is determined by the composition, structure, and defect concentration within the oxide layer. Additionally, the solution pH plays a critical role in shaping the growth dynamics. Chemical reactions at both the metal-oxide and oxide-electrolyte interfaces are also significant in determining the growth behavior. This presentation will provide a detailed analysis of the growth mechanisms of anodic oxide films under high electric field, with particular emphasis on ionic migration and the chemical and electrochemical reactions that occur in various electrolytes under high electric field.

Keywords: anodic oxide film; growth mechanism; migration of ions; electrochemical reactions

Frequency-Dependent Effect of Cathodic Polarization on Spark Anodizing of Zr Alloy

Zhang Qianna, Wu Jiale, Cheng Yingliang*

College of Materials Science and Engineering, Hunan University, Changsha 410082, China

E-mail address: chengyingliang@ hnu. edu. cn (Cheng Yingliang)

Abstract: : Cathodic polarization is a key factor influencing the behavior of spark anodizing (SA). This study systematically investigates its effects on the SA of a Zirlo alloy in a silicate electrolyte under two frequencies (100 Hz and 1000 Hz). Cathodic-to-anodic current density ratios of R = 0, 0.7, 1.1, 1.6 are used for the investigation. At 100 Hz, soft sparking occurred when R > 1, leading to coating delamination, reduced interface temperatures, and coatings dominated by monoclinic ZrO_2 (m-ZrO_2). In contrast, at 1000 Hz, soft sparking was suppressed even at R = 1.6, resulting in uniform, thicker, and more stable coatings. The 1000 Hz coatings with cathodic polarization showed higher Si content as well as more t-ZrO_2, reaching ~37 wt. % at R = 1.1. These results highlight that both frequency and cathodic polarization are critical in tailoring the microstructure, composition, and energy efficiency of SA coatings on Zirlo alloy.

Keywords: spark anodizing; cathodic polarization; frequency; phase transition; t-ZrO_2

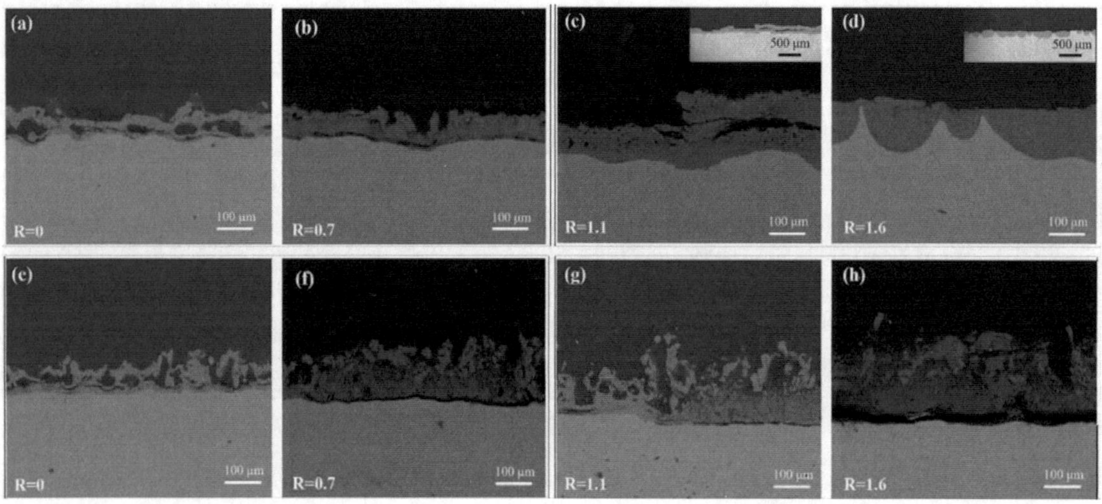

Fig. 1 SEM images showing the cross sections of the PEO coatings on Zirlo alloy formed under different R ratios at different frequencies: (a-d) 100 Hz; (e-f) 1000Hz

Biocompatibility of the Coatings on Tantalum Formed by Spark Anodizing in NaCl Solution

Wu Jiale, Zhang Qianna, Cheng Yingliang*

College of Materials Science and Engineering, Hunan University, Changsha 410082, China

E-mail address: chengyingliang@ hnu. edu. cn (Cheng Yingliang)

Abstract: : Sodium chloride (NaCl) is a well-known corrosive medium that is commonly employed in corrosion tests. Conventional valve metals tend to undergo rapid anodic dissolution when exposed to NaCl solution under applied anodic potentials, making NaCl as an electrolyte unsuitable for anodization processes. In this study, we found that spark anodizing coatings with enhanced biocompatibility can be formed on tantalum in NaCl electrolytes of varied concentrations (1 M, 0. 1 M, and 0. 01 M). The resulting oxide layers were systematically characterized by scanning electron microscopy (SEM) in conjunction with energy-dispersive X-ray spectroscopy (EDS) and X-ray diffraction (XRD). The biocompatibility of the coatings was evaluated by assessing the proliferation and morphology of mouse fibroblasts (L929) using Cell Counting Kit-8 (CCK-8) assays and SEM imaging. Results demonstrated that NaCl, despite its corrosive nature, can successfully initiate spark discharges on tantalum surfaces and lead to the formation of oxide coatings. These coatings exhibited distinct microstructural morphologies depending on the NaCl concentration, with all containing crystalline Ta_2O_5 phases. All coatings enhanced fibroblast proliferation, with the 0. 1 M NaCl-derived coating supporting the highest cell density by day seven. Furthermore, the porous structures formed at 0. 1 M and 1 M, as well as the nodular morphology observed at 0. 01 M, were found to be favorable for fibroblast adhesion, indicating strong potential for biomedical surface applications.

Keywords: tantalum; sodium chloride; plasma electrolytic oxidation; biocompatibility

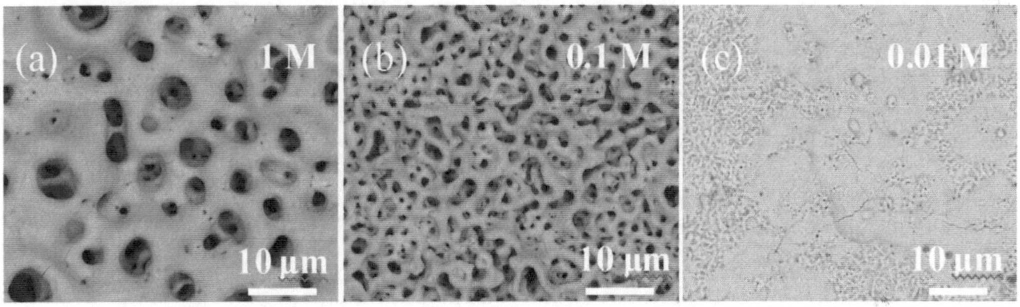

Fig. 1 SEM images showing the surface morphologies of oxide coatings formed on tantalum in NaCl electrolytes of varying concentrations

Anodizing Behavior of the Cast Magnesium Alloys in Ammonium Borate Solution

Wang Lishi *

Hubei Provincial Key Laboratory of Green Materials for Light Industry, School of Materials and Chemical Engineering, Hubei University of Technology, Wuhan 430068, China

E-mail address: lswang@ hbut. edu. cn (Wang Lishi)

Abstract: The anodizing processes of magnesium and its alloys often related multi-component electrolyte composition, such as commercial treatment HAE, DOW17, Anomag, Tagnite, Magoxid and Keronite etc. The formed anodic coating usually contains various pore structure. The pore-type structure is originated from the multi-function of electrolyte solution, the initial barrier layer and local electric field. The formation of barrier layer can employ single component electrolyte. This study investigated the coating formation and growth for pure magnesium and as-cast magnesium alloy under the conditions of an 0. 1 M ammonium borate electrolyte solution. The voltage-time curves were recorded with the constant current density $10mA/cm^2$ and $50mA/cm^2$. By virtue of ultramicrotomy technique for magnesium specimen preparation, this paper observed the cross-section structure and barrier layer morphology of the anodic oxide coatings. The second phase adjacent the surface region inhibited the coating growth.

Keywords: pure magnesium; cast magnesium alloy; anodizing; ammonium borate solution; coating growth; barrier layer

Electrochemical Anodization of TiAl Alloy and Its Oxidation Performance

Wu Liankui*, Yan Haojie, Tao Fan, Sun Qingqing, Cao Fahe

School of Materials, Sun Yat-sen University, Shenzhen 518107, China

E-mail address: wulk5@mail.sysu.edu.cn (Wu Liankui)

Abstract: The long-term isothermal oxidation and cyclic oxidation performance of the Ti48Al2Nb2Cr alloy anodized in NH_4F containing electrolyte were investigated. Results showed that the preliminarily generated Al_2O_3 and TiO_2 components at the anodization process benefit the formation of the α-Al_2O_3-based oxide scale. During thermal exposure, Al_2O_3 acts as the crystal nucleus and promotes the thermal nucleation of α-Al_2O_3, and TiO_2 promotes the direct growth of Al_2O_3 to α-Al_2O_3 owing to their well-matched oxygen sublattices. After isothermal oxidation at 1000 ℃ and cyclic oxidation at 900 ℃ for 1000 h, the weight gains of the anodized Ti48Al2Nb2Cr alloys are 1.12 mg cm^{-2} and 0.48 mg cm^{-2}, respectively. The excellent oxidation performance is ascribed to the fluorine effect. Raman and photo-stimulated luminescence spectra revealed that anodization impedes the formation of rutile and promotes the generation of α-Al_2O_3 at high temperatures. When exposed to air at high temperatures, TiF_4 evaporates, while AlF_3 and CrF_3 transform to their corresponding oxides. The transformation of Cr_2O_3 from chromium fluorine promotes the generation of Al_2O_3.

Keywords: TiAl alloy; anodization; high temperation oxidation; fluorine effect; raman spectra

Synergistic Protective Mechanisms of Micro-Arc Oxidation and Sealing Treatments on Cold-Spray-Repaired 2024Al and ZM6 alloys

Zhong Yuan, Huang Yueyu, Wang Qun *

College of Materials Science and Engineering, Hunan University, Changsha, China

E-mail address: wangqun@hnu.edu.cn (Wang Qun)

Abstract: Cold spray (CS) repair of damaged aluminum and magnesium alloy components often results in interfaces that are either "homologous" (similar composition and structure) or "heterologous" (different structure or composition). To enhance the corrosion and wear resistance of such bonded materials, thick 2024Al and 6061Al coatings were deposited on rolled 2024Al-T3 and cast ZM6-Mg substrates, respectively, using CS technology. Subsequently, carbon nanotube (CNT)-reinforced ceramic composite coatings were fabricated via micro-arc oxidation (MAO) on the 2024Al-T3/2024Al-CS ("homologous" interface) and ZM6-Mg/6061Al-CS ("heterologous" interface) systems. The ceramic coatings were further modified through polymer sealing treatment. The surface and cross-sectional morphologies, mechanical performance, corrosion resistance, and wear behavior of the base materials, unsealed MAO (U-MAO), and sealed MAO (S-MAO) samples were systematically characterized using X-ray diffraction (XRD), Raman spectroscopy, scanning electron microscopy (SEM), transmission electron microscopy (TEM), potentiodynamic polarization (PDP), electrochemical impedance spectroscopy (EIS), and sliding wear tests. A comprehensive analysis was conducted to elucidate the formation mechanisms and failure behaviors of the MAO coatings. The results demonstrated that complete ceramic coatings were successfully formed on both types of interfaces using CNT-doped silicate electrolytes. The MAO + sealing treatment significantly reduced corrosion current density—by over two orders of magnitude for the 2024Al-T3/2024Al-CS system and over five orders for the ZM6-Mg/6061Al-CS system. Furthermore, wear rates were reduced by more than two orders of magnitude and one order of magnitude for the respective systems. This study presents a viable strategy for producing corrosion-and wear-resistant coatings on CS-repaired aluminum and magnesium alloy structures.

Keywords: cold spray; micro-arc oxidation; spark anodizing; corrosion; wear

Failure Analysis and Atmospheric Corrosion Behavior of Tin Plating Layer

Wu Tangqing*

School of Materials Science and Engineering, Xiangtan University, Xiangtan, China

E-mail address: tqwu@ xtu. edu. cn (Wu Tangqing)

Abstract: Two works will be presented in this speak. The first one was the failure cause of a tinned copper wire clamp in marine atmosphere. The main components of the green rust on the surface of the damaged wire clamp are SnO_2, CuO, $Cu_2Cl(OH)_3$ and $CuCO_3(OH)_2$. Much of green rust distributes at the platform edge along the axial direction on the crimp connection, and severe corrosion and corrosion pits occur at the platform edge zone along the axial direction. The enriching Cl-at the marine atmosphere and the existing O2 in air collectively enhance the corrosion process of the tin coating and the copper matrix. Finite element model results show that the residual stress and strain of the tin coating are the largest at the platform edge along the axial direction on the crimp connection, and the corresponding electrode potential of the tin coating at this zone drops significantly. The above results indicate that the residual strain increases the driving force of the corrosion electrochemical reactions and accelerates the corrosion rate and the pit corrosion of the tin coating at this zone. The second work was a comparative study of Atmospheric corrosion of tin coatings on H62 brass and T2 copper. The microstructure and atmospheric corrosion of the tin coatings on the H62 brass and T2 copper were compared by XRD, SEM/EDS and electrochemical measurements. The results showed that the preferred planes structure of the tin coatings were (101) and (112). More SiO2 inclusions existed in the tin coating on the brass than on the copper. In the atmospheric environment, the corrosion rates of tin coatings on the brass and copper were 1. 427 and 0. 578 $\mu m \cdot y^{-1}$, respectively. The local corrosion sensitivity of the tin coatings on the brass was higher than that on the copper. No matter on the copper or the brass, the local corrosion sensitivity and R of the skyward surface of the tin coatings were greater than that of the earthward surface. The preferential corrosion of the (101) oriented crystal particles was the dominated reason for the higher corrosion sensitivity of the tin coatings. The preferential wetting, the more suspended particles and the intense scouring of rainwater affected the corrosion process of the tin coating as well.

Keywords: failure analysis; atmospheric corrosion; tin plating layer

Tailored Fabrication of Versatile Nano-structured Coatings with Enhanced Electrical/Thermal Conductivity and Wear Resistance for Automotive Connectors and Power Electro-/Electrical Devices

Shochiku Kure, Song-Zhu S. Kure-Chu *

Jiacheng Liu, Shunsuke Mori, Yumi Hattori, Peng Wang, Erli Lin, Kouhei Okumura, Hidetoshi Miyazaki, Takehiko,

Hihara Materials Function and Design, Nagoya Institute of Technology, Nagoya, 466-8555 Japan

E-mail address: kure. shochiku@ nitech. ac. jp (Song-Zhu S. Kure-Chu)

Abstract: Surface finishing technology of materials plays an essential role in ensuring safety and performance in versatile fields including power electronic devices, automobiles and mechanical equipment, battery and energy conversion, as well as aircraft and aerospace parts. Various electrochemical methods, including electroplating, anodizing, electro-etching or electro-polishing, alternating current plating, as well as electroless plating, are usually used to fabricate versatile metallic and oxide films on various materials for broad industrial applications. We introduce several novelelectrochemical approaches to fabricate functional nanocomposite coatings/films on various non-ferrous materials through electroplating anodizing, direct and/or alternating current electrodeposition, e. g. to fabricate Ag (Sn, Cu, Ni)-Graphene composite coatings on copper and aluminum alloys with high electrical and thermal conductivities for automotive terminals/connectors and power electrical/electronic devices (Fig. 1a), Al_2O_3/Sn (Ni)-MoS_2 composite coatings on Al-Si-Cu alloys with enhanced lubricity/wear resistance for engine/transmission parts, TiO_2-TiN (-Sn, MoO_x, MoS_2, Ni-W-Mo) composite films on titanium for high-safety anode materials of lithium-ion batteries with large capacity and for super-hard engine parts in aircrafts (Fig. 1b), and Al (Ti, Mg)-Ni-W-Mo-O-based composite oxide coatings on Ti, Al, and Mg alloys for various transportation/aircraft parts with enhanced mechanical strength and excellent corrosion resistance.

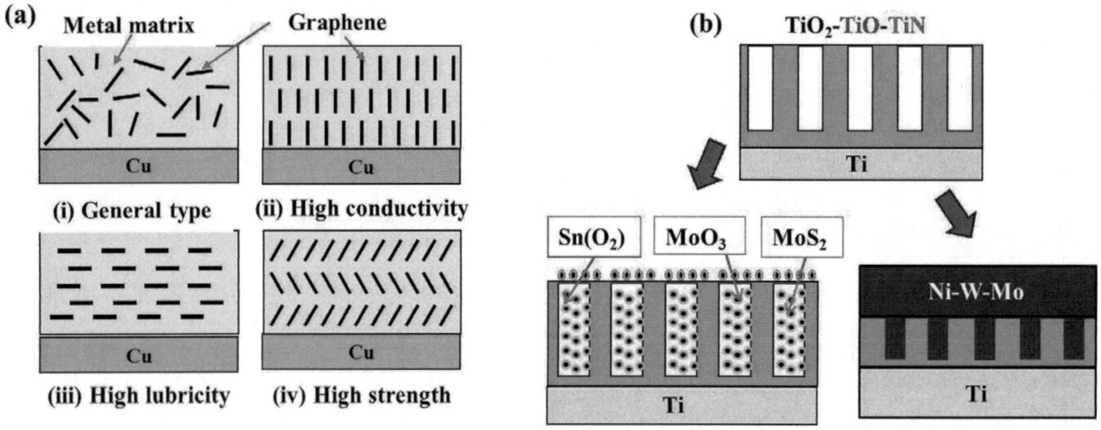

Fig. 1 Schematic drawing of cross-sections of functional coatings: (a) Ag-Graphene composite electroplating films on Cu alloys; (b) nanoporous TiO_2-TiO-TiN composite oxide films and TiO_2-TiO-TiN/($Sn(O_2)$, MoO_x, MoS_2, Ni-W-Mo) composite films on Ti sheets

Keywords: nano-structured coatings; contact resistance; thermal conductivity; wear resistance; super-hardness; capacity

Influence of Microstructural Evolution in Al-Mg-Si (-Cu) Aluminum Alloys on the Composition, Structure, and Corrosion Resistance of Anodic Oxide Films

Yu Mei*, Guo Zixin, Shi Haobo, Wu Yanlin, Liu Jianhua

School of Materials Science and Engineering, Beihang University, Beijing, China

E-mail address: yumei@buaa.edu.cn (Yu Mei)

Abstract: Aluminum alloy anodizing of is a necessary and commonly used surface treatment method to improve the corrosion resistance of aluminum alloy, and its process and the corrosion resistance of the formed film are affected by the grain size, grain boundary characteristics, and particle differences between metals. Therefore, the influence of microstructure on oxide film formation and corrosion resistance has been widely concerned. AA6061 is a commonly used Al-Mg-Si aluminum alloy, and its excellent properties make it widely used in the fields of vehicles, ships and aerospace. In this paper, AA6061 aluminum alloy was taken as the research object, and the commercial AA6061 aluminum alloy and discharge plasma sintering (SPS) AA6061 aluminum alloy were selected as the matrix, and the influence mechanism of the microstructure changes caused by alkali washing time, grain size, process path and closed process on the formation behavior of oxide film was discussed, and the corrosion resistance of the film was evaluated and the parameters were optimized. The large grain size reduces the grain boundary density, the matrix surface is more continuous, the current distribution during anodizing is more uniform, and the quality of the resulting film is better. The grain size is too small, resulting in a strong dissolution reaction in the matrix during anodic oxidation, and the oxidation efficiency is low. Increasing the distribution of large-angle grain boundaries is conducive to increasing the ion migration rate and increasing the growth rate of the oxide film. However, too high grain size will also make the potential distribution on the surface of the matrix too dispersed, which is not conducive to the stable formation of oxide film.

Keywords: aluminum alloy; anodizing; corrosion resistance; microstructure

Mechanisms of Cr Precursor Film "Catalyzing" Selective Oxidation of low Cr alloys

Xie Yun
NanchangHangkong University, Nanchang, China
E-mail address: yun. xie@ nchu. edu. cn (Xie Yun)

Abstract: For structural metallic materials aiming for high temperature applications, their Cr and Al contents have to be maintained at low levels so as to obtain excellent mechanical properties. As a result, they are incapable to achieve selective oxidation, i. e. thermally grow protective chromia or alumina scales, and the resistance to high temperature oxidation is unsatisfactory. According to theoretical analysis based on classical Wagner's oxidation theory, the critical Cr content ($N_{Cr}^{(1)}$) required for transition of Cr_2O_3 from internal precipitation to external scaling is lower than that ($N_{Cr}^{(2)}$) for sustaining the steady growth of an external Cr_2O_3 layer. This implies that the alloys with the Cr content lying between $N_{Cr}^{(2)}$ and $N_{Cr}^{(1)}$ have a marginal capability to from a protective Cr_2O_3 layer. Based on this finding, we propose a novel concept to "catalyze" the selective oxidation of low Cr alloys relying on their relatively low Cr content, by preparing a Cr precursor film (below 1 μm thick) which can assist the initial establishment of an Cr_2O_3 layer on the alloy surface. Therefore, the high Cr content threshold required by $N_{Cr}^{(1)}$ is overcome, and subsequent steady growth of the formed Cr_2O_3 layer can be sustained relying the alloys' relatively low Cr content. Eventually, selective oxidation of Cr can be achieved in low Cr alloys.

Keywords: low Cr alloys; Cr precursor film; selective oxidation; catalyzing

Selective Emitter Based on Photonic Structure for Terrestrial and Space Radiative Cooling

Lv Jinpeng

Nanjing University of Aeronautics and Astronautics, Nanjing, China

E-mail address: jinpenglv@ nuaa. edu. cn (Lv Jinpeng)

Abstract: Passive radiative cooling offers a promising solution to achieve *refrigeration* and global carbon neutrality without energy consumption. Herein, *photonic radiative cooler* based on fluoride-metal multilayer stack structure are developed and their outdoor cooling performances are demonstrated. The as-designed selective emitters exhibit ideal spectral properties with reflectance up to 97. 2% and atmospheric window emissivity up to 0. 83. Field tests illustrate that the selective emitters have excellent all-day sub-ambient radiative cooling performance. The MgF2/Ag emitter shows more than 19 ℃ all-day sub-ambient cooling ability despite the season, demonstrating its superior cooling performance. Furthermore, the real-world application test shows that the MgF2/Ag cube cooler has a maximum of 14 ℃ cooling effect under sunlight and an average cooling effect of 2 ℃ during night. The porotype cooler provides an innovative strategy for achieving deep sub-ambient cooling and broad energy saving applications.

Keywords: selective emitter; photonic structure; radiative cooling; sub-ambient cooling; multilayer stack

Effect of Slurry Solid Content and Sintering Temperature on Dielectric Breakdown Strength in Nanostructured Alumina Coatings During Feedstock Powder Preparation

Zhang Xiaodong

School of Materials Science and Engineering, Harbin Institute of Technology, China

E-mail address: zhangxiaodong@hit.edu.cn (Zhang Xiaodong)

Abstract: In this study, the effects of slurry solid content and sintering temperature of the reconstituted nano-Al_2O_3 feedstock powder on the phase composition, microstructure, and breakdown strength of the deposited coatings were investigated. The results indicated that the slurry solid content primarily influences the morphology of feedstock powders, while the sintering temperature affects the phase composition and grain size. The phase structure of the as-sprayed coatings was composed of α-Al_2O_3 and γ-Al_2O_3, regardless of whether the phase of the feedstock powder used was composed of θ and α phases or pure α phase. The highest dielectric breakdown strength of 10.32 ± 0.56 kV/mm was achieved when the feedstock powder preparation parameters were set at a slurry solid content of 60 wt% and a sintering temperature of 1100 ℃. A simplified model was proposed to describe the relationship between the dielectric breakdown strength of the deposited coating and the α phase and porosity. The dielectric breakdown strength of the deposited coatings is approximately proportional to the α phase content and inversely proportional to the porosity.

Keywords: slurry solids content; sintering temperature; feedstock powder; dielectric breakdown; Al_2O_3 coatings; Nanostructured coating

Design and Realization of Interface Strengthening to GNPs/YSZ Nanocomposite Coating

Liu Huixian, Wang You*, Zhang Xiaodong*
Harbin Institute of Technology, Harbin, China
E-mail address: wangyou@hit.edu.cn (Wang You)
E-mail address: zhangxiaodong@hit.edu.cn (Zhang Xiaodong)

Abstract: This study enhances the radiation resistance of GNPs/YSZ nanocomposites through interface engineering. To address the phase instability and radiation hardening issues of YSZ, functionalized graphene nanosheets (GNPs) were introduced to optimize the interface bonding. The 0.1-5.0 vol% composites were prepared using the colloidal treatment method and discharge plasma sintering (SPS). The results show that 0.5 vol% GNPs increase the fracture toughness by 15% (4.42 MPa · $m^{1/2}$) and the hardness by 22% (16.1 GPa), attributed to the extraction of GNPs and the deflection of cracks; higher concentrations lead to performance degradation due to agglomeration. The radiation test indicates that GNPs inhibit the phase transformation and charge accumulation of YSZ, reducing surface bubble formation by 40% and promoting defect elimination at the grain boundary interface. Interface functionalization enhances adhesion and reduces thermal stress. This study provides a solution with both toughness and stability for aerospace coatings. Future in-orbit verification will be conducted.

Keywords: graphene nanosheets; YSZ nanocomposites; interface engineering; radiation resistance; aerospace coatings

Proceedings of IFHTSE 2025

The 30th World Congress of International Federation for Heat Treatment and Surface Engineering
August, 19-22, 2025, Suzhou, China

Design and Performance Study of SiC-Reinforced Nanostructured Thermal Barrier Coatings Based on First-Principles Calculations

Han Xu, Wang You*, Zhang Xiaodong*
Harbin Institute of Technology, Harbin, China
E-mail address: wangyou@hit.edu.cn (Wang You)
E-mail address: zhangxiaodong@hit.edu.cn (Zhang Xiaodong)

Abstract: This work presents the development and characterization of nanostructured thermal barrier coatings (TBCs) reinforced with SiC for high-temperature applications. The aim is to improve toughness and service life compared to conventional YSZ coatings. The composite coating system was designed starting from atomic-scale calculations using density functional theory (DFT) to evaluate the surface energy and electronic structure of ZrO_2 and SiC, identifying stable heterogeneous interface characteristics. Composite feedstock powders were prepared using nano spray granulation, followed by plasma spheroidization to improve powder morphology and flowability. Coatings were deposited on GH4169 Ni-based alloy substrates via thermal spraying. Three types of coatings—nano YSZ, nano SiC/YSZ, and plasma-spheroidized nano SiC/YSZ—were evaluated. Microstructural analysis showed enhanced nanostructure and reduced internal crack propagation with the addition of SiC. Plasma spheroidization also reduced porosity and improved melting efficiency. This paper presents results from tensile and microstructural characterization, comparing the effects of different coating compositions and processing routes. The findings demonstrate that the optimized nanostructured coatings show improved mechanical behavior and dimensional stability under high-temperature service conditions.

Keywords: plasma spraying; nanometer powder regranulation; SiC; YSZ; first principles

Continuously Innovating High-End Thermal Protection Coatings

Wang You, Zhang Xiaodong

School of Materials Science and Engineering Harbin Institute of Technology, Harbin, China

E-mail address: wangyou@ hit. edu. cn (Wang You)

Abstract: Advanced aircraft urgently require high-performance aircraft engines. High performance gas turbines are urgently needed for ships, power equipment, and other applications. The effective protection of high-temperature hot end components of aircraft engines, ground heavy gas turbines, and ship gas turbines cannot be achieved without thermal protection coatings. This report combines the research progress of our research group in the fields of nanostructured thermal barrier coatings, nanostructured environmental barrier coatings, and nano modified base coatings in the past two decades. It introduces the research progress of thermal protection coatings from the following four levels, emphasizing the innovative development ideas in the researchprocess: 1. Nanostructured ceramic coating has been successfully applied. 2. Pain points still exist in the thermal protection coatings. 3. Innovation in Materials and Micro/Nano Structures of Coatings. 4. Concept of New Structural Thermal Protection Coatings. Finally, the broad application prospects of nanostructured thermal protective coatings on hot end components of aircraft engines and gas turbines were discussed.

Keywords: nanostructured ceramic coatings; thermal protection coatings; micro/nano structures; new structural thermal protection coatings

Fabrication and Tribological Performance of TiWCrSiN Nanocoatings via Multi-Arc Ion Plating

Jiang Mingquan, Li Yang

Yantai University, China

Abstract: Transition metal nitride coatings have been extensively applied to address friction and wear challenges in critical components across mechanical manufacturing, transportation, marine equipment, and aerospace industries, owing to their outstanding mechanical properties and wear resistance. Multi-arc ion plating (MAIP) technology is widely utilized for fabricating these coatings due to its superior film quality, high deposition rate, and precise control over chemical composition and microstructure. In this study, aTiWCrSiN composite coating system was successfully fabricated using MAIP with magnetic filtration assistance. The synergistic strengthening mechanisms of coating structure regulation, multi-component design, and plasma nitriding pretreatment were systematically investigated, elucidating the regulatory effects of microstructural evolution and interface optimization on coating performance. Furthermore, the influence of nitriding pre-treatment temperature on coating properties was analyzed, leading to optimized ion nitriding/coating composite processing techniques and exploration of intrinsic relationships among phase composition, mechanical properties, and tribological behavior in composite modified layers. (1) TiN coatings with varying structures were deposited on quenched and tempered 42CrMo low-alloy steel substrates. The microstructure, phase composition, mechanical properties, and tribological performance were characterized. HRTEM analysis revealed that increasing turntable rotation speed promoted the transformation of TiN grains into nanocrystalline structures, with coating hardness rising from 1122.2 $HV_{0.05}$ to 1532.8 $HV_{0.05}$ (3-5 times higher than the substrate). The composite structure of columnar TiN + nanocrystalline TiN exhibited enhanced toughness, as evidenced by a 48.12% increase in H^3/E^{*2} ratio compared to columnar TiN and an 8.41% improvement over nanocrystalline TiN. This structural optimization significantly improved coating adhesion and tribological performance. (2) TiWCrSiN nanogradient coatings with varying W content (0-12.1 wt%) were synthesized on 42CrMo substrates. Experimental results demonstrated that coating hardness and toughness initially increased and then decreased with rising W content, achieving hardness values of 1131.2-1431.7 $HV_{0.05}$, maximum adhesion of 72 N, and stable friction coefficients of 0.14-0.17. Optimal mechanical and tribological performance was observed at 8.94 wt% W, where in-situ formation of Fe_3O_4 and WO_2 lubricious phases during friction effectively reduced friction coefficient and wear rate. (3) TiN, TiSiN, TiCrSiN, and TiWCrSiN coatings were deposited on plasma nitrided substrates to investigate tribological performance under increasing loads (2-10 N). The TiWCrSiN nanogradient coating exhibited superior performance, maintaining stable average friction coefficients of 0.133-0.155 (11.28%-16.67% lower than TiN). Comparative studies on 42CrMo substrates subjected to low-temperature (480 ℃) and high-temperature (520 ℃) plasma nitriding revealed distinct interface characteristics: low-temperature nitriding produced α-Fe (N) diffusion layers, while high-temperature treatment generated additional compound layers (Fe_3N, Fe_4N). These interfacial layers enhanced coating/substrate bonding, providing superior mechanical support. The high-temperature nitrided composite sample achieved 93 N adhesion (29.2% higher than single-layer coatings) and maintained a stable friction coefficient of 0.127 under high-speed conditions (10.5% reduction compared to non-nitrided counterparts).

Keywords: transition metal nitride coatings; multi-arc ion plating; plasma nitriding; duplex treatment; friction and wear

Investigationof Thermal Barrier Coatings Modified via Pulsed Electron Beam Technology

Cai Jie

Jiangsu University

Abstract: : Thermal barrier coatings (TBCs) are a critical thermal protection technology for the advanced of modern aero-engine. However, their long-term performance is often compromised by poor service stability under high-temperature and complex environments, commonly manifesting as interfacial cracking and premature spallation. Controlling the structural stability of TBC interfaces and suppressing uncontrolled oxide growth remain unresolved challenges that critically affect coating durability. This study proposes a novel approach utilizing pulsed electron beam technology to synergistically regulate the morphology and microstructure of the metal/ceramic interface in TBCs, aiming to achieve steady thermally grown oxide (TGO) growth and prolong coating lifespan. Specifically, irradiation process parameters were optimized to polish and purify the bond coat surface, enabling targeted modulation of its microstructure and compositional distribution at the microscale for precise control over TGO growth behavior. Subsequently, an electron beam physical vapor deposition (EB-PVD) technique was employed to deposit ceramic layers on the modified bond coat. By investigating the influence of bond coat microstructural evolution on the conformal growth of ceramic layers, interfacial adhesion, and thermal cycling lifetime, this work establishes the structure-property relationship of TBCs under pulsed electron beam treatment and unveils the underlying mechanisms of TGO growth regulation and lifespan extension. The proposed method ultimately achieves controllable fabrication of TBCs with enhanced interfacial reliability, oxidation resistance, and longevity. The large beam diameter of pulsed electron beams, coupled with multi-pass overlapping irradiation, demonstrates compatibility with complex blade geometries, highlighting its potential for improving the service life of thermal protection coatings in engineering applications.

Preparation and Applications of Feather-Shaped High-Entropy Thermal Barrier Coatings

Lv Bowen*, Mao Xiye, Li Weihai, Mao Jie, Deng Chunming, Deng Changguang, Liu Min, Zhou Kesong

National Engineering Laboratory for Modern Materials Surface Engineering Technology, Institute of New Materials, Guangdong Academy of Sciences, Guangzhou, PR China

E-mail address: lvbowen@gdinm.com (Lv Bowen)

Abstract: With the increasing service temperature in aero-engines and gas turbines, thermal barrier coatings (TBCs) face significant challenges regarding their high-temperature performance. Conventional yttria-stabilized zirconia undergoes phase transformation and sintering during prolonged exposure to temperatures above 1200 ℃, leading to premature fracture and failure. This situation necessitates the development of novel ultra-high-temperature TBC material systems. In this study, we first prepared high-entropy ceramic powders suitable for plasma spraying through solid-state reaction and spray drying techniques. High-temperature stability tests at 1500 ℃ were conducted to elucidate the sintering resistance and phase stability. TBCs and thermal/environmental barrier coatings (T/EBCs) were subsequently fabricated via plasma spray-physical vapor deposition (PS-PVD). The results demonstrate that the high-entropy ceramics exhibit superior mechanical and thermal properties, as well as high-temperature stability. The feather-shaped columnar microstructure, characterized by high strain tolerance, significantly enhances thermal shock resistance and water-oxygen corrosion resistance, indicating great potential for engineering applications.

Development of Novel Environmental Barrier Coatings for SiC-based Ceramics to Enable Temperatures above 1450 ℃

Meng Yifan, Yuan Shuai, Zhang Zhenya*, Zhang Shihong*

Laboratory of Green Fabrication and Surface Technology of Advanced Metal Materials, Ministry of Education,

Anhui University of Technology, Maanshan City, 243000, Anhui Province, PR China

E-mail address: zzhenya@ahut.edu.cn (Zhang Zhenya)

Abstract: Environmental barrier coating (EBC) has been developed to protect SiC-based ceramics exposed to harsh environments, such as air and steam conditions. The conventional EBCs are composed of the Si bond coat and rare earth disilicate topcoat layers. However, as the operating temperature of advanced gas turbines typically exceeds 1400 ℃, it has been required to replace conventional Si bond coat in EBC with materials possessing higher thermal stability. The HfO_2-SiO_2 was developed as the novel bond coat deposited on SiC substrate using atmospheric plasma spray in EBC, and the service temperature of EBC was improved to 1475 ℃. Furthermore, novel $Y_xYb_{(2-x)}Si_2O_7$ (x = 0.3, 0.5, 0.8, and 1.0) topcoat layer were deposited using atmospheric plasma spray as the topcoat layer in EBC to reduce the thermal conductivity in EBC. The novel EBC of HfO_2-SiO_2/$Y_xYb_{(2-x)}Si_2O_7$ was conducted in air and steam conditions at 1475 ℃ for various times. The results showed that the novel EBC could effectively reduce the oxidation of EBC-coated SiC and improve the service temperature of EBC in air and steam conditions, which is a promising candidate for the next generation of EBC.

Keywords: environmental barrier coating; atmospheric plasma spray; service temperature; oxidation

Preparation and Thermal Shock Behaviors of the novel Hf6Ta2O17/YSZ Thermal Barrier Coating

Cheng Chunyu, Yang Li, Zhou Yichun

Abstract: Single-phase$Hf_6Ta_2O_{17}$ powder was synthesized via a solid-state reaction method. The powder was subsequently deposited onto a metallic substrate using supersonic atmospheric plasma spraying (SAPS) to fabricate the $Hf_6Ta_2O_{17}$ thermal barrier coating. Microstructural characterization revealed that the as-sprayed coating exhibited a multiphase architecture. To investigate the thermal stability of $Hf_6Ta_2O_{17}$, experiment and ab initio molecular dynamics simulation (from 1400 to 1700 K) were employed. First-principles calculations demonstrated that the Hf-O bonds on the outermost layer of the $Hf_6Ta_2O_{17}$(0 0 1) surface at ambient temperature exhibit superior stability compared to Ta-O bonds. This stability disparity suggests that Ta-O bonds are more prone to rupture under high temperatures. Heat treatment experiments conducted at elevated temperatures (1400, 1700, and 2000 K) further revealed that while both Ta-O and Hf-O bond populations and strengths progressively decreased with increasing temperature, the Hf-O bonds retained significantly higher bond density and strength than their Ta-O counterparts. These findings collectively demonstrate the enhanced thermal stability of surface Hf atoms relative to Ta atoms in the $Hf_6Ta_2O_{17}$ system, which critically governs the coating's resistance to thermally induced degradation. This result indicates that the topmost Hf atoms have stronger stability than Ta atoms after heat treatment. To suppress the decomposition of$Hf_6Ta_2O_{17}$ during thermal spraying, Ta_2O_5 was introduced into $Hf_6Ta_2O_{17}$ powders, and a series of $Hf_6Ta_2O_{17}$ powders with varying stoichiometric ratios were synthesized. The results indicate that increasing the Ta_2O_5 content progressively reduces the HfO_2 phase fraction in the $Hf_6Ta_2O_{17}$ coating, ultimately leading to its complete elimination, while the Ta_2O_5 concentration increases from 1% to 17%. The highest $Hf_6Ta_2O_{17}$ phase content in the coating was achieved at an HfO_2-to-Ta_2O_5 precursor ratio of 6:2. The low-melting-point Ta_2O_5 infiltrated the coating in a liquid state during spraying, which not only expanded the molten regions on the coating surface but also enhanced its structural densification. Furthermore, the inherent low hardness and reduced elastic modulus of Ta_2O_5 induced a non-monotonic evolution in the mechanical properties of the $Hf_6Ta_2O_{17}$ coating, characterized by an initial increase followed by a gradual decline in both hardness and elastic modulus with rising Ta_2O_5 content. The thermal shock resistance of the structurally optimized$Hf_6Ta_2O_{17}$/YSZ TBCs was evaluated under 1200 ℃ simulated service conditions using a gas thermal shock testing apparatus. Experimental results demonstrate that the 6Hf-1Ta/YSZ, 6Hf-2Ta/YSZ, and 6Hf-3Ta/YSZ coatings exhibit significantly higher thermal cycling lifetimes compared to the YSZ coating. Among these, the 6Hf-3Ta/YSZ coating achieved the maximum gas thermal shock resistance, which is attributed to the reduced elastic modulus induced by the incorporation of low-modulus Ta_2O_5, effectively lowering the crack driving force. Real-time monitoring of the TBC failure process via acoustic emission (AE) nondestructive testing revealed distinct fracture evolution stages: vertical surface cracks dominated the initial failure signals, followed by rapid propagation of interfacial shear cracks during intermediate and final stages. The synergistic interaction of these crack modes, coupled with substrate deformation, ultimately led to coating spallation. Additionally, high-temperature chemical reactions between the decomposed $Hf_6Ta_2O_{17}$ and HfO_2 phases further accelerated the spallation process.

Preparation of Hafnium Tantalum Based Carbide Solid Solution Coating and Study on Its Ablation Resistance Properties

Zhu Xiaofei*
Xidian University, Xian, China
E-mail address: zhuxiaofei@ xidian. edu. cn (Zhu Xiaofei)

Abstract: The single-phase solid solution of hafnium tantalum carbide is expected to meet the urgent demand for thermal protection materials for the hot end components of remote hypersonic aircraft due to its ultra-high temperature melting point (above 3900 ℃) and erosion resistance. However, the single-phase solid solution coating of hafnium tantalum carbide prepared by traditional methods has low solid solubility and poor density. To address this issue, this paper first studied the effects of three powder preparation processes on the structural properties of spray coatings, revealed the microstructure evolution law and solid solution mechanism inside the particles, determined the step-by-step plasma preparation process of IPS-VPS, and obtained HfTaC solid solution coatings with improved density and solid solubility. Explored the effect of WC doping on the mechanical and ablative properties of ceramics and coatings. The results showed that WC doping can further improve the solubility, density, and mechanical properties of ceramic coatings. After 120 seconds of oxyacetylene ablation at a heat flux density of 2. 38mW/m2, the hafnium tantalum tungsten carbon solid solution coating exhibited excellent anti ablation performance compared to the non WC doped hafnium tantalum carbon coating. The evolution law of coating ablation structure performance was analyzed, and the oxidation ablation protection mechanism of hafnium tantalum based carbide coating was elucidated.

Keywords: hafnium tantalum based carbide solid solution; coating; IPS-VPS; ablation resistance properties

The Effect of Porosity on the Mechanical and Thermal Insulation Properties of Ultra-Thick Thermal Barrier Coatings and Its Synergistic Optimization

Song Peng*

Faculty of Materials Science and Engineering, Kunming University of Science and Technology, Kunming 650093, China

Faculty of Civil Aviation and Aeronautics, Kunming University of Science and Technology, Kunming 650500, China

E-mail address: songpeng@ kust. edu. cn (Song Peng)

Abstract: Extreme hot-end components impose stringent requirements on thermal protection coatings, with the primary challenge being the coordinated optimization of mechanical properties and thermal insulation performance. This study focuses on the development of a high-performance structural ceramic and investigates its potential application in ultra-thick thermal barrier coatings (TBCs). The results reveal that the ultra-thick 8YSZ coating with high porosity exhibits a low thermal conductivity of 0.77 W/(m · K) at 1100 ℃, while demonstrating excellent thermal shock resistance with a lifetime exceeding 530 cycles under the same temperature, indicating its outstanding performance in high-temperature service environments. To further improve the bonding strength, a gradient porosity architecture was constructed by optimizing the microstructure of the ceramic layer. As a result, the bonding strength was successfully enhanced to over 20 MPa, while the coating retained excellent thermal shock resistance. This structural ceramic integrates efficient thermal insulation with reliable service performance, offering a novel strategy for the design of advanced thermal protection coatings for extreme environments.

Keywords: ultra-thick thermal barrier coatings; particle size; porosity; thermal shock resistance; gradient coatings

Invited Speakers in Functional Surfaces and Coatings

Huang Wenzhi

Science and Technology on Advanced Ceramic Fibers and Composites Laboratory, College of Aerospace Science and Engineering, National University of Defense Technology

E-mail address: 18173123353, huangwenzhi_kd@163.com (Huang Wenzhi)

Abstract: Thermal barrier coatings (TBCs), as a ceramic coating deposited on metallic hot-section components of aeroengines, significantly enhance operational temperature and oxidation resistance. However, corrosion induced by molten environmental deposits (CaO-MgO-Al_2O_3-SiO_2, CMAS) severely threate the service life of the aeroengine. To address the insufficient CMAS resistance of conventional 8 wt% Y_2O_3-stabilized ZrO_2 (8YSZ) coatings, this study proposes an innovative optimization strategy combining second-phase composite modification and high-concentration doping modification. $LaPO_4$-YSZ composite coatings and 56YSZ coatings with elevated Y_2O_3 doping were fabricated via atmospheric plasma spraying, followed by a systematic investigation of their corrosion behaviors and mechanisms at different temperatures. Results reveal that the aggregation of second-phase particles at YSZ grain boundaries forms weak interfaces, thereby enhancing the fracture toughness. Simultaneously, the second phase rapidly reacts with CMAS to generate a continuous and dense apatite-based barrier layer, effectively inhibiting CMAS infiltration. In 56YSZ coatings, sufficient Y^{3+} ions react with CMAS to form an apatite layer, achieving corrosion self-inhibition. This work reveals that the enhancement of CMAS resistance originates from the regulation of interfacial reaction pathways, where rapid crystallization of barrier layers blocks the penetration of molten salt. These findings provide critical insights for the composition design and performance optimization of next-generation TBCs.

Fabrication and Performance of Electrothermal/Superhydrophobic Active-Passive Synergistic Anti-Icing Coatings

Zhou Xin[1,2*], Feng XinJie[1,2], Kang Xiaoli[3], Zhang Hao[3]

[1] Institute for Advanced Studies in Precision Materials, Yantai University, Yantai 264005, China

[2] Shandong Key Laboratory of Advanced Structural Materials Genome Engineering, Yantai University, Yantai 264005, China

[3] School of Materials and Energy, Jiangxi Science and Technology Normal University, Nanchang 330038, China

E-mail address: zhouxin@ytu.edu.cn (Zhou Xin)

Abstract: Ice formation is a common natural phenomenon. With the rapid development of China's wind energy industry and aviation sector, icing on wind turbine blades and aircraft has become increasingly problematic. Superhydrophobic coatings inspired by the "lotus leaf effect" in nature represent an ideal anti-icing/de-icing approach. However, they suffer from limitations such as failure upon ice accumulation, poor weather resistance, and low durability. This study proposes the fabrication of high-performance biomimetic anti-icing structures with superior mechanical strength using thermal spraying technology, integrating them with thermally sprayed electrothermal coatings to create low-energy-consumption superhydrophobic electrothermal coatings. High-throughput preparation of Al-Si-M (M = Mn, Fe, Co, Ni) alloys via multi-target co-sputtering establishes the relationship between alloy composition and electrical resistivity, enabling the screening of electrothermal coating materials suitable for diverse anti-icing/de-icing requirements. Using nano-CeO_2-coated micron alloy powders or polyurethane-nano-Al_2O_3 suspensions as feedstocks, micro-nano composite structured surfaces are fabricated to investigate cross-scale coordination control methods and mechanisms governing the composition-process-structure-property relationships of thermal spray coatings. Combining simulation and experimental validation, this research reveals the active-passive collaborative anti-icing/de-icing mechanisms of superhydrophobic electrothermal coatings.

Keywords: anti-icing; electrothermal coating; superhydrophobic coating; active-passive synergistic; thermal spray

Formation Mechanisms of Columnar/Dense Structure for $Gd_2Zr_2O_7$ Ceramics Coatings with Superior CMAS corrosion resistance

Abstract: Corrosion failure induced by calcium-magnesium-alumina-silicate (CMAS) deposits is a predominant degradation factor in advanced thermal barrier coating (TBCs) systems. This study systematically investigates the surface laser modification strategy for reconstructing the microstructure near GZO ($Gd_2Zr_2O_7$) coatings surface prepared by air plasma spraying (APS), with aims to enhance TBCs CMAS corrosion resistance. Numerical simulations revealed that laser power thresholds govern microstructural transformations, a fully dense layer forms near GZO surface at 4 W, while micro columnar crystallization (honeycomb crystals first form at the bottom of the bath, growing from the bottom to the top of the bath with a slight bias towards the laser scanning direction, and then forming columnar dendritic crystals in the middle and upper part of the bath) initiates at 12 W. A parameter set of 14 W power, 100 mm/s scan speed, 0.02 mm line spacing and 0.5 mm spot diameter is filtered to form a hybrid textured like surface, vertically-aligned microcolumns (10 μm) overlying a dense sublayer (15 μm). Superior CMAS corrosion resistance was achieved through the low wettability combine infiltrating (columnar micro-nanostructures increase the apparent contact angle, resulting in super-hydrophobic properties) to CMAS of this composite structure. The synergistic effects of surface texturing and subsurface structural engineering provide new insights into designing corrosion-resistant TBCs through laser-induced phase-selective recrystallization.

Keywords: CMAS corrosion; thermal barrier coatings (TBCs); laser surface modification; columnar/dense structure; gadolinium zirconate; air plasma spray

"High-end Equipment, Green Manufacturing" Advanced Materials and Modern Surface Engineering

Lv Wenquan

Lanzhou Institute of Chemical Physics, Chinese Academy of Sciences

E-mail address: wqlv@licp.cas.cn (Lv Wenquan)

Abstract: In the context of industrialization, optimizing material properties is of paramount importance. Emerging industries demand advanced surface characteristics such as wear resistance, corrosion resistance, and heat resistance, posing dual challenges of cost and technology to traditional material development. Surface engineering, an emerging interdisciplinary field, employs advanced techniques including electron beams, ion beams, and laser beams, integrated with nanotechnology and micro-nano structures, to economically enhance material surface properties. Its applications span multiple industries, such as aerospace, defense, automotive, petrochemical, and mechanical manufacturing. Notably, laser additive manufacturing and thermal spraying are utilized for protecting critical components and repairing/reinforcing failed parts. Chemical plating technology improves wear and corrosion resistance of components. Physical vapor deposition (PVD) technology addresses wear and corrosion issues in aerospace and petrochemical sectors by developing high-performance wear-resistant and lubricating thin-film materials. These advancements enhance product performance and quality, fulfilling the needs of high-end equipment manufacturing and national projects, while fostering sustainable development and innovation in the materials industry.

The Collaborative Treatment of Al Current Collector by Surface Roughness and ZIF-8-derived Carbon Coating for High Electrochemical Performance Lithium Ion Battery

Jia Yongqiang[a,b,c], Han Maoyi[b], Wang Xunjie[b], Chen Mingxi[b], Zhang Lingling[b], Liu Ti[d], Wu Mingfang[a*], Ba Zhixin[b,c*]

[a] School of Materials Science and Engineering, Jiangsu University of Science and Technology, Zhenjiang, 212100, Jiangsu, China

[b] School of Materials Science and Engineering, Nanjing Institute of Technology, Nanjing 211167, Jiangsu, China

[c] Jiangsu Key Laboratory of Advanced Structural Materials and Application Technology, Nanjing Institute of Technology, Nanjing 211167, Jiangsu, China

[d] Jiangsu Dingsheng New Energy Materials Co., Ltd

E-mail address: wu_mingfang@163.com (Wu Mingfang)

E-mail address: bzhx@njit.edu.cn (Ba Zhixin)

Abstract: Surface roughness and carbon coatings are the effective technique for achieve the high adhesion and battery performance of the Al current collector in the field of lithium-ion battery. Henin, the Al current collector was etched by the simple electrochemical method using the acid-free, environmentally friendly ferric nitrate-based electrolyte, also with the variable etching voltages from 2.0 V ~ 3.6 V and durations from 1 min ~ 9 min. Furthermore, the N-doped porous carbon was prepared by calcining the zeolitic imidazole framework (ZIF)-8 precursors with the temperature of 600 ℃ and 800 ℃, and employed the ZIF-8-derived carbon as the coatings on the etched Al current collector. The etched Al current collector showed the uniform nano-rough structure with the parameter of 2.8 V, 3 min. The ZIF-8-derived carbon achieved from calcining at 800 ℃ have the lower degree of graphitization, larger specific surface area and more defective active sites. With synergistic effects of the roughened surface having nano-pit structure and the well-defined structure, porous nature of ZIF-8-derived porous carbon, the modified Al current collectors exhibit the better interface adhesion and the smaller charge transfer impedance. The LIB cell based on the modified Al current collector exhibits the excellent discharge specific capacity (164.99 mAh · g^{-1}) at 0.1 C rate, excellent cycle stability with the average coulombic efficiency of 99.81%.

Keywords: lithium-ion battery; electrochemical etching; ZIF-8-derived carbon; aluminum current collector; electrochemical performance

Research on Processing Technology of Biomedical Cobalt-Chromium-Molybdenum Alloys

Li Yawen

Yantai University, China

E-mail address: lyw18264583242@163.com (Li Yawen)

Abstract: Cobalt-chromium-molybdenum (Co-Cr-Mo) alloys are widely used in medical implants such as artificial joints due to their excellent biocompatibility, wear resistance, and mechanical properties. However, China's forged Co-Cr-Mo alloy production heavily relies on imports, and breakthroughs are urgently needed in processing technologies, particularly in hot rolling forming techniques. This study aims to achieve domestic production of Co-Cr-Mo alloys by optimizing hot rolling processes and regulating heat treatment systems, successfully preparing alloy materials that meet performance standards. The research first benchmarked the composition, microstructure, and mechanical properties of imported products, confirming their compliance with ASTM F799 standards. Focusing on the processing characteristics of high-nitrogen Co-Cr-Mo alloys, the study systematically investigated the effects of hot rolling parameters (temperature, deformation amount, rolling speed) on alloy microstructure and properties. The rolling path was optimized to avoid cracking issues commonly encountered in traditional free forging processes. Different heat treatment regimes are planned to further regulate grain size and precipitate distribution to enhance comprehensive properties. Results demonstrate that the optimized hot rolling process enables the samples to meet the mechanical property requirements for implant materials (tensile strength ≥ 1172 MPa, elongation ≥ 12%). This study provides a feasible processing solution for the industrial production of domestic forged Co-Cr-Mo alloys, holding significant importance for promoting the localization of high-end biomedical materials in China.

Keywords: CoCrMo alloy; hot rolling process; heat treatment regime; microstructure and properties

Preparation of NiCrAlY-CeO2@PDMS Superhydrophobic Coating by Atmospheric Plasma Spraying

Kang Xiaoli[1], Zhang Hao[1*], Zhou Xin[2*]

[1] Jiangxi Science & Technology Normal University, Jiangxi, China

[2] Yantai University, Shandong, China

E-mail address: zhanghao@jxstnu.edu.cn (Zhang Hao)

E-mail address: zhouxin@ytu.edu.cn (Zhou Xin)

Abstract: Superhydrophobic coatings, characterized by water contact angles exceeding 150°, have emerged as what seems to be a pivotal research area due to their versatile applications in anti-fouling, anti-icing, drag reduction, and self-cleaning. What these properties suggest is a synergy of hierarchical surface roughness and low surface energy materials, effectively mimicking what appears to represent natural superhydrophobic surfaces like those of lotus leaves. However, the predominant reliance on traditional organic polymers, such as fluoropolymers and silicones, tends to impose what might be characterized as significant limitations. These materials ostensibly degrade rapidly under mechanical abrasion, high-temperature environments, or chemical corrosion, rendering them unsuitable for demanding applications in aerospace, power transmission, and marine engineering. What appears particularly significant about these findings is that developing superhydrophobic coatings with enhanced mechanical robustness and environmental resilience has apparently become a critical research frontier. In this study, inspired by what seems to constitute the multi-layered architecture of thermal barrier coatings, what the investigation indicates is that a novel composite coating system was developed. This coating integrated rare earth oxide (REO), specifically CeO_2, with a NiCrAlY alloy through atmospheric plasma spraying (APS). What tends to emerge from these findings is that APS allows for what might be characterized as precise control over coating microstructure by adjusting parameters such as spraying power, gas flow rate, and particle feed rate. Subsequently, polydimethylsiloxane (PDMS) was applied to impart low surface energy. What the research seems to point toward, given the complexity of these theoretical relationships, is a systematic investigation of the effects of varying CeO_2 doping ratios (ranging from approximately 5% to 15%) and spraying power (typically 30-50 kW) on the coating's wettability, characterized by contact angle measurements and surface roughness analysis using scanning electron microscopy (SEM) and 3D profilometer. Experimental results suggest what seems to be a negative correlation between spraying power and superhydrophobicity. As power increased, enhanced particle melting apparently led to a somewhat smoother surface, which in turn tends to suggest a decrease in the contact angle from approximately 162° to around 148°. What appears particularly significant about these findings is that among the tested samples, the NiCrAlY-10% CeO_2@PDMS coating seems to demonstrate what might be characterized as optimal performance. Mechanical stability tests, including sandpaper abrasion, water impact test, and adhesive tape peeling, indicate only a marginal decrease in contact angle to about 143°, which suggests what appears to be substantial resistance to physical damage. Within this broader analytical framework, electrochemical impedance spectroscopy and 264-hour salt spray tests seem to lend support to what may represent exceptional corrosion protection, ostensibly achieving an efficiency of roughly 82.74% for 316 stainless steel substrates. What the data suggest that the coating typically delayed ice formation for approximately 480 seconds at −10 ℃ and around 285 seconds at −15 ℃, and it apparently facilitated rapid ice melting within about 100 seconds by heating to roughly 42 ℃ using a low-voltage (2 V) electrical current, outperforming conventional electrothermal coatings in terms of what appears to be energy efficiency. What this investigation indicates is not only a practical approach for fabricating what seems to be high-performance anti-icing composite coatings but also what tends to illuminate the complex relationship between APS parameters and coating properties. Given the multifaceted nature of this evidence, what these findings point toward is a contribution to the theoretical work of thermal-sprayed functional coatings and what offers what might be characterized as innovative strategies for material surface engineering.

Keywords: hydrophobic coatings; mechanical durability; water contact angle; NiCrAlY-CeO_2

Design of Multifunctional Rare-earth Silicates Coatings

Zhang Jie

Shenyang National Laboratory for Materials Science, Institute of Metal Research, Chinese Academy of Sciences, Shenyang 110016, China

Abstract: Environmental barrier coatings (EBCs) are needed to protectSiCf/SiC ceramics from degradation advanced modern gas turbines. Multifunctional requirements were established for EBC, including matched coefficient of thermal expansion (CTE), low thermal conductivity as well as corrosion resistance to both steam and calcium-magnesium-aluminum-silicate (CMAS) deposits. In the current talk, tailoring thermal expansion behavior of rare earth silicates by means of local disordered structure was illustrated. Coordinative local disturbances introduced by strategic high-entropy design have been proposed as the key method for CTE regulation. To achieve thermomechanical robustness against CMAS deposits, thermochemical equilibrium between rare-earth oxides and calcium-magnesium aluminum-silicate deposit was investigated. The variation of rare earth constituents results in the transformation of crystalline products from apatite/silicocarnotite to garnet/diopside, which accelerates the consumption of CMAS melt and facilitates corrosion mitigation. A pseudo-ternary phase diagram was established, which had a great potential to describe phase equilibrium in coating-deposit systems and could provide guidance for compositional design of corrosion-resistant coatings.

Keywords: functional design; environmental barrier coating

Preparation and Characterization of Novel Bond Coat Materials for High-Entropy Thermal Barrier Coatings

Feng Xinjie, Zhou Xin

Yantai University, Yantai, China

E-mail address: fxj60122023@163.com (Feng Xinjie)

E-mail address: zhouxin@ytu.edu.cn (Zhou Xin)

Abstract: To accelerate the exploration and screening of high-performance oxidation-resistant bond coat materials for thermal barrier coatings (TBCs), this study employed magnetron co-sputtering to deposit compositionally graded AlxCoCrFeNi high-entropy alloy (HEA) films onto single-crystal Si (100) and CM247 Ni-based superalloy substrates. An innovative rapid screening strategy was proposed, integrating short-term cyclic oxidation (1100 ℃, 15 min-2 h) with high-throughput characterization. By precisely controlling oxidation time to trigger initial reactions, this method minimizing film degradation and substrate interdiffusion effects. Scanning electron microscopy/energy-dispersive spectroscopy (SEM/EDS) was systematically used to analyse the surface morphology and composition distribution of the oxide layers, while micro-area X-ray diffraction (μ-XRD) or Raman spectroscopy identified the phases of oxidation products, with a focus on the formation of protective α-Al_2O_3. By integrating data on original composition, microstructure, post-oxidation morphology, and phase composition, a comprehensive database was established, revealing clear composition-structure-oxidation resistance correlations. Focusing on early-stage oxidation behavior and leveraging high-throughput surface analysis techniques, this study overcame the research limitations of thin-film systems, achieving efficient screening and data-driven correlations. It provides a rapid evaluation framework for developing novel oxidation-resistant TBC bond coats.

Keywords: magnetron co-sputtering; high-entropy alloy (HEA) films; high-throughput; Short-term oxidation behavior; alumina

Microstructure and High-Temperature Oxidation Behavior of Y_2O_3 Strengthened Inconel 625 Alloy Manufactured by Laser Direction Energy Deposition

Wang Shouqiang, Zhou Xin

Yantai University, China

E-mail address: armstrongwsq@163.com (Wang Shouqiang)

E-mail address: zhouxin@ytu.edu.cn (Zhou Xin)

Abstract: To address the insufficient application of additive manufacturing in Oxide Dispersion Strengthened (ODS) nickel-based superalloy research, this study proposes a composite fabrication strategy. Initially, high-energy resonant acoustic mixing was employed to achieve uniform dispersion of 1 wt. % Y_2O_3 nanoparticles within Inconel 625 powder. Subsequently, defect-free ODS-IN625-Y_2O_3 alloy was successfully fabricated via Laser Directed Energy Deposition (LDED). The principal conclusions are as follows: At low laser power, prominent Y-Al-Si-O slag were observed. Increasing the power to 850 W eliminated these inclusions by enhancing melt pool sptter, while excessive power exacerbated Y_2O_3 loss and promoted Ostwald ripening. The REO effect of Y_2O_3 increased the melt pool energy absorption rate (increasing depth to 700 μm and width to 2500 μm), leading to coarser average grain size (249 μm) and microstructural homogenization. Grain coarsening and homogenization increased room-temperature and 700 ℃ elongation by 25.7% and 18.2%, respectively. However, the low Y_2O_3 retention rate (< 1 wt. %), Nb/Mo segregation, and weakened solid solution strengthening collectively resulted in reduced strength. Y_2O_3-derived phases formed diffusion barriers at grain boundaries, inhibiting internal oxygen diffusion (thinning the internal oxidation layer by 53.7%) and outward Cr diffusion. This promoted the formation of a dense Cr_2O_3 protective layer, reducing the 1000 ℃ oxidation weight gain by 40.9%.

Keywords: oxide dispersion strengthening; laser directed energy deposition; nickel-based superalloy; resonant acoustic mixing; high-temperature oxidation

The Corrosion Resistance of YTZ-Al_2O_3 Ceramics to CMAS and CMAS + V_2O_5 (9:1) at 1250 ℃ and 1400 ℃

Abstract: Calcium magnesium aluminosilicate (CMAS) corrosion is one of the main failure forms of thermal barrier coatings (TBCs), and the deposition of molten salt V_2O_5 in low-grade fuels is also one of the reasons for the failure of thermal barrier coatings. The effects of Al_2O_3 content in YTZ on the corrosion resistance of CMAS and CMAS + V_2O_5 (9:1) were investigated by methods such as VHX-500 ultra-depth of field microscope, X-ray diffractometer (XRD), scanning electron microscope (SEM) and wetting test. The research results show that: When the Al_2O_3 content was 5 wt%, 10 wt%, and 20 wt%, high-melting-point feldspar was formed. The 20 wt% Al_2O_3-modified YTZ had the best resistance to CMAS corrosion and the poorest wettability. Under two corrosion conditions of 1250 ℃ for 20 hours, the corrosion depths were only 15 μm and 19 μm respectively, demonstrating good resistance to CMAS and CMAS + V_2O_5 (9:1) corrosion. Among them, due to the mechanism of V_2O_5 accelerating the corrosion and penetration of CMAS, this composite ceramic has a better anti-corrosion performance against CMAS.

Keywords: Al-YTZ; CMAS corrosion; CMAS + V_2O_5 corrosion; ceramic material

Ultrafast Laser-Induced TBC/EBC Performance Enhancement

Wu Yang

Beijing University of Technology, Beijing, China

E-mail address: wuyang@ bjut. edu. cn (Wu Yang)

Abstract: Thermal barrier coatings (TBCs) are extensively applied in the hot-section components of engines to enhance their high-temperature resistance. During service, these coatings often suffer from calcium-magnesium-alumina-silicate (CMAS) corrosion. Inspired by the hydrophobic properties of lotus leaves, we employed ultrafast laser direct writing to fabricate a triple-scale micro-nano grid structure on the TBC surface, aiming to improve its resistance to molten CMAS infiltration. First, the triple-scale structures with varying laser scanning spacings (70, 80, and 90 μm) were fabricated on ceramic bulks. After 10 min of volcanic ash corrosion, the ceramic with a scanning spacing of 70 μm exhibits the largest contact angle (115°). Subsequently, a triple-scale micro-nano structure with a scanning spacing of 70 μm was prepared on the coating surface. Following 10 min of volcanic ash exposure, this structure demonstrates superior performance, with a lárger contact angle (116°) and enhances resistance to CMAS penetration compared to the as-sprayed coating (105°). With the extension of corrosion time, the structured coating surface reacts with molten CMAS and forms a reaction layer composed of ZrO2 and apatite phase, which inhibits further infiltration. The laser-ablated micro-nano structure increases surface roughness, resulting in a higher friction coefficient (0. 280) compared to conventional coatings (0. 214).

Keywords: thermal barrier coatings (TBCs); EBCs

$CrTaO_4$ and $CrNbO_4$: Dual Functional Scales to Protect RHEAs From Oxidation and Thermal Attack

Zhou Yanchun, Zhang Shuang

Zhengzhou University, Zhengzhou, China

Suzhou Laboratory,Suzhou City Industrial Park, Suzhou, China

Abstract: CrTaO4 and $CrNdO_4$ have been unexpectedly found to play decisive roles in improving the oxidation resistance of Cr and Ta-or Nd containing refractory high-entropy alloys (RHEAs). These rarely encountered complex oxides can effectively prevent the outward diffusion of metal cations from the RHEAs. However, they hav been ignored and their mechanical and thermal properties have yet to be studied. To fill this research gap and explore the untapped potential for their applications, here we report for the first time the microstructure, mechanical and thermal properties of CrTaO4 and $CrNdO_4$ prepared by hot-press sintering of solid-state reaction synthesized powders. Using the HAADF and ABF-STEM techniques, rutile crystal structures were confirmed and short range ordering was directly observed. Intriguingly, CrTaO4 and $CrNdO_4$ exhibit elastic/mechanical properties similar to those of yttria stabilized zirconia (YSZ), which spurred inquiries to lucrative leverage them as a new thermal barrier material. The results show that they have high melting point, low thermal conductivity, and close thermal expansion coefficients to refractory metals and RHEAs. From the perspective of matched thermal expansion coefficient, CrTaO4 and $CrNdO_4$ pertain to eligible thermal barrier materials for refractory metals and RHEAs. As such, they can be regarded as novel dual functional scales on top of RHEAs to protect them from oxidation and thermal attack.

Oxidative Wear Behaviors of Inconel 625 Alloy in a Simulated Rotary Calcination Furnace Environment at 600 and 800 ℃

Wang Chengxu

Yantai University, China

E-mail address: cxwang@ ytu. edu. cn (Wang Chengxu)

Abstract: This study investigates the oxidative wear behaviors of Inconel 625 alloy in a simulated rotary calcination furnace environment at 600 and 800 ℃. Results show that both the coefficient of friction and wear loss increase with temperature from 600 ℃ to 800 ℃. At 600 ℃, the alloy exhibits strong resistance to oxidative wear, due to the formation of an intact oxide scale composed of Ni-Cr-O and Cr-rich layers. However, the oxide scale formed at 800 ℃ features a coarser grain structure, coupled with increased wear debris, leads to higher surface roughness and coefficient of friction. The oxide scale at 800 ℃ has poor effectiveness in protecting the substrate, being more prone to spallation and contributing to greater wear loss.

Keywords: Inconel 625; oxidative wear; microstructures; vitrification

Mechanistic Study on the Influence of Pre-Corrosion on the Oxidation Behavior ofFeCrAl coating under Accident Conditions

Shen Zhao, Zeng Xiaoqin

School of Materials Science and Engineering, Shanghai Jiao Tong University, 200240 Shanghai, China

Abstract: Experimental kinetic dataindicate that FeCrAl coatings subjected to prior uniform corrosion treatment exhibit significantly higher mass gains under identical high-temperature steam oxidation conditions (1000 ℃ for 2 h, 1200 ℃ for 2 h, and 1400 ℃ for 2 h) compared to untreated samples. Comparative tests show that the untreated samples maintain structural integrity in steam environments at 1400 ℃ for 2 hours, whereas the pre-corroded samples undergo destructive oxidation under the same conditions. This performance degradation is attributed to the detrimental effects of the iron(III) oxide layer formed during the pre-corrosion process on the high-temperature oxidation resistance of the FeCrAl coating. Microstructural analysis reveals that during high-temperature steam oxidation, the phase transformation of the Fe_2O_3 particles generated by uniform corrosion into iron-aluminum spinel inhibits the formation of the protective α-Al_2O_3 layer, leading to a dramatic deterioration in oxidation resistance. This study, by simulating more realistic nuclear fuel cladding service conditions, elucidates the mechanism by which pre-corrosion affects the oxidation behavior of FeCrAl coatings under accident scenarios.

Design, Composition Optimization, Oxidation/Ablation Resistance of Multi-Component Ceramic Coating

Dong Shun*, Yu Yue

Science and Technology on Advanced Composite in Special Environments Laboratory, Harbin Institute of Technology, Harbin, China

E-mail address: dongshun@hit.edu.cn (Dong Shun)

E-mail address: 25B318027@stu.hit.edu.cn (Yu Yue)

Abstract: With the fast development of advanced aerospace vehicles, higher requirement is demanded for thermal protection materials. Multi-component ceramic coating has been acknowledged to be one of the most promising thermal protection materials, due to the excellent high-temperature physical/chemical stability and the good oxidation/ablation resistance. In order to further improve the service performance of thermal protection materials, it is crucial to focus on the study of composition design and structural optimization, preparation process. In our recent research, the $TaSi_2$-$MoSi_2$-SiB_6 coating with the addition of SiC_w was prepared by spraying and flash sintering, expected to improve the oxidation/ablation resistance and fracture toughness. Due to composition optimization, the mass ablation rate and line ablation rate were 1.03×10^{-3} g/s and 1.2×10^{-3} mm/s after ablation at 1600 ℃ for 1000 s. Furthermore, we synthesized the ZrB_2-SiC-TaB_2 ceramic in order to clarify the effect of Ta addition on the oxidation resistance. Ta elements were dissolved into the surface glassy layer forming Ta-O-Si network structure, contributing to the improvement of anti-oxidation ability. Subsequently, the thermal protection coating will be prepared based on this optimized result. With the development of composition optimization and structural design, advanced multi-component ceramic is expected to meet the service requirement for thermal protection materials and contribute to the application of advanced aerospace vehicles in extreme environment.

Keywords: Ablation; oxidation; composition design; structural optimization

Format Requirements for Extended Abstract to the 30th World Congress of International Federation for Heat Treatment and Surface Engineering Development and Performance Testing of Environmental Barrier Coatings for Silicon Carbide Ceramic Matrix Composites

Wang Weize

School of Mechanical and Power Engineering, East China University of Science and Technology, Shanghai, China

E-mail address: wangwz@ ecust. edu. cn (Wang Weize)

Abstract: Silicon carbide ceramic matrix composites (SiC/SiC CMCs) exhibit low density and excellent high-temperature thermomechanical properties, but they are susceptible to corrosion by high-temperature water vapor generated from gas combustion, leading to rapid surface degradation and significant strength decline. This study focuses on improving the performance of environmental barrier coating (EBC) systems. Through first-principles calculations, thermodynamic property tests, a rapid screening method for high-entropy rare-earth disilicate materials, and high-temperature CMAS corrosion tests, a low thermal conductivity high-entropy (Sc0. 2Y0. 2Er0. 2Yb0. 2Lu0. 2) 2Si2O7 (5RESO) EBC material was developed. Compared with traditional EBC materials, this material has comparable thermal expansion coefficient, lower thermal conductivity, and enhanced CMAS corrosion resistance. Using this material as the top coat, the water vapor corrosion resistance of the coating system has been significantly improved compared to SiC/SiC CMCs.

Keywords: SiC/SiC CMCs; environmental barrier coating; rapid screening method; high-entropy rare-earth disilicate; thermal expansion coefficient; thermal conductivity; CMAS and water vapor corrosion resistance

High-Temperature CMAS Corrosion-Resistantbehaviours of Rare-earth Hafnates Thermal/Environmental Barrier Coatings

Yiming Wu, Xin Zhong, Yaran Niu, Xuebin Zheng, Chuanxian Ding

Key Laboratory of Inorganic Coating Materials CAS, Shanghai Institute of Ceramics, Chinese Academy of Sciences, China

E-mail address: wuyiming211@ mails. ucas. ac. cn (Yiming Wu)

Abstract: With the continuous advancement of the aerospace technology toward high performance and thrust-to-weight ratio, the service environment of advanced aero-engines has become increasingly severe. High-temperature corrosion, particularly caused by CMAS (CaO-MgO-Al_2O_3-SiO_2) molten salts, has emerged as a critical factor limiting the lifespan of engine hot-section components. Rare-earth hafnates, distinguished by its high melting points (>1500 ℃) and low thermal conductivity, have become prime candidates for thermal/environmental barrier coatings (T/EBCs) in recent years. In this work, a series of rare-earth hafnate coatings (RE = La, Gd, Yb, Lu) were fabricated vias plasma sprayed method and the influence of rare-earth ionic radii and temperature variations on the CMAS corrosion resistance of rare-earth hafnates were systematically investigated. Results indicated that rare-earth hafnates coatings demonstrated superior corrosion resistance compared to widely-studied 8YSZ or rare-earth zirconates coatings. The reaction with CMAS resulted in the formation of a layer composed of apatite, garnet, and fluorite phases, which effectively mitigated further corrosion. The study also shown that temperature exerted a significant effect on the CMAS corrosion behaviours of rare-earth hafnates in the aspects of corrosion depth, the types of corrosion products, and the rare-earth ion content within the corrosion products in hafnates. Besides, hafnates with smaller rare-earth ion radii (RE = Yb, Lu) demonstrated superior corrosion resistance across different temperatures, which can be attributed to the differences in reaction rates with CMAS and the variation in corrosion products. Due to the favorable substitution of Ca^{2+} by larger ions promoting apatite phase crystallization, the reaction rate between the coating and CMAS increased with the radius of the rare earth ion at all tested temperatures. Furthermore, garnet and cuspidine products formed by Yb^{3+} and Lu^{3+} (with smaller ionic radii) at different temperatures, respectively, consuming significant amounts of Ca, Mg, Al, and Si elements, which also effectively alleviated CMAS corrosion. This work may provide theoretical support for the material design and application of next-generation high-temperature multilayer T/EBC systems.

Keywords: CMAS corrosion; rare-earth hafnates; thermal/environmental barrier coating

Research on Design and Anti-Oxidation/Ablation Property of ZrB_2-Based UHTC Coatings

Niu Yaran

Key Laboratory of Inorganic Coating Materials CAS, Shanghai Institute of Ceramics, Chinese Academy of Sciences, Shanghai, China

E-mail address: yrniu@ mail. sic. ac. cn (Niu Yaran)

Abstract: Ultra-high temperature ceramics (UHTCs) are potential thermal protective materials due to their extremely high melting points. It is well accepted that the application of UHTC coatings on the surface of fiber reinforced composites, such as C_f/C and C_f/SiC composites, is an effective way to realize the demands of combination of structural and surface demands. One key problem is to improve the high temperature oxidation/ablation resistant property of mono UHTC. In this presentation, the research progress on the design, oxidation/ablation behavior and failure mechanism of ZrB_2-based composite coatings in our lab are reported. Firstly, the oxidation behaviors of ZrB_2, ZrB_2-SiC and ZrB_2-$MoSi_2$ coatings are compared, Secondly, the influence of the amount and distribution of the second phase on the oxidation behavior of ZrB_2-based coatings are reported. Thirdly, we fabricated the ZrB_2-based ternary composite coating and evaluated the influence of the third phase on the oxidation/ablation behavior and products. Lastly, the effect of heating-cooling processes on the oxidation behavior and products are briefly introduced. It is hoped that this work could attributed to the development and application of UHTC coatings.

Keywords: Ultra-high temperature ceramics coating; high temperature oxidation/ablation resistant; ZrB_2-based composite coatings; amount and distribution of second phases; effect of heating-cooling process

Investigation on the Relationship Between CMAS Corrosion Mechanism and Microstructure of Multicomponent Monosilicates

Zeng Yi

Inorganic Materials Analysis and Testing Center, Shanghai Institute of Ceramics, Chinese Academy of Sciences, Shanghai, China

E-mail address: zengyi@ mail. sic. ac. cn (Zeng Yi)

Abstract: Understanding corrosion mechanisms and enhancing material corrosion resistance are of significant importance for reducing global corrosion costs. However, uncertainties regarding the primary mechanisms of CMAS corrosion in extreme environments have limited the enhancement of anti-CMAS corrosion performance in monosilicates through mixing and matching critical rare earth components. In this report, we propose a methodology utilizing correlated electron microscopy techniques combined with theoretical calculations to elucidate the competing mechanisms governing CMAS corrosion in monosilicates. Our findings reveal a temperature-dependent competition between thermodynamics and kinetics during corrosion. Based on this, we developed novel multicomponent monosilicates that exhibit exceptional corrosion resistance even at 1,500 ℃. This achievement was realized through precise regulation of rare earth ion radii in monosilicates to establish a delicate balance between thermodynamic and kinetic competition. After 50-hour and 100-hour corrosion tests, the thinnest reaction layers measured were merely 28. 8 μm and 35. 4 μm, respectively. Furthermore, we proposed a descriptor for evaluating the solid solubility of multicomponent rare earth monosilicates. Based on this descriptor and corrosion assessment results from over twenty monosilicates, we established composition design guidelines for multicomponent monosilicates capable of predicting optimal rare earth compositions for superior corrosion resistance at different operating temperatures. These findings provide crucial theoretical support for accelerating the development of multicomponent rare earth monosilicates, enabling customized compositions with outstanding anti-corrosion performance under specific service temperatures.

Keywords: Environmental barrier coating; anti-CMAS corrosion performance; rare-earth monosilicates; thermodynamic and kinetic competition; rare earth ion radii; composition design guidelines

Format Requirements for Extended Abstract to the 30th World Congress of International Federation for Heat Treatment and Surface Engineering High Entropy Design and Thermophysical Properties of Rare Earth Cerate Thermal Barrier Coating Materials

Zhao Xiaobing

School of Materials Science and Engineering, Changzhou University, Jiangsu, China

E-mail address: zhaoxiaobing00@ 163. com (Zhao Xiaobing)

Abstract: The performance of current thermal barrier coating materials is increasingly unable to meet the growing demands for high-temperature protection. Therefore, the exploration of new types of thermal barrier coating materials has become an urgent need in both research and application in this field. Rare-earth cerates are considered potential candidates for the next generation of thermal barrier coating materials due to their excellent high-temperature stability, relatively high thermal expansion coefficient, and low thermal conductivity. However, their thermal contraction in the low-temperature range (473-673 K) limits their application scope. In recent years, high-entropy materials have attracted widespread attention due to their unique high-entropy effects, providing a new research direction for improving the performance of rare-earth cerates. Thispresentation introduces our recent research progress in the field of rare-earth cerate thermal barrier coating materials. Specifically, we adopted a "high-entropy" strategy to design rare-earth cerate ceramic materials and investigated the effects of configurational entropy, disorder parameter, and element valence variation on their structure and thermophysical properties. We prepared high-entropy rare-earth cerate thermal barrier coatings using atmospheric plasma spraying (APS) technology and investigated the influence of spray power on the element valence, structure, and thermophysical properties of the coatings. The results show that through high-entropy design, the low-temperature thermal contraction problem of rare earth cerates has been significantly improved and even eliminated, and their thermophysical properties have also been greatly enhanced. The plasma-sprayed coating exhibits excellent thermophysical properties, with a thermal expansion coefficient of up to 11.69 $\times 10^{-6} \cdot K^{-1}$ and a thermal conductivity as low as 0.54 $W \cdot m^{-1} \cdot K^{-1}$ at 1273 K. This material has a promising application potential in the field of thermal barrier coatings.

Keywords: Thermal barrier coating; rare-earth cerates; high-entropy strategy; low-temperature thermal contraction; thermophysical properties

Microstructural Control Strategies and Long-Term Performance of Protective Nitride Coatings in Harsh Environments

Gao Zhaohe

Shanghai University, Shanghai, China

E-mail address: z_gao@ shu. edu. cn (Gao Zhaohe)

Abstract: Ti alloys blades in aero engine have applied surface coatings to increase their performance against oxidation/corrosion, erosion and even fire, which is the critical foundation for the safe application of aero engine in demanding environments. This talk will focus on the design of the thermal stable and low thermal conductivity dual-phase nitride coatings. This talk also uncovers the interfacial microstructural modification techniques, e. g. interfacial diffusion bonding, interfacial mechanical twinning, interfacial roughness, between nitride coatings and Ti alloys substrate.

Keywords: Nitride coating; Aero engine; Ti alloys

Fabrication of High-entropy Boride Powders for Plasma Spray by Inductive Plasma Spheroidization (IPS)

Yu Leyangyang, Niu Yaran*, Zheng Xuebin*, Ding Chuanxian

Shanghai Institute of Ceramics, Chinese Academy of Sciences, Shanghai 200050, China

E-mail address: yrniu@ mail. sic. ac. cn (Niu Yaran)

E-mail address: xbzheng@ mail. sic. ac. cn (Zheng Xuebin)

Abstract: The fabrication of feedstock powders is a critical technology that directly influences the microstructure and performance of plasma-spray coatings. In this work, an induction plasma spheroidization (IPS) process was employed to fabricate ($Zr_{1/4}Hf_{1/4}Ta_{1/4}Ti_{1/4}$) B_2 high-entropy powders for plasma spray and compared with the other two traditional powder preparation routes meanwhile. The morphology, internal structure, particle size distribution, density, and other fundamental properties of powders were systematically characterized. The effects of different powder fabricating processes were researched. The results demonstrate that using commercial micron-sized boride powders as precursors, a hybrid process combining mixing, spray drying, sintering with IPS enables the fabrication of high-entropy powders with homogeneous elemental distribution. The powders exhibit spherical morphology, smooth surfaces, high internal density, and high apparent/tap density. Further experiments synthesizing different high-entropy borides with varied compositions confirm the broad applicability of this method. The formation mechanism of high-entropy solid solutions was elucidated through first-principles calculations combined with the unique characteristics of IPS process. This work proposes a promising method for fabricating high-entropy ceramic powders suitable for plasma-spray coatings.

Keywords: Induction plasma spheroidization; high entropy ceramics; borides; powders for spray

Design, Fabircation and Characterization of Self-Healing Thermal Barrier Coatings Fabricated by Atmospheric Plasma Spraying

Zhao Weiling, Wang Liang*

Shanghai Institute of Ceramics, Chinese Academy of Sciences, Shanghai, China

E-mail address: L. Wang@ mail. sic. ac. cn (Wang Liang)

Abstract: Compared with conventional thermal barrier coatings (TBCs), intelligent self-healing thermal barrier coatings can initiate an intelligent self-healing mechanism under high-temperature service conditions. This mechanism could effectively inhibit the kinetic process of crack propagation, thereby extending the service life of the coatings at elevated temperatures. In the current study, a thermo-mechanic coupled finite element simulation combined with the ideal point method was used to optimize the thickness of the atmospheric plasma sprayed intelligent self-healing thermal barrier coatings (YSZ/20wt. % TiC + 10wt. % Al_2O_3 + 70wt. % YSZ (TAZ) /BC/SUB). The optimization aims to balance the high thermal insulation performance with minimal residual stresses within the coating. Results indicate that an optimal thickness ratio of the self-healing layer and top-coat layer is 1: 4, assuming fixed thicknesses for the bare substrate and bond-coat. Given that residual stresses from the coefficient of thermal expansion (CTE) mismatch of materials are crucial in causing spallation and delamination failure of the coatings, the study has revealed the influence of interface morphology on the residual stresses systematically. The effects of different interface geometries on the residual stress distributions at the YSZ/TAZ and TAZ/BC interfaces were systematically investigated using the ideal cosine curve interface and the real interface morphology model based on computational micro-mechanics (CMM). It is found that the morphology of the YSZ/TAZ interface can affect the stress distribution along the TAZ/BC interface, and cracks are more inclined to initiate at the TAZ/BC interface. In addition, the research provides a deep analysis of the coating failure mechanisms due to residual stresses, providing comprehensive theoretical guidance for the interface regulation of intelligent self-healing thermal barrier coatings and the optimization design of the preparation process. The self-healing thermal barrier coatings have been fabricated based on the results of the thickness optimization, and the self-healing mechanism during high temperature service process has been characterized systematically in the current work.

Keywords: thermal barrier coatings (TBCs); intelligent self-healing; thickness optimization; interface topography; finite element simulation; self-healing mechanism

YSZ Composite Thick Thermal Barrier Coatings with Improved Thermal Cycle Life

Zou Binglin[a]

[a] State Key Laboratory of Rare Earth Resources Utilization, Changchun Institute of Applied Chemistry, Chinese Academy of Sciences, Changchun 130022, Jilin, China

E-mail address: zoubinglin@ ciac. ac. cn (Zou Binglin)

Abstract: The adoption of YSZ thick thermal barrier coatings (TTBC) with thickness of above 500 mm is an effective method to achieve high thermal insulation performance, however, the low thermal cycle life of traditional YSZ TTBC greatly restricts its high-temperature application. This study designed a new type of mollusk-like "mud-brick" laminated structure YSZ composite TTBC. In-situ intralamellar cracks were produced at high temperatures, instead of through cracks, and the cracks propagation stopped at the interface of "mudbrick" . The interlamellar cracks formed in "mud layer" not only released the residual thermal stress caused by the thermal expansion mismatch, but also released the thermal stress caused by the YSZ phase transformation, sintering at high temperature. In addition, the interlayer cracks can effectively block the heat flow from diffusing directly through the cracks to the substrate compared with the through cracks. The imitation shell structure design could effectively solve the problems of high-temperature resistance, high insulation and long thermal cycle life of YSZ TTBC. which could find attractive potential applications in the high-temperature thermal protection of the hot end components of new generation of aircraft engines, heavy-duty gas turbines and hypersonic vehicles.

Keywords: YSZ, thick thermal barrier coating; thermal cycle; thermal protection

Research on the Design, Preparation, and High-Temperature Service Performance of Ultra-High-Temperature Iridium and Modified Iridium Coatings

Zhu Lian

National University of Defense Technology

E-mail address: mr_ zla@ 163. com (Zhu Lian)

Abstract: : With the increase in the flight speed of high-speed vehicles, the key hot-end components of their propulsion and thermal protection systems are imposing increasingly higher temperature requirements on ultra-high-temperature thermal protection materials. There is an urgent need to develop ultra-high-temperature thermal protection coating materials with a service temperature exceeding 2000 ℃. Iridium is the material with the lowest high-temperature oxygen permeability among all known materials. When used as a thermal protection coating on the surface of refractory metals or carbon-based materials, it can significantly enhance the service performance of metal/carbon-based composite materials in high-temperature, oxygen-rich environments. This report focuses on the research progress of our team in the design, fabrication, and high-temperature service performance of iridium and its alloyed/ceramized modified coatings. It is hoped that this work will provide an ultra-high-temperature, zero-ablation, and wide-temperature-range thermal protection coating solution for thermal structural components serving in extreme thermal environments.

Keywords: iridium; high temperature coatings; zero ablation

Study on the CMAS Corrosion Resistance and Thermal Cycle Properties of Nanostructured Lutetium Silicate Environmental Barrier Coatings

Xu Baosheng

Beijing Insitute of Technology, China

E-mail address: xubsh@ bit. edu. cn (Xu Baosheng)

Abstract: Ceramic matrix composites (CMCs) are considered the preferred material for high-temperature components of the new generation of high thrust to weight ratio aircraft engines. However, CMCs still face severe environmental corrosion issues during the service. Environmental barrier coatings (EBCs) can effectively improve the calcium-magnesium-aluminumsilicates (CMAS) corrosion resistance and service life of CMCs. In order to enhance the high-temperature CMAS corrosion and long-term service performance of EBCs, Lu_2SiO_5 and $Lu_2Si_2O_7$ spherical feedstocks with micron size but still nanostructure inside for atmospheric plasma spraying (APS) were firstly prepared via a nanopowder granulation technology in this work. Secondly, the nanostructured $Lu_2SiO_5/Lu_2Si_2O_7$ double-ceramic coating was prepared on the SiC substrate coated with a Si bond layer using APS. Finally, the CMAS corrosion resistance and thermal cycling performance of nanostructured $Lu_2SiO_5/Lu_2Si_2O_7$ double-ceramic coating were studied, respectively. The results indicated that the reaction layer and TGO layer thickness of the nanostructured $Lu_2SiO_5/Lu_2Si_2O_7$ double-ceramic coating after corrosion at 1400℃ for 50 h were 82. 5 μm and 2. 27 μm, respectively. Due to the formation of the apatite dense layer inside the coating, the CMAS melt did not penetrate the Lu_2SiO_5 top layer, that is, did not penetrate into $Lu_2Si_2O_7$ intermediate layer, and the coating remained intact. In addition, no new phases were generated on the surface of nanosturctured $Lu_2SiO_5/Lu_2Si_2O_7$ double-ceramic coating after 1200 cycles at 1350℃. The grain size of the coating after thermal cycling is 1. 02 ± 0. 47 μm. The coating structure remained intact. In short, the nanosturctured $Lu_2SiO_5/Lu_2Si_2O_7$ coating exhibited superior CMAS corrosion resistance and thermal cycling resistance.

Keywords: environmental barrier coatings; CMAS; thermal cycle

Proceedings of IFHTSE 2025

The 30th World Congress of International Federation for Heat Treatment and Surface Engineering
August, 19-22, 2025, Suzhou, China

High-Entropy ($Y_{0.2}La_{0.2}Nd_{0.2}Sm_{0.2}Eu_{0.2}$)$_2Zr_2O_7$ Thermal Barrier Coatings with Improved Sintering Resistance and Phase Stability

Mao Xiye[1], Xu Baosheng[1], Lv Bowen[2]
[1] Beijing Institute of Technology, China
[2] Institute of New Materials, Guangdong Academy of Sciences, China
E-mail address: 3120246437@ bit. edu. cn (Mao Xiye)
E-mail address: xubsh@ bit. edu. cn (Xu Baosheng)

Abstract: Thermal barrier coatings (TBCs) are a key thermal protection technology that insulates turbine and combustor components from the high-temperature gas, thereby improving the durability and efficiency of aero-engines. With the continuous development of aero-engines towards high thrust-to-weight ratio and high efficiency, the inlet gas temperature is constantly increasing. Conventional TBC materials, such as yttria-stabilized zirconia, can no longer meet the demands of high-temperature resistance and high thermal insulation. It has therefore become an urgent need to find novel TBC materials that can replace YSZ. In recent years, high-entropy ceramics (HECs), especially high-entropy rare-earth zirconates (HE-REZs), have been widely researched due to their superior thermal properties and corrosion resistance. However, the relatively low fracture toughness of HE-REZs leads to short thermal cycling life when used as coatings, severely restricting their practical application. Plasma spray-physical vapor deposition (PS-PVD) is a novel spraying technique based on low pressure plasma spraying, which can prepare coatings with feather-like columnar structure. This high strain tolerant structure can mitigate the influence of low fracture toughness and thus improve the thermal cycle life of coatings. To the best of our knowledge, there are no reports on the performance of PS-PVD HE-REZ coatings. Based on this, the HE-REZ coatings with a composition of ($Y_{0.2}La_{0.2}Nd_{0.2}Sm_{0.2}Eu_{0.2}$)$_2Zr_2O_7$ (YLNSE) were first prepared by PS-PVD technique in this work. The sintering resistance and phase stability of the YLNSE coatings at 1500 ℃ were investigated and compared to the referential gadolinium zirconate (GZO) coatings. Scanning electron microscopy, X-ray diffraction and Raman spectra confirmed that the as-sprayed YLNSE and GZO coatings had the typical PS-PVD feather-like columnar structure, and were composed of defective fluorite phases with uniform elements distribution. At 1500 ℃, the microstructures and mechanical properties of the two coatings evolved with heat treatment, showing a general trend of significant change in the early stage and stabilization in the later stage. The evolution of grain size, porosity and Young's modulus of the YLNSE coatings as a function of heat treatment duration was slower than that of the GZO coatings, indicating that better sintering resistance. After heat treatment, both coatings underwent a phase transformation from the disordered defective fluorite structure to the ordered pyrochlore structure. Compared with the GZO coatings, the YLNSE coatings exhibited better phase stability at 1500 ℃, as evidenced by their later phase transformation. The improved sintering resistance and phase stability of high-entropy YLNSE TBCs can be attributed to the sluggish diffusion effect of HECs. Thermal cycling test results demonstrated that the PS-PVD YLNSE coatings exhibited obviously improved thermal cycle life. This work demonstrated the feasibility of HE-REZ TBCs with high strain tolerant structure and deepened the understanding of high-temperature stability in terms of microstructure, mechanical property, as well as phase composition. In addition, these findings provided a solid foundation for the application of strain tolerant HE-REZ TBCs in advanced aero-engines.

Keywords: high-entropy rare-earth zirconates; plasma spray-physical vapor deposition; thermal barrier coatings; sintering resistance; phase stability

Investigation of Heat Accumulation and Cooling Control Mechanism in Internal Diameter Atmospheric Plasma Spray

Li Hongchen, Wang Weize*

School of Mechanical and Power Engineering, East China University of Science and Technology, Shanghai, China

E-mail address: Y11220024@ mail. ecust. edu. cn (Li Hongchen)

E-mail address: Wangwz@ ecust. edu. cn (Wang Weize)

Abstract: Using internal diameter atmospheric plasma spray (ID-APS) to prepare coatings on the internal wall is an effective approach to extend the life of the critical equipment. However, the heat accumulation and poor particle deposition during ID-APS adversely affect the safety of the gun and coating performance. The introduction of a cooling process could mitigate these negative effects, but it may influence the plasma jet. Therefore, the mechanism of heat accumulation and effect of cooling airflow need to be further studied. A more effective cooling method with less impact on the jet needs to be investigated. This study analyzes the temperature, velocity and particle distribution fields during spraying inside cylinders. Different cooling airflow processes are introduced around the cylinder, at the side of the gun and near the wall and around the jet, compared with the spray process without cooling. The effects of cylinder diameter and cooling airflow on temperature distribution, trajectory of the wall jet and particle behaviour in ID-APS were investigated. As the cylinder diameter decreases, the restricted heat dissipation capacity results in a significant increase in environment temperature and plasma jet temperature within 5 mm of the substrate. The confined space and internal wall with curvature enclose the high temperature wall jet, forming stronger backflow inside cylinders. Due to the relative position between the enclosed wall jet and the gun, the temperature is higher at the sides of the gun. As the cylinder diameter decreases from 130 mm to 70 mm, the highest temperature area moves from the back side to the front side of the gun. The increase of jet temperature near the substrate leads to a significant increase of temperature during the deposition of 1-20 μm particles. Particles around 10 μm are more easily deflected from the plasma jet due to the increased edge velocity of the plasma jet. Cooling airflow introduced around the entire cylinder causes the wall jet to flow in a " sine wave" pattern, carrying the heat far away from the gun and completely out. However, it also leads to jet deflection and a shift in particle deposition position, particularly affecting particles smaller than 10 μm. Cooling airflow introduced at the side of the gun and near the wall can effectively interrupt the wall jet and protect the gun with less influence on the plasma jet, though its cooling effectiveness is highly dependent on the inlet position and less in the front area. Appropriate rate of radial cooling airflow on both sides of the jet can make the wall jet behave similarly to that on a flat substrate and cooling the front of the gun effectively. However, excessive flow rate can significantly disturb the jet near the substrate, disrupting particle deposition positions and temperature-velocity distributions. Therefore, a combination of sidewall and radial cooling methods can minimize the impact on the jet and achieve an optimal wall jet trajectory, achieving better cooling effect in ID-APS.

Keywords: internal diameter atmospheric plasma spray; plasma jet simulation; temperature distribution; cooling airflow optimization

Proceedings of IFHTSE 2025

The 30th World Congress of International Federation for Heat Treatment and Surface Engineering
August, 19-22, 2025, Suzhou, China

Research on Composition Design and Dopant Element Selection for CMAS-Resistant Medium-and High-Entropy Thermal Barrier Coating Materials

Zhang Wenkang, Wang Weize*

East China University of Science and Technology, Shanghai, China

E-mail address: Zhangwenkang2022@163.com (Zhang Wenkang)

E-mail address: wangwz@ecust.edu.cn (Wang Weize)

Abstract: Thermal barrier coatings (TBCs) are advanced surface treatments applied to aircraft engine components to insulate against heat, lower substrate temperatures, prolong service life, and improve overall engine efficiency. Conventional yttria-stabilized zirconia (YSZ), despite its widespread application, is nearing its performance ceiling and is no longer capable of satisfying the escalating demands of next-generation engines. This limitation is especially pronounced in its insufficient resistance to the ever-growing threat of CMAS (CaO-MgO-Al_2O_3-SiO_2) corrosion. With the rapidly growing performance demands for high-temperature components in critical fields such as aerospace and energy power systems, there is an increasingly urgent need to design and develop advanced thermal barrier coating materials with superior performance characteristics. Medium-and high-entropy ceramics based on YSZ were synthesized via the solid-state reaction method by incorporating rare-earth elements of varying types and quantities. With an increasing number of doped elements, the ceramic materials exhibit enhanced resistance to CMAS corrosion, improved phase stability, as well as superior mechanical and thermodynamic properties. YSZ materials that are doped with more than three elements can effectively suppress elemental diffusion at the CMAS/YSZ interface under high-temperature conditions, which reduces material degradation in the CMAS environment. When modifying YSZ through rare earth element doping, the choice of dopant has a significant impact on the material's performance. To identify the rare earth elements that are most effective in enhancing the overall properties of the material. This study employs density functional theory (DFT) to construct crystal models of rare-earth-doped ZrO_2 (ReSZ; Re = Sc, Y, and lanthanides) and a Resz/CMAS interface model. It aims to clarify how different dopants influence the mechanical properties, thermodynamic stability, and CMAS corrosion resistance of ZrO_2. Crystal stability criteria, electronic density of states (DOS), surface and interfacial stability, dipole moment, and key parameters such as thermal expansion coefficient, thermal conductivity, and Young's modulus are calculated. These data are then integrated using the entropy weight method to perform a comprehensive multi-dimensional evaluation of material performance. The results show that the LaSZ and YSZ systems exhibit the best overall performance. The LaSZ demonstrates significant advantages in thermodynamic properties, including reduced thermal conductivity and increased thermal expansion coefficient. The SmSZ, DySZ, and PmSZ systems show improved interfacial stability, with Griffith rupture work increasing by more than 50%. The CeSZ system excels in mechanical performance, with a 20% increase in elastic modulus, a 50% increase in hardness, and a 10% improvement in fracture toughness. This study establishes a quantitative criterion for selecting rare earth dopants, providing a theoretical basis for designing high-entropy ceramic thermal barrier coatings (TBCs) with enhanced CMAS corrosion resistance. Additionally, a "function-oriented" doping strategy is proposed, aiming to balance overall performance optimization with the specific requirements of targeted application scenarios.

Keywords: YSZ; CMAS; Density functional theory (DFT); Rare earth element doping; Mechanical properties

Study of Oxidation Behavior and Mechanism of Non-Equimolar (Zr, Ti, Ta, W) B_2 at 1500 ℃

Fan Dong[1], Huang Jian[2], Niu Yaran[1], Zheng Xuebin[1], Ding Chuanxian[1]

[1] Key Laboratory of Inorganic Coating Materials CAS, Shanghai Institute of Ceramics, Chinese Academy of Sciences

[2] Materials Genome Institute, Shanghai University

E-mail address: fandong1998@126.com (Fan Dong)

E-mail address: yrniu@mail.sic.ac.cn (Niu Yaran)

Abstract: Ultra-high temperature ceramics (UHTCs), particularly transition metal borides, are essential for applications in extreme high-temperature environments due to their high melting points and excellent mechanical properties. However, improving their oxidation resistance remains a significant challenge. This study investigates the oxidation behavior of non-equimolar $(Zr_{0.5}Ti_{0.3}Ta_{0.1}W_{0.1})B_2$ and equimolar $(Zr_{0.25}Ti_{0.25}Ta_{0.25}W_{0.25})B_2$ ceramics at 1500 ℃, comparing them with single-component ZrB_2. The results show that $(Zr_{0.5}Ti_{0.3}Ta_{0.1}W_{0.1})B_2$ exhibited superior oxidation resistance, with an oxide layer thickness reduced by approximately 25% compared to ZrB_2 and minimal mass loss. Characterizations like XRD, SEM, and EDS reveal that a dense multi-element oxide layer formed on $(Zr_{0.5}Ti_{0.3}Ta_{0.1}W_{0.1})B_2$ during oxidation, effectively blocking oxygen penetration. Molecular dynamics simulations indicate that the oxygen diffusion coefficient in the multi-component oxide (Zr, Ti, Ta, W) O_2 is significantly lower than that in ZrO_2 and decreases further with increasing Ti content. Moreover, the lower content of Ta and W elements in $(Zr_{0.5}Ti_{0.3}Ta_{0.1}W_{0.1})B_2$ reduces the formation of low melting point phases at high temperature, slowing down mass loss and structural degradation. These findings suggest that the optimized elemental composition and synergistic effects of multiple elements significantly enhance the oxidation resistance of the ceramics. This research is expected to provide theoretical support for the material design and application of UHTCs with better performance in high-temperature environments.

Keywords: ultra high temperature ceramics; high entropy ceramic; oxidation resistance; transition metal diborides

Influence of Chemical Disorder on Mechanical and Thermal Properties of Multi-Component Rare-Earth Monosilicates RE_2SiO_5 (RE = Y、Er、Yb、Lu)

Zhang Shuxin[1], Huang Jian[2]

[1] Shanghai Institute of Ceramics, Chinese Academy of Sciences, China

[2] Shanghai University China

E-mail address: a728965935@163.com (Zhang Shuxin)

E-mail address: huangj@shu.edu.cn (Huang Jian)

Abstract: To explore the effects of chemical disorder or high entropy effect on mechanical and thermal properties of rare-earth monosilicates, six two-component, three three-component and one four-component rare-earth monosilicates, RE_2SiO_5 (RE = Y, Er, Yb, Lu), were studied by using first principles calculations. The applicability of the special quasi-random structure method in the study of multi-component X2-RE_2SiO_5 systems was also validated. The results indicate that chemical disorder has a weak influence on the crystal structure as well as the mechanical and thermal properties. The lengths of RE-O bond and Si-O bond increases linearly with the radius of RE^{3+}, regardless of the degree of chemical disorder. The mechanical properties of RE_2SiO_5 are mainly influenced by the types of rare-earth elements. The Young's modulus of RE_2SiO_5 can be enhanced by the addition of a small amount of Lu, and $(Yb_{15/16}Lu_{1/16})_2SiO_5$ has the largest Young's modulus. Compared with single-component RE_2SiO_5, the ductility of multi-component RE_2SiO_5 decreases, while the thermal conductivity approximately follows the " mixing rule" . $(Y_{1/4}Yb_{1/4}Er_{1/4}Lu_{1/4})_2SiO_5$ exhibit similar phonon vibration modes and comparable thermodynamic properties to those of single-component rare-earth silicates. This work also analyzes the gap between the calculated results of first principles and experimental results, suggesting that the low thermal conductivity caused by the high entropy effect may not be solely attributed to increased phonon anharmonicity due to enhanced lattice distortion. Instead, greater consideration should be given to the influence of high entropy effect on grain boundaries, defects, and other factors.

Keywords: environmental barrier coatings; high-entropy rare-earth monoslicates; density functional theory; thermal conductivity; mechanical properties

High Entropy Ceramic Coating on C/C Composite: From AI Design to HT Performance

Sun Jia
Northwestern Polytechnical University, China
E-mail address: j. sun@ nwpu. edu. cn (Sun Jia)

Abstract: Carbon/Carbon (C/C) composites have demonstrated extensive application prospects in aerospace thermal structural components due to their low density, excellent thermal stability, and high-temperature mechanical properties. However, these materials are prone to oxidation reactions in oxidative environments above 400 ℃, which can lead to a significant decline in mechanical properties, thus severely limiting their practical applications. To address this challenge, ultra-high temperature ceramic materials, with their extremely high melting points and good oxidation resistance, are widely used as surface coatings for C/C composites to enhance their protection against high-temperature oxidation. This presentation focuses on three aspects: i) Machine learning driving the composition selection and high temperature performance of ultra-high temperature ceramic coating; ii) AI design of high-entropy carbides and the prediction of single-phase formation, and iii) Ablation behavior investigation on rare-earth-containing high-entropy carbide coatings. This report will show our research group's recent progress on coating composition selection, preparation, and their ultra-high temperature performance under data-driven artificial intelligence models, and also provide an outlook on future work.

Keywords: composite; design; fabrication; performance

Proceedings of IFHTSE 2025

The 30th World Congress of International Federation for Heat Treatment and Surface Engineering
August, 19-22, 2025, Suzhou, China

Nanostructured $Yb_4Hf_3O_{12}$ Thermal/Environmental Barrier Coatings via Plasma Spraying: Microstructure and Properties

Wang Shun*, Xu Baosheng

Institute of Advanced Structure Technology, Beijing Institute of Technology, Beijing, China

E-mail address: wangshun@ bit. edu. com (Wang Shun)

Abstract: Ceramic matrix composites (CMCs) are promising high-temperature structural materials for next-generation high thrust-to-weight ratio aeroengines, but they face challenges such as water-oxygen corrosion and CMAS attack. Thermal and environmental barrier coatings (TEBCs) are essential for protecting CMCs, with nanostructured coatings gaining attention due to their superior thermomechanical properties. This study compares the microstructure and nanomechanical properties of nanostructured and conventional $Yb_4Hf_3O_12$ coatings deposited by atmospheric plasma spraying. The nanostructured coating exhibited a bimodal microstructure and, after annealing, showed a more significant improvement in elastic modulus and nano-hardness than its conventional counterpart. Additionally, nanostructured coatings demonstrated lower thermal conductivity, while both coatings maintained compatible thermal expansion behavior despite phase transitions (δ to F phase), ensuring practical applicability.

Keywords: nanostructured; rare earth hafnate; thermal/environmental barrier coating; physical properties

Thermal Insulation Performance and Thermal Shock Resistance of Plasma Sprayed TiAlCrNb/$Gd_2Zr_2O_7$ tHermal Barrier Coatings Sprayed on TC4 Alloy

Shen Tinghuan, Hong Du, Niu Yaran, Zheng Xuebin, Ding Chuanxian

Shanghai Institute of Ceramic, Chinese Academy of Sciences, Shanghai, China

E-mail address: 2262515855@ qq. com (Shen Tinghuan)

E-mail address: yrniu@ mail. sic. ac. cn (Niu Yaran)

Abstract: Titanium alloys have been widely utilized in the aerospace industry due to their exceptional specific strength. However, their insufficient oxidation resistance and creep performance above 600℃ significantly limit high-temperature applications. Conventional MCrAlY coatings exhibit severe interdiffusion with titanium alloy substrates during service, while typical oxidation-resistant TiAlCr coatings often contain high Cr content, resulting in inherent brittleness and proneness to cracking. In this study, a low-Cr-content Ti-48Al-2Cr-2Nb (at. %) bond coat with balanced properties was prepared on a TC4 substrate via vacuum plasma spraying (VPS), followed by atmospheric plasma spraying (APS) deposition of Gd2Zr2O7 thermal barrier coatings (TBCs) with two thicknesses (350 μm and 1 mm). The thermal insulation performance and thermal shock resistance of the coatings were systematically investigated. Experimental results demonstrate that under service conditions of 600-800 ℃, the temperature reductions achieved by 350 μm and 1 mm-thick Gd2Zr2O7 coatings were 221-338 ℃ and 284-399 ℃, respectively, representing approximately 50% improvement compared to YSZ coatings of equivalent thickness. Water-quenching thermal shock tests were conducted to evaluate the failure behavior of the coatings at 900 ℃. The 350 μm coating exhibited approximately 10% spallation after 150 thermal cycles at 900 ℃, whereas the 1 mm coating showed rapid failure at the same temperature, with an average spallation area of 20% after merely 20 cycles. The remarkably enhanced thermal insulation performance is primarily attributed to the low thermal conductivity of the Gd2Zr2O7 layer. Finite element simulation was employed to analyze the thermal insulation characteristics, stress distribution, and failure mechanisms of the TiAlCrNb/Gd2Zr2O7 coating system. The well-matched coefficients of thermal expansion among the Ti-48Al-2Cr-2Nb bond coat, Gd2Zr2O7 top coat, and TC4 substrate effectively mitigated residual stresses, thereby endowing the coating with durable thermal shock resistance. Nevertheless, the intrinsically low fracture toughness of Gd2Zr2O7 caused thick coatings (1 mm) to experience drastic reduction in stress tolerance during thermal cycling, leading to accelerated system failure.

Keywords: TC4 alloy; TiAlCrNb; TBCs; thermal insulation performance; thermal shock resistance

High Temperature Thermal Shock Resistance of Plasma-Sprayed High-Entropy (Y0. 2Yb0. 2Ho0. 2Er0. 2Tm0. 2) 2Hf2O7 Coating

Zhang Jinyuan, Zhu Jingpeng

Zhengzhou University, Zhengzhou, China

E-mail address: jinyuanzhang@ gs. zzu. edu. cn (Zhang Jinyuan)

E-mail address: zhujinpeng@ zzu. edu. cn (Zhu Jingpeng)

Abstract: Rare earth hafnate is considered to have great application potential in the field of thermal protection due to their relatively higher melting point and lower thermal conductivity. Based on multi-principal element high-entropy design concept, a novel high-entropy $(Y_{0.2}Yb_{0.2}Ho_{0.2}Er_{0.2}Tm_{0.2})_2Hf_2O_7$ (HE-YYb) as top coating was deposited by air plasma spraying. This study investigates the feasibility of preparing high-entropy rare earth hafnate (HE-YYb) coating plasma spraying and exploring the thermal shock resistance behaviors of the HE-YYb coating carried out by the plasma flame shock test. Results indicate that the as-sprayed powders with good sphericity, good flowability and uniform particle size distribution are obtained by exploring the spraying process. By optimizing the spraying power, a typical layered structure HE-YYb coating with a smooth and dense surface, tight bonding boundaries between layers, and few pores and cracks can be prepared. During plasma flame shock at 1200 ℃, 1300 ℃, and 1400 ℃, the sluggish diffusion effect significantly improves the high-temperature phase stability and shock resistance of the HE-YYb coating. The HE-YYb coating showed no spalling at 1300 ℃, only impact craters and spherical particles appeared, and most areas still maintained their original appearance. However, at 1400 ℃, severe over-burning occurred at the surface edges of the coating. The newly formed structure after melting exhibited a fragmented and uneven morphology but still retained a defective fluorite structure. The unique high-entropy effect makes material exhibit good phase stability that ensures better thermal protection during high-temperature work. During the shock process, the thermal insulation temperature of coating was 450 ℃, 420 ℃, and 360 ℃, respectively. The low thermal conductivity of the HE-YYb material, which is attributed to lattice distortion, ensures the excellent thermal insulation performance of the HE-YYb coating. Additionally, the thermal cycling resistance experiment shows that the HE-YYb coating only exhibited edge delamination and cracking after 13 cycles of the furnace cycle test at 1200 ℃-300 s, which effectively alleviates thermal stress concentration, indicating the coating possesses thermal shock resistance. In the thermal cycle test at 1200 ℃-300 s, the sluggish diffusion effect and higher CTE is the crucial factor in enhancing the cyclic lifespan of the HE-YYb coating. The main mode of coating failure under thermal cycling is the shear and radial stress concentration causing edge position coating spallation. This study clearly shows that the novel HE-YYb coating expands the application prospects of high-entropy rare earth oxide coatings in thermal protection coatings.

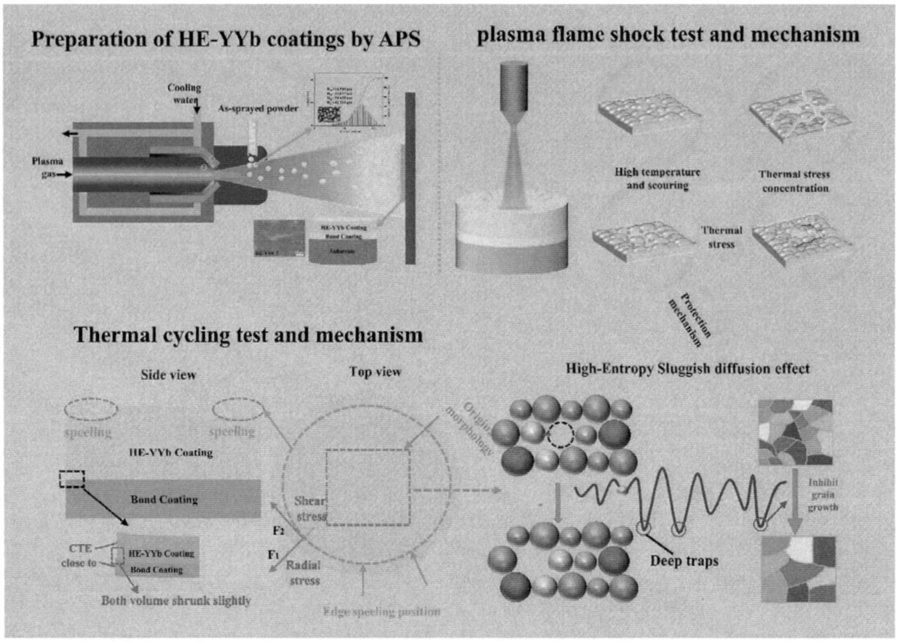

Fig. 1 SEM images showing the cross sections of the PEO coatings on Zirlo alloy formed under different R ratios at different frequencies: (a-d) 100 Hz; (e-f) 1000Hz

High-Temperature Thermal Protective Performance of High-Entropy Rare-Earth Tantalate ($Nd_{0.2}Dy_{0.2}Ho_{0.2}Y_{0.2}Er_{0.2}$) TaO_4 Coating Deposited by Atmospheric Plasma Spraying

Zhang Yulong, Zhu Jingpeng

Zhengzhou University, Zhengzhou, China

E-mail address: yulongzhang@ gs. zzu. edu. cn (Zhang Yulong)

E-mail address: zhujinpeng@ zzu. edu. cn (Zhu Jingpeng)

Abstract: High-entropy rare-earth tantalates have potential for use in thermal barrier coatings, but current research has focused mostly on powder or bulk materials instead of coatings. In this work, atmospheric plasma spraying technology was employed to prepare novel high-entropy rare-earth tantalate ($Nd_{0.2}Dy_{0.2}Ho_{0.2}Y_{0.2}Er_{0.2}$) TaO_4 (HE-$RETaO_4$) coatings on the surface of nickel-based alloys. At a power of 36 KW, the HE-$RETaO_4$ powder can be fully melted, and the prepared coating exhibits excellent adhesive strength (35. 21 MPa), which is higher than that of 8YSZ (28 MPa) . The as-sprayed HE-$RETaO_4$ powders undergo a ferroelastic phase transition from the monoclinic (m) phase to the metastable tetragonal (t′) phase during the spraying process. Owing to the influence the cationic radii of multispecies HE-$RETaO_4$, its phase transition temperature is lower than that of $YTaO_4$, leading to the formation of a ferroelastic domain toughened structure at lower temperatures. Based on the high-entropy effect and ferroelastic phase-transformation mechanism of HE-$RETaO_4$, the HE-$RETaO_4$ coating in this work showed excellent mechanical properties. Specifically, the Vickers hardness (Hv) of the HE-$RETaO_4$ coating (8. 05 ±0. 92 GPa) is greater than that of the YSZ coating (6. 51 ±0. 29 GPa), the Young's modulus (E) is 137 GPa, which is lower than that of YSZ (250 GPa) and 7YSZ (210 GPa), and the fracture toughness of the HE-$RETaO_4$ coating (1. 3 MPa · $m^{1/2}$) is similar to that of the 8YSZ coating (1. 31 MPa · $m^{1/2}$) . These excellent mechanical properties enable the coating to effectively resist damage caused by thermal cycling, thermal mismatch, and thermal shock. The significant RE^{3+} radius difference in HE-$RETaO_4$ materials caused considerable lattice distortion, greatly enhanced phonon scattering, and caused HE-$RETaO_4$ ceramics to exhibit excellent thermophysical properties with a thermal expansion coefficient of $10.3 \times 10^{-6}\ K^{-1}$ (1000 ℃) and thermal conductivity of 1. 43 $W^{-1} \cdot m^{-1}$ (800 ℃) . Under the plasma flame assessment at 1200 and 1300 ℃, the insulation temperatures of the HE-$RETaO_4$ coatings are 220 and 200 ℃, respectively. The center of the HE-$RETaO_4$ coating is subjected to concentrated thermal load, resulting in grain growth compared to the edge region. The sluggish diffusion of HE-$RETaO_4$ hinders diffusion of elements, ultimately engendering a profusion of equiaxed grains with comparable aspect ratios and a substantial divergence in energy between intragranular regions and boundaries. During the thermal shock process, micro-cracks occur within the coating due to the large difference in the coefficient of thermal expansion. In addition, the microcracks generated within the HE-$RETaO_4$ coatings rapidly expand along the high-energy grain boundaries of the equiaxed grains, eventually leading to mechanical delamination and coating failure. Meanwhile, during the cooling process of the coating system, the coating edges are spalled due to the influence of stress. This work provides a powerful theoretical basis for the application of HE-$RETaO_4$ as a high-temperature protective material.

Keywords: high-entropy rare-earth tantalate; Thermal barrier coating; atmospheric plasma spraying; ablation behaviour

Design, Composition Optimization, Oxidation/Ablation Resistance of Multi-Component Ceramic Coating

Dong Shun*, Yu Yue

Science and Technology on Advanced Composite in Special Environments Laboratory, Harbin Institute of Technology, Harbin, China

E-mail address: dongshun@hit.edu.cn (Dong Shun)

E-mail address: 25B318027@stu.hit.edu.cn (Yu Yue)

Abstract: With the fast development of advanced aerospace vehicles, higher requirement is demanded for thermal protection materials. Multi-component ceramic coating has been acknowledged to be one of the most promising thermal protection materials, due to the excellent high-temperature physical/chemical stability and the good oxidation/ablation resistance. In order to further improve the service performance of thermal protection materials, it is crucial to focus on the study of composition design and structural optimization, preparation process. In our recent research, the $TaSi_2$-$MoSi_2$-SiB_6 coating with the addition of SiC_w was prepared by spraying and flash sintering, expected to improve the oxidation/ablation resistance and fracture toughness. Due to composition optimization, the mass ablation rate and line ablation rate were 1.03×10^{-3} g/s and 1.2×10^{-3} mm/s after ablation at 1600 ℃ for 1000 s. Furthermore, we synthesized the ZrB_2-SiC-TaB_2 ceramic in order to clarify the effect of Ta addition on the oxidation resistance. Ta elements were dissolved into the surface glassy layer forming Ta-O-Si network structure, contributing to the improvement of anti-oxidation ability. Subsequently, the thermal protection coating will be prepared based on this optimized result. With the development of composition optimization and structural design, advanced multi-component ceramic is expected to meet the service requirement for thermal protection materials and contribute to the application of advanced aerospace vehicles in extreme environment.

Keywords: Ablation; oxidation; composition design; structural optimization

Air Plasma-Sprayed High-entropy Rare Earth Oxide Coating With High Thermal Protection Performance

Zhu Jinpeng

School of Material Science andEngineering , Zhengzhou University, Zhengzhou 450001, China

Abstract: High-entropy rare-earth aluminate ($Y_{0.2}Yb_{0.2}Lu_{0.2}Eu_{0.2}Er_{0.2}$)$_3$ Al_5O_{12} (HE-$RE_3Al_5O_{12}$) has been considered as a promising thermal protection coating (TPC) material based on its low thermal conductivity and close thermal expansion coefficient to that of Al_2O_3. However, such a coating has not been experimentally prepared, and its thermal protection performance has not been evaluated. To prove the feasibility of utilizing HE-$RE_3Al_5O_{12}$ as a TPC, HE-$RE_3Al_5O_{12}$ coating was deposited on a nickel-based superalloy for the first time using the atmospheric plasma spraying technique. The stability, surface and cross-sectional morphologies, as well as the fracture surface of the HE-$RE_3Al_5O_{12}$ coating were investigated, and the thermal shock resistance was evaluated using the oxyacetylene flame test. The results show that the HE-$RE_3Al_5O_{12}$ coating can remain intact after 50 cycles at 1200 ℃ for 200 s, while the edge peeling phenomenon occurs after 10 cycles at 1400 ℃ for 200 s. This study clearly demonstrates that HE-$RE_3Al_5O_{12}$ coating is effective for protecting the nickel-based superalloy, and the air plasma spraying is a suitable method for preparing this kind of coatings.

Keywords: high-entropy rare earth oxide; thermal protection coating; air plasma spraying; thermal protection performance

Continuously Innovating High-End Thermal Protection Coatings

Wang You

Harbin Institute of Technology, China

E-mail address: wangyou@hit.edu.cn (Wang You)

Abstract: Advanced aircraft urgently require high-performance aircraft engines. High performance gas turbines are urgently needed for ships, power equipment, and other applications. The effective protection of high-temperature hot end components of aircraft engines, ground heavy gas turbines, and ship gas turbines cannot be achieved without thermal protection coatings. This report combines the research progress of our research group in the fields of nanostructured thermal barrier coatings, nanostructured environmental barrier coatings, and nano modified base coatings in the past two decades. It introduces the research progress of thermal protection coatings from the following four levels, emphasizing the innovative development ideas in the research process: 1. Nanostructured ceramic coating has been successfully applied. 2. Pain points still exist in the thermal protection coatings. 3. Innovation in Materials and Micro/Nano Structures of Coatings. 4. Concept of New Structural Thermal Protection Coatings. Finally, the broad application prospects of nanostructured thermal protective coatings on hot end components of aircraft engines and gas turbines were discussed.

Keywords: thermal protection coatings; nanostructured ceramic coating; high-temperature hot end components

Synthesis Pathways of (HfZrTiCe/La/Y) O_{2-x} Nanoparticles via Benzyl Alcohol Route at Critical Temperature

Guo Fangwei

Shanghai Jiao Tong University, China

E-mail address: fwguo2014@ sjtu. edu. cn (Guo Fangwei)

Abstract: High entropy oxides (RE-HEO) containing rare earth elements can provide excellent phase stability and oxidation resistance. However, the controllable synthesis of RE-HEO at low temperature is a challenge due to poor understanding on nucleation. We synthesized the (HfZrTiLn) -HEO nanoparticles with 15% ionic radius mismatch via benzyl alcohol route at 220 ℃-5 min in presence of $PtCl_4$ and Fe (acac)$_3$. Nucleation pathways of the 5HEO at the critical temperature were elucidated by using a comparison study of conventional heating and microwave irradiation heating. Consistency of XRD patterns and STEM-EDX observation indicate that the resultant Hf-OBn monomers acted as the nucleation center of the 5HEO, determined by diffusion kinetics. The nucleation rate depended on the metal monomers assembly and esterification reaction, which was accelerated by water vapor pressure produced in-situ by $PtCl_4$ catalyst. The Fe-metal organic cages served as the structure stabilizer of Zr/Ti monomers, and prevented early hydrothermal reaction route.

Keywords: high-entropy fluorite oxide; rare earth elements; Nucleation pathway; metal monomers, Fe-metal organic cages

Microstructure Evolution and Mechanical Property Degradation Mechanism of Micro/Nanostructured Zirconate Thermal Barrier Coatings in CMAS Environment

Zhou Feifei

Beijing Institute of Technology, China

E-mail address: zhoufeifei@ bit. edu. cn (Zhou Feifei)

Abstract: Focusing on the structure-property relationship between the microstructure and mechanical properties of thermal barrier coatings subjected to calcium-magnesium-aluminum-silicate (CMAS) corrosion, this project designs La2 (Zr0. 75Ce0. 25) 2O7 (LCZ) coating materials based on " defect" engineering. On the basis of achieving low thermal conductivity and high toughness, this project intends to prepare nanostructured La2 (Zr0. 75Ce0. 25) 2O7 (LCZ) spherical powder feedstock through nano-powder re-granulation technology, construct micro/nanostructured LCZ thermal barrier coatings using atmospheric plasma spraying, and investigate the CMAS corrosion behavior of micro/nanostructured LCZ thermal barrier coatings at 1300 ℃. The aim is to clarify the rules and mechanisms of CMAS corrosion on the microstructure and mechanical property degradation of nano-zirconate thermal barrier coatings, reveal the structure-property relationship between the microstructure and mechanical properties of micro/nanostructured LCZ thermal barrier coatings, and lay a theoretical foundation for the development and engineering application of micro/nanostructured rare earth zirconate thermal barrier coatings.

Keywords: thermal barrier coatings (TBCs); nanostructure; CMAS corrosion

Design and Preparation of Ultra-High-Temperature Thermal Barrier and Environmental Barrier Coatings on Turbine Blades

He Jian*, Guo Hongbo, Gong Shengkai*

School of Materials Science and Engineering, Key Laboratory of High-Temperature Structural Materials and Coatings Technology (Ministry of Industry and Information Technology), Beihang University, Beijing, China

E-mail address: hejian511690@buaa.edu.cn (He Jian)

E-mail address: gongsk@buaa.edu.cn (Gong Shengkai)

Abstract: High-pressure turbine blades are the core components in aero-engines which suffer from the highest temperature and the most complex load. The application of blades, including the in-service single-crystal superalloy blades and the SiCf/SiC ceramic matrix composite (CMC) blades under research, requires advanced high-temperature protective coatings. For example, the single-crystal superalloy blades rely on thermal barrier coatings (TBCs) for thermal insulation and oxidation resistance, while CMC blades need environmental barrier coatings (EBCs) to prevent water vapor corrosion. These coatings are critical technologies for enhancing the thrust-to-weight ratio, service life and operational safety of high-performance engines. However, As the thrust-to-weight ratio of engines continually rises, the operating temperature of turbine blades increases correspondingly, and as a result, the conventional TBCs and EBCs can't meet the application requirements for higher temperatures. Developing ultra-high-temperature TBCs (≥1300 ℃) and EBCs (≥1400 ℃) becomes critical for achieving ultra-high-temperature application of turbine blades in aero-engines with high thrust-to-weight ratio. In viewing of this, the report introduces the research on the design and preparation of ultra-high-temperature TBCs and EBCs by the High-Temperature Protective Coating Group of Beihang University, which includes the design of rare-earth modified zirconate TBC materials and the corresponding columnar structure with interstitial particles, the modified hafnium silicate EBC bond coat materials and high-entropy rare-earth silicate EBC topcoat materials as well as their thermal spraying technologies. These studies provide important support for developing ultra-high-temperature protective coatings on advanced aero-engine turbine blades.

Keywords: thermal barrier coatings; environmental barrier coatings; thermal spraying; ultra-high temperature

Optimization of Bonding Strength and Microstructure in Atmospheric Plasma-Sprayed Y_2O_3 Coatings via Response Surface Methodology

Zhang Shitao, Li Wenge*, Zhao Yuantao

School of Merchant Marine, Shanghai Maritime University, Shanghai 201306, China

E-mail address: zstcameron@ sina. com (Zhang Shitao)

E-mail address: wgli@ shmtu. edu. cn (Li Wenge)

Abstract: Atmospheric plasma spraying (APS) is one of the most effective techniques for the preparing Y_2O_3 ceramic coatings. To investigate the effect of spraying parameters on the microstructure and the bonding strength of Y_2O_3 coatings, a response surface methodology (RSM) has been adopted to design the plasma spraying parameters. The results showed that the optimized bonding strength of the coating (23. 72 MPa) was obtained at the current of 610. 61 A, Ar flow rate of 48. 23 L/min, spraying distance of 15. 06 cm, which was supported by both RSM and experimental results. The optimized Y_2O_3 coating possessed the small grain size of 37. 01 nm and the low porosity of 1. 80%, which brought in the optimized bonding strength of the coatings. Meanwhile, the pore sizes of the Y_2O_3 coatings also affected the bonding strength of the coatings, in which pore size less than 10 μm were found to support a higher bonding strength. Therefore, the microstructure of the coatings strongly affected the bonding strength including the adhesive strength and cohesive strength of the coatings, which consequently dominated the fracture failure mechanism of the Y_2O_3 coatings. With decreasing the bonding strength of the coatings, the fracture failure mechanisms were that the fracture mainly appeared within the coating, the fracture appeared at the coating/substrate interface and within the coating, the fracture mainly appeared at the coating/substrate interface, respectively.

Keywords: atmospheric plasma spraying; Y_2O_3; Response Surface Methodology; bonding performance

Research Progress of Self-Healing Thermal Barrier Coatings Modified by Laser Surface Treatment

Zhang Panpan[1,2,3], Yao Jianhua[1,2,3*]

[1] Institute of Laser Advanced Manufacturing, Zhejiang University of Technology, Hangzhou 310023, China

[2] School of Mechanical Engineering, Zhejiang University of Technology, Hangzhou 310023, China

[3] Collaborative Innovation Center of High-end Laser Manufacturing Equipment (National "2011 Plan"), Zhejiang University of Technology, Hangzhou 310023, China

E-mail address: panpanzhang166@163.com (Zhang Panpan)

E-mail address: lam@zjut.edu.cn (Yao Jianhua)

Abstract: Due to thermal protection from extreme temperatures and reduction in fuel consumption, thermal barrier coatings were extensively applied in hot components of gas turbine engines for aircraft/marine/vehicle propulsion and power generation. In order to improve the microstructure and thermal service performance of plasma-sprayed thermal barrier coatings, self-healing materials were introduced by laser surface treatment to obtain the self-healing thermal barrier coatings. Nickel superalloy substrates were sprayed with NiCrAlY bond coating and then with yttria-stabilized zirconia (YSZ) top coating by atmospheric plasma spraying. The microstructure, phase composition, and thermal service performance of the plasma-sprayed and self-healing coatings were examined. The results indicated that the laser-treated zone was characterized by the segmented microcracks and dense columnar crystal structure. The thermal service performance of the self-healing thermal barrier coatings was better than that of the as-sprayed coating. On the one hand, the columnar grains and segmented cracks in the laser-treated zone could increase the strain tolerance of coatings. On the other hand, the microcracks healing delayed the connection of vertical cracks and horizontal cracks during thermal cycling, which further improved the thermal service performance.

Keywords: thermal barrier coatings; laser surface treatment; self-healing materials; thermal service performance

Stability of Thermal Barrier Coatings in Natural Environment

Cao Xueqiang

Wuhan University of Technology, Wuhan, China

E-mail address: xcao@ whut. edu. cn (Cao Xueqiang)

Abstract: Thermal barrier coatings (TBCs) are usually composed of NiCrAlY as bond coat (BC) and yttria stabilized zirconia (YSZ) as top coat (TC) . During storage of TBCs in natural environment, TBCs may be seriously corroded, and the corrosion mechanism is studied. On one hand, BC would be corroded by the residual Cl^- in YSZ powder. The nano-powder of YSZ contains a lot of $ZrOCl_2$ and YCl_3, HCl is formed in moisture, leading to the corrosion of BC. The BC is also NiCrAlY alloy, with Al and Cr content higher than the substrate Ni-based high-temperature alloys, leading to about 50 mV lower corrosion potential for BC compared to the substrate, therefore, resulting in galvanic couple with the BC. The influence of different Al contents in NiAl alloys on their corrosion behavior in dilute H_2SO_4, simulating the corrosive environment was studied. The research demonstrates that a high Al content in the BC is the primary cause of corrosion in conditions of natural environment. On the other hand, the structural stabilizer Y^{3+} in YSZ would react with H_2O in moisture, leading to the phase transformation of YSZ from the tetragonal to the monoclinic phase. A core/shell structure of ZrO_2/Y_2O_3 in YSZ crystalline grains has been proved. Calcination of YSZ above 1100 ℃ can prevent the degradation.

Keywords: stability; thermal barrier coatings; corrosion mechanism; phase transformation; degradation

AI-Assisted Visualization of the Service-Induced Damage Process in Thermal Barrier Coatings

Wang Liang

Shanghai Institute of Ceramics, Chinese Academy of Sciences, Shanghai, China

E-mail address: L. Wang@ mail. sic. ac. cn (Wang Liang)

Abstract: Thermal barrier coatings (TBCs) inevitably face failure during practical service, and predicting their failure modes and service lifespan remains a common challenge in the scientific community. To address this, we have developed an AI-driven, dual-modal (" ear-listening" and " eye-seeing") visualization framework for the full-process monitoring of TBCs' service-induced damage. By dynamically analyzing in-situ acoustic emission (AE) signals from TBCs under mechanical loading and actual service conditions—including filtering, fast Fourier transform (FFT), wavelet (packet) analysis, K-Means unsupervised machine learning clustering, wavelet energy coefficient calculation, and neural network analysis (BP and RBF neural networks) —we designed a software system that establishes real-time correlations between internal crack propagation modes and AE signal features, enabling rapid " ear-listening" identification of coating failure modes. Additionally, leveraging batch and automated labeling of microstructural features in cross-sectional images of as-prepared and early-service TBCs, combined with deep learning and long short-term memory (LSTM) neural networks, we achieved dynamic prediction of microstructural evolution, particularly tracking the crack propagation patterns over extended service durations. This integrated approach realizes AI-powered dual-modal (auditory and visual) visualization of TBCs' failure behavior, bridging the gap between real-time damage monitoring and predictive lifespan assessment.

Keywords: AI; damage process; thermal barrier coatings; acoustic emission; machine learning; deep learning

Proceedings of IFHTSE 2025

The 30th World Congress of International Federation for Heat Treatment and Surface Engineering
August, 19-22, 2025, Suzhou, China

Fabrication of GaN (Gallium Bitride) Coatings on Stainless Steel via Kinetic Spray Method

Zhou Shaoyun

Nagoya University, Nagoya, Japan

E-mail address: zhou. shaoyun. d4@ f. mail. nagoya-u. ac. jp (Zhou Shaoyun)

Abstract: Ceramic materials are inherently characterized by high melting points and brittleness, which pose manufacturing challenges in diverse applications. Kinetic spray methods, such as Cold Spray (CS) and Aerosol Deposition (AD), have emerged as promising techniques for depositing coatings without significant chemical reactions or phase transformations. However, it has been established that native oxide layers impede metallurgical bonding in kinetic spraying of metallic materials. Due to inherent challenges and the limited research on brittle ceramic particles, a consensus on the impact of oxide layers on ceramic coating bonding remains elusive. In this study, agglomerated GaN powders, featuring primary particle sizes of tens of nanometers and agglomerate diameters of several microns, were used to successfully fabricate nanoporous GaN coatings on stainless steel substrates via Low-Pressure Cold Spray (LPCS). During high-velocity particle impact, native oxide films on both particles and substrates were removed, followed by the generation of a new oxide layer at the particle-substrate interface. X-ray photoelectron spectroscopy (XPS) and high-resolution transmission electron microscopy (HRTEM) were employed to characterize interface oxidation, clarifying chemical stability and oxidation types. Local heteroepitaxy near the interface, associated with newly formed gallium oxide, promoted coating formation. The GaN coatings exhibited boundary amorphization and high dislocation density, attributed to high strain rates during particle impact. Thus, bonding mechanisms between nitride ceramic particles and stainless steel substrates can be attributed to newly formed interfacial oxidation and boundary amorphization. To mitigate environmental interactions, the Aerosol Deposition (AD) technique was also explored as an alternative kinetic spraying method for GaN coatings. Preliminary results indicate that successful GaN deposition through AD significantly depends on optimized process parameters, particularly those achieving high particle velocities. Consequently, dense and crack-free GaN layers were successfully fabricated at room temperature using helium as the process gas. Ongoing studies aim to investigate internal interfaces and clarify underlying bonding mechanisms. In conclusion, this research demonstrates the feasibility and effectiveness of kinetic spray methods for fabricating GaN coatings. The insights gained not only validate this novel approach for GaN deposition but also provide valuable guidance for kinetic spraying of a broader range of ceramic materials.

Keywords: GaN (gallium nitride) coatings; kinetic spray method; Low-Pressure Cold Spray (LPCS)

Design and Preparation of Novel Plasma Sprayed Coatings for Anti-Laser Application

Gao Lihong

School of Materials Science & Engineering, Beijing Institute of Technology, Beijing, China

E-mail address: gaolihong@bit.edu.cn (Gao Lihong)

Abstract: As a light source, laser has unique advantages, high brightness, good directivity and great monochromatic. Many countries develop laser technology and try to use it in different fields. So laser has attracted much attention. It's widely used in many fields, such as additive manufacturing, laser cutting, medical treatment, etc. Of course we try to use it as well as possible, but if we use high energy laser, we should pay attention, especially for continuous wave laser, it could cause serious thermal and mechanical damage. So how to protect our human body and equipments? This is a new challenge.

Keywords: Plasma spray; coatings; anti-laser; high energy laser

Oxidation Behavior and Failure Mechanisms of a HVOF-sprayed NiCoCrAlY Coating in High-temperature Water Vapor Environment

Zhao Xiaofeng

Shanghai Jiao Tong University, Shanghai 200240, China

E-mail address: xiafoengzhao@ sjtu. edu. cn (Zhao Xiaofeng)

Abstract: Hydrogen-firing gas turbines are developed to reduce CO_2 emission and improve operating efficiency. One potential material challenge is that hydrogen combustion produces a much higher water vapor concentration in the exhaust gas compared to natural gas combustion. MCrAlY (M = Ni, Co or both) alloys are widely applied as the overlay coatings or bond coats in thermal barrier coatings to protect the hot components in gas turbines (e. g., turbine blades and combustors) against high-temperature oxidation. The high-water-vapor service environment in hydrogen-firing gas turbines is expected to undermine the oxidation resistance of MCrAlY coatings and potentially accelerate their degradation. Thus, it is of necessity to elucidate the effect of high-temperature water vapor on the oxidation behavior and failure mechanisms of MCrAlY coatings. We present an in-depth study on the oxidation behavior and failure mechanisms of a HVOF-sprayed NiCoCrAlY coating in high-temperature water vapor. Compared to air oxidation during which an α-Al_2O_3 layer is quickly established, oxidation in water vapor promotes the growth of metastable γ-Al_2O_3 and delays its transformation to α-Al_2O_3, which is attributed to the surface hydroxylation originating from high OH^- concentration generated through H_2O dissociation. Both the fast outward growth of γ-Al_2O_3 and the low formation energies of Al and O vacancies due to the interstitial protons deriving from H_2O dissociation and the substitution of base metal cations at the Al^{3+} sites in γ/α-Al_2O_3 accelerate the outward Al diffusion, and thus increase the oxide growth rate. This result in turn generates a high counter current vacancy flux at the oxide/metal interface, consequently resulting in the formation of numerous interface pores, which significantly reduces the interface toughness by diminishing contact between the metal and the oxide scale. Ultimately, the fast oxide growth and reduced interface toughness lead to premature spallation failure of the oxide scale.

Keywords: HVOF; Coatings; Water vapor; High-temperature oxidation; alumina scale

New Environmental Barrier Coating for Aircraft Engines

Zhang Xiaofeng

Institute of New Materials, Guangdong Academy of Sciences, Guangdong, China

E-mail address: zxf200808@126.com (Zhang Xiaofeng)

Abstract: Compare the dual blade 7YSZ thermal barrier coatings prepared by emerging plasma spraying physical vapor deposition (PS-PVD) and existing electron beam physical vapor deposition (EB-PVD), and study the deposition mechanisms of both. Cut open the twin blades coated with 7YSZ thermal barrier coating, take samples around each surface of the twin blades, and use SEM, thickness measurement and other detection methods to characterize the microstructure and thickness evolution of the thermal barrier coating. The PS-PVD feather columnar coating grows perpendicular to the substrate, with almost no angular tilt, and the decrease in thickness in the shaded area is relatively small. Overall, the coating thickness is relatively uniform. The EB-PVD needle shaped coating shows severe tilted growth in the shaded area, accompanied by a sharp decrease in coating thickness. The thickness difference between the directly deposited area and the shaded area is significant and overall uneven. The high energy and high powder feeding rate of PS-PVD provide a greater driving force for coating growth during deposition compared to EB-PVD gas-phase particles. The energy supply, deposition units (elements), and their arrangement in the coating preparation process result in PS-PVD being a mixed element deposition, while EB-PVD is a single element deposition. The non line of sight deposition characteristics of PS-PVD result in direct deposition of the coating on the front side, while a small amount of deposition is also observed on the shadow surface. The deposition efficiency of the shadow surface is related to the back angle of the workpiece shape. The PS-PVD process, due to its higher power and plasma flow, can achieve better deposition in shaded areas, especially in engineering applications where it can deposit uniformly thick thermal barrier coatings on complex shaped twin blades.

Keywords: Thermal barrier coating; plasma spraying physical vapor deposition; electron beam physical vapor deposition; sedimentary mechanism; double blade

Dot-Matrix Diffusion Barrier Lamellae Redefining Anti-Sintering Topcoat in Thermal Barrier Coatings

Dong Hui, Feng Yukun, Bai Ruxue, Wang Fei, Yi Jing

Xi'an Key Laboratory of High Performance Oil and Gas Field Materials, School of Materials Science and Engineering, Xi'an Shiyou University, Xi'an China

E-mail address: donghui@ xsyu. edu. cn (Dong Hui)

Abstract: To address the premature spallation caused by the yttria-stabilized zirconia (YSZ) sintering, a novel anti-sintering topcoat of thermal barrier coatings (TBC) was designed and deposited by the core-shell powder (the YSZ core partly covered by $La_2Zr_2O_7$ (LZ) shell). The anti-sintering coating (NAS) included of YSZ skeleton and dot-matrix LZ lamellae. LZ lamellae in YSZ skeleton was designed as the diffusion barrier. The microstructural evolution, sintering behavior and adhesion work of NAS coating and YSZ coating were comparatively elaborated by ultra-high temperature exposure test (≥1400 ℃). The results showed that a spherical YSZ@ LZ core-shell powder with mechanical bonding interface, i. e. YSZ core partly covered by LZ shell, could be produced by dual-feed atmospheric plasma spraying (APS). The NAS coating deposited by YSZ@ LZ powder included the YSZ matrix and uniformly dispersed LZ lamellae. The lamellar gap areas decreased with thermal exposure duration, but the area decreasing velocity between YSZ lamellae and LZ lamellae was dramatically lower than that between YSZ lamellae. In terms of the contribution of the area reduction ratio, the LZ lamellae was only about 45% of YSZ lamellae. The fracture toughness of YSZ coating was ideally maintained under addition of 13wt. % LZ in YSZ topcoat. The propagation path of indentation cracks and the adhesion work indicated that the micro-cracks induced by the sintering stress tended to initiate and propagate in LZ lamellae or the ZrO_2/LZ interface and tended to close once the crack tips arrived at ZrO_2 region. Therefore, the NAS coating (i. e. inserting anti-sintering lamellae in YSZ coating) in TBC could not only improve the ultra-high temperature sintering resistance, but relieve the stress caused by YSZ sintering.

Keywords: thermal barrier coatings; anti-sintering topcoat; core-shell powder; heterogeneous interface; anti-sintering mechanism

Durable Dual-State Duplex Si-HfO2 with Excellent Oxidation and Cracking Resistance

Chen Lin[1], Luo Jingchuan[1*], Yang Wenqi[2], Li Changjiu, Yang Guanjun

[1] National Key Laboratory for Mechanical Behavior of Materials, School of Materials Science and Engineering, Xi'an Jiaotong University, China

[2] Xi'an Aerospace Engine Company Limited in CASC, China

E-mail address: chenlinxjtu@ xjtu. edu. cn (Chen Lin)

E-mail address: luojc123@ stu. xjtu. edu. cn (Luo Jingchuan)

Abstract: The lifetime of Si bond coatings in environmental barrier coatings is constrained by phase-transition-induced cracking of the SiO_2 scale. In this study, Si-HfO_2 dual-state duplex composite materials are proposed to address this issue by partially forming $HfSiO_4$ and minimizing the SiO_2 content. The as-prepared composite exhibited a structure comprising discrete HfO_2 " bricks" embedded in a continuous Si " mortar", while the oxidized state transformed into discrete $HfSiO_4$ " bricks" within continuous thin SiO_2 " mortars" . The results indicate that continuous thin SiO_2 contributes to reducing the oxidation rate to a level comparable to that of pure Si, and discrete $HfSiO_4$ particles aid in relieving phase transition-induced stress and inhibiting crack propagation, thereby enhancing oxidation and cracking resistance simultaneously. Consequently, the composite with 20 mol% HfO_2 and a mean particle size of ~500 nm at 1370 ℃ exhibited a service lifetime 10 times greater than that of pure Si. This research provides valuable insights for designing Si-based bond coatings with improved service lifetimes.

Keywords: environmental barrier coatings; bond coatings; oxidation; phase transition; cracking

Small-Angle Al_2O_3 Grain Boundaries Achieve Ultralow Oxygen Diffusion

Wang Wenjing, Yang Guanjun, Chen Lin*
Xi'anJiaotong University, Xi'an 710049, China
E-mail address: chenlinxjtu@ xjtu. edu. cn (Chen Lin)

Abstract: Alumina (Al_2O_3) is extensively utilized as an oxidant barrier in aeroengines and gas turbines at elevated temperatures, where small-angle grain boundaries significantly reduce its oxygen permeation. However, experimentally observing and regulating these small-angle grain boundaries poses considerable challenges. In this study, we search for these small-angle Al_2O_3 grain boundaries via first-principles calculations. We identify two families of ultralow-oxygen-diffusion grain boundaries, namely, Gb (xxx) / (001) and Gb (xxx) / (110), whose diffusion rates are 2 to 4 orders of magnitude lower than those of conventional grain boundaries. On the basis of these findings, we summarize and analyze the influence mechanisms and relative importance of various factors affecting grain boundary permeation. Our results indicate that, in addition to the grain boundary angle, the average bond length and band gap play primary roles in impeding oxidant transport. Finally, the [0001] selective nucleation growth of alumina was achieved by stepwise sintering, and ultralow-oxygen-diffusion Gb (104) / (110) was successfully prepared, which was consistent with the simulation results. These findings provide valuable insights for the design of ultralow-oxygen-diffusion Al_2O_3 grain boundaries, thereby significantly enhancing the oxidation resistance of aluminum-containing superalloys.

Keywords: Al_2O_3; oxidant diffusion; grain boundary angle; bond length; band gap

Structural Evolutions and Failure Mechanisms of APS Tantalate High-Entropy Ceramics Coatings Response to Thermal Cycle up to 1500 ℃

Chen Lin*, Wang Jiankun, Zhang Luyang, Chen Xunlei, Feng Jing*

Faculty of Materials Science and Technology, Kunming University of Science and Technology, Kunming 650093, China

E-mail address: 11068466502qq. com (Chen Lin)

E-mail address: jingfeng@ kust. edu. cn (Feng Jing)

Abstract: Thermal barrier coatings (TBCs) are widely used in gas turbine engines to provide thermal insulation for high-temperature alloy blades, and rare-earth tantalates are extensively studied and applied in recent years. To further boost the service performance of tantalate TBCs, high-entropy ceramic coatings of tantalates (HECs tantalates) are designed and synthesized via air plasma spraying (APS) in this work. The typical microstructures and failure mechanisms of HECs tantalates coatings are analyzed after thermal cycle at temperatures up to 1500 ℃, and it is found that the formation and thickening of thermal growth oxides (TGO) lead to the failure of HECs tantalates coatings. The evolutions of crystal structure, microstructures, and interfacial thermal stress of the synthesized HECs tantalates TBC systems are investigated in detail, which are used to further reveal the failure mechanisms. It is believed that the service life of HECs tantalates coatings can be further optimized by improving the APS parameters and designing novel bond coat (BC) compositions. In summary, HECs tantalates coatings can be used at temperatures up to 1500 ℃, and they can provide excellent thermal insulation for high-temperature components.

Keywords: rare-earth tantalates; TBCs; high-entropy ceramics; thermal cycle; failure mechanisms

High Hydrothermal Corrosion Resistance of Semi-Crystalline Enamel Coating Modified by Self-healing Sib_6

Feng Xinchang[1], Guo Fangwei[1*], Zhang Wenchen[1], Zeng Qingfeng[2], Li Junning[3], Zhao Xiaofeng[1]

[1] Shanghai Key Laboratory of Advanced High-temperature Materials and Precision Forming, School of Materials Science and Engineering, Shanghai Jiao Tong University, China

[2] Tianmushan Laboratory, China

[3] Science and Technology on Advanced Functional Composite Laboratory, Aerospace Research Institute of Materials & Processing Technology, China

E-mail address: fengxinchang0815@163.com (Feng Xinchang)

E-mail address: fwguo2014@sjtu.edu.cn (Guo Fangwei)

Abstract: Enamel possesses excellent thermal resistance and low oxygen permeability. However, defects like pores and cracks introduced during enamel coating preparation can significantly compromise corrosion resistance in high-temperature water vapor environments. This study demonstrates that the modified semi-crystalline enamel coating with a trace amount of SiB_6, sintered at a low temperature, yields promising hydrothermal corrosion resistance. Hydrothermal corrosion behaviors of the modified enamel coating at 80 ~ 600 ℃ and electrochemical impedance were investigated systematically. The modified enamel coating with SiB_6 content of 1 wt.% exhibited an increased degree of amorphization and a significantly low mass change of ~0.01% after 480h of hydrothermal corrosion. A significant amount of flocculent products filled the residual pores of the modified enamel coating during hydrothermal corrosion, preventing the permeation of water vapor and oxygen. The residual stress of the modified enamel coating was lower than that of the undoped enamel coatings, eliminating the through-cracks. Both the equivalent coating resistance and the equivalent charge transfer resistance increased by 2-3 times compared to the undoped enamel coatings. The enhanced performance of the modified enamel coating was attributed to the synergistic optimization of the amorphous phase and residual stress, dynamically regulating hydrothermal corrosion resistance through the interplay between composition and microstructure.

Keywords: semi-crystalline enamel coating; SiB_6; porosity; hydrothermal corrosion; electrochemical impedance spectroscopy (EIS)

Densification and Wear Resistance of High-Content SiC/Al Composite Coatings: The Role of Doped Ni@SiC Core-Shell Particles

Liang Taoding[1,2], Gong Chunzhi[1], Hang Chunjin[1,2], Tian Xiubo[1]

[1] State Key Laboratory of Precision Welding & Joining of Materials and Structures, Harbin Institute of Technology, Harbin 150001, P. R. China

[2] Zhengzhou Research Institute, Harbin Institute of Technology, Zhengzhou 450000, P. R. China

E-mail address: 18693405727@163.com (Liang Taoding)

E-mail address: chunzhigong163.com (Gong Chunzhi)

Abstract: β-silicon carbide particles (β-SiCp), distinguished by their unique cubic crystal structure and exceptional hardness, exhibit outstanding potential for precision tribological applications. In this study, the core-shell type Ni-SiCp composite powder was constructed by electroplating coating technology, which increased the deposition threshold of ceramic particles, effectively solved the problem of kinetic mismatch of Al/SiC two-phase deposition, mechanical locking effect and inhibition of interface fracture tendency during the cold sprayed process, and further enhanced the cohesion of the coating. The deposition efficiency of the doped SiCp-Ni/Al composite coating is increased by 31.1 %. The core-shell structure effectively reduces the fragmentation rate of SiCp, increases the retention of SiCp by 19.1 %, improves the uniform dispersion state of SiCp, and finally obtains a densified coating with a porosity of less than 1 %. The coefficient of friction the doped SiCp-Ni/Al composite coating is reduced by 29.7 % compared with the traditional system, and the wear rate is reduced by 57.7 %. The performance improvement mechanism can be attributed to the synergistic effects of three aspects: the buffering effect of the intermediate Ni layer dissipates the impact energy, the interface bonding inhibits the shedding of abrasive particles, and the dense structure delays the failure process of the oxide layer.

Keywords: Cold spraying; Electroplating Ni; Core-shell SiCp-Ni; SiC-Ni/Al composite coating; Tribological properties

Gas Pressure and Soft-Phase Tuning in Cold Spray: Achieving High-Performance Al/7075 Al Composite Coatings with Ultra-Low Porosity and Enhanced Mechanical Characters

Liu Qiming[1,2], Gong Chunzhi[1,2], Zhou Changzhuang[1], Liu Yuan[1], Tian Xiubo[1]

[1] State Key Laboratory of Precision Welding & Joining of Materials and Structures, Harbin Institute of Technology, Harbin 150001, P. R. China

[2] Zhengzhou Research Institute, Harbin Institute of Technology, Zhengzhou 450000, P. R. China

E-mail address: m15638522225@163.com (Liu Qiming)

E-mail address: chunzhigong@163.com (Gong Chunzhi)

Abstract: As a ultra-high-strength wrought aluminum alloy, 7xxx series aluminum alloys are widely used in aerospace and other fields. However, due to harsh service environments involving high loads, high stresses, and complex corrosive conditions, they inevitably develop various surface defects. Therefore, timely and effective surface repair of 7xxx aluminum alloys is of significant importance. Cold spray (CS) has emerged as a viable method for repairing aluminum alloy damage due to its advantages of low process temperature, high deposition rate, and flexible structural repair capabilities. In this study, a novel low-cost and rapid cold spray repair method for 7075 aluminum alloy surfaces was achieved for the first time by utilizing soft-phase Al-assisted densification of cold-sprayed 7075 Al deposits. The microstructure and properties of the repair coatings were characterized using scanning electron microscopy (SEM), X-ray diffraction (XRD), friction and wear testing, microhardness testing, and tensile testing. The research first investigated the effect of varying gas pressures (3.5 MPa, 4 MPa, 4.5 MPa, 5 MPa) on pure 7075 Al coatings. Results indicated that as the gas pressure increased, the density and overall performance of the coatings initially improved and then deteriorated. The optimal performance was attained at 4.5 MPa gas pressure. At this pressure, the porosity of the 7075 Al deposit decreased from 3.9% to 0.89%, and its Vickers hardness increased from 108.55 $HV_{0.5}$ to 122.26 $HV_{0.5}$. The average friction coefficient of the coating reduced from 0.5431 to 0.4713, and the wear rate decreased from 3.31×10^{-3} $mm^3/N \cdot m$ to 1.25×10^{-3} $mm^3/N \cdot m$, exhibiting characteristics of adhesive and oxidative wear. Subsequently, at the optimized gas pressure of 4.5 MPa, the influence of different soft-phase Al contents (5 wt.%, 10 wt.%, 15 wt.%, 20 wt.%) on Al/7075 Al composite coatings was examined. The results demonstrated that the density of the Al/7075 Al coatings significantly improved due to the excellent plastic deformation capability of Al, with porosity reduced to below 0.1% from the baseline 0.89%. As the Al content increased under the same 6-layer "S"-shaped deposition path, the deposition thickness of the composite coatings markedly increased from 3.27 mm to 7.14 mm, indicating substantially enhanced deposition efficiency. Benefiting from the reduction in void defects, the composite coating with 5 wt.% Al content achieved optimal overall performance, exhibiting a surface Vickers hardness of 126.23 $HV_{0.5}$, which surpassed that of the pure 7075 Al coating. However, due to the intrinsic softness of Al, excessive Al content detrimentally affected the composite coating properties. All composite coating cross-sections exhibited a gradient increase in Vickers hardness from the surface towards the substrate interface, attributed to the tamping effect during deposition. With increasing Al content, the tensile strength of the Al/7075 Al coatings exhibited a declining trend. The maximum tensile strength of 135.9 MPa was achieved at 5 wt.% Al. SEM analysis revealed that fracture surfaces of all composite coatings displayed distinct brittle fracture characteristics. Furthermore, nitrogen gas was employed as the process gas. The incorporation of soft-phase Al was found to further mitigate wear on the powder feeding system during deposition, thereby significantly reducing operational costs and equipment maintenance requirements.

Keywords: cold spray; Al/7075 Al; gas pressure; repair coatings; mechanical characters

Enhanced Resistance to Environmental Deposit Penetration via Synergistic Design of Multi-Rare-Earth Co-Stabilized Zirconia

Zan Yaoxu*, Liu Meijun, Yang Guanjun

School of Materials Science and Engineering, Xi'anJiaotong University, China

E-mail address: yx_ zan@ stu. xjtu. edu. cn (Zan Yaoxu)

E-mail address: liumjun@ xjtu. edu. cn (Liu Meijun)

Abstract: The durability of thermal barrier coatings (TBCs) in modern turbine engines is severely compromised by infiltration from molten environmental deposits, such as calcium-magnesium-alumino-silicates (CMAS). To address this critical failure mode, we have developed a new class of TBC material engineered to actively combat such attacks. Our approach leverages the synergistic interaction of multiple rare-earth co-stabilizers (Y, Gd, Yb) to precisely control the in-situ formation of reaction products upon exposure to corrosive melts. This strategy led to the synthesis of a unique defect-fluorite ceramic, $Zr_{3/7}Y_{8/21}Yb_{2/21}Gd_{2/21}O_{12/7}$ (denoted as YYbGdZO). When tested against the conventional material, gadolinium zirconate (GZO), YYbGdZO demonstrated exceptional performance. After an intensive 100-hour exposure at 1400 ℃, the penetration depth of the molten environmental deposits was dramatically minimized, measuring only 2% of that observed in the GZO sample. The source of this outstanding resistance lies in the cooperative action of the Y, Gd, and Yb dopants. Upon contact with the melt, these elements trigger a selective reaction, precipitating a dense and highly protective composite layer. This in-situ formed barrier, composed of multi-rare-earth disilicate alongside an apatite phase within a rare-earth-stabilized zirconia matrix, effectively arrests further melt infiltration. By preventing degradation from environmental deposits, this mechanism ensures the long-term structural integrity and stability of the coating. Our findings establish a new design approach for creating next-generation materials capable of withstanding high-temperature corrosion, with broad applicability for diverse and harsh operating conditions.

Keywords: multicomponent rare earth; non-equimolar cationic design; dense barrier layer; CMAS resistance; corrosion of environmental deposits

Research on 2219 Al Alloy Coatings Deposited by Plasma-Assisted Cold Spraying Method

Lin Chaoqun, Gong Chunzhi, Liu Yuan, Zhou Changzhuang, Tian Xiubo

State Key Laboratory of Precision Welding & Joining of Materials and Structures, Harbin Institute of Technology, Harbin 150001, P. R. China

E-mail address: 17614661866@163.com (Lin Chaoqun)

E-mail address: chunzhigong@163.com (Gong Chunzhi)

Abstract: This work presents a process for cold spraying assisted by a plasma heat source. Cold spraying technology is a solid-state powder deposition process, and the bonding mechanism is mainly based on plastic deformation and mechanical embedding of high-speed solid particles. Inadequate plastic deformation, mechanically embedded bonding, and stress accumulation during layer-by-layer deposition can easily lead to poor bonding, which may cause cracking or spalling of the coating. These problems can significantly shorten the service life of components and can lead to catastrophic failures, especially in harsh environments such as aerospace and energy equipment. Utilizing a plasma-assisted heat source can effectively mitigate these problems and improve the prospects for the application of cold spray technology. Plasma-assisted cold spraying uses high-temperature plasma flame flow to heat the substrate and powder to improve the surface state, soften the particles, enhance plastic deformation, and improve the bonding strength; at the same time, the high temperature prompts the particles to recrystallize, optimize the grain structure and reduce the residual stress. The purpose of this paper is to deposit cold sprayed coatings of 2219 Al alloy. By adding a plasma heat source and varying the spray gas pressure, this paper compares the microstructure and macroscopic mechanical properties of the coatings. Among them, the highest coating-substrate bond strength is up to 43.32 MPa, the highest hardness of the coating is 107.98 HV, and the lowest average coefficient of friction is 0.4221.

Keywords: Cold spraying; Plasma source; 2219 Al alloy coating; microstructure; mechanical properties

Study on the Strengthening Mechanism of High Pressure Cold Spray Coating for Surface Repair of 6061 Aluminum Alloy

Gong Chunzhi, Tian Xiubo

State Key Laboratory of Precision Welding & Joining of Materials and Structures, Harbin Institute of Technology, Harbin 150001, P. R. China

E-mail address: chunzhigong@163.com (Gong Chunzhi)

E-mail address: xiubotian@163.com (Tian Xiubo)

Abstract: Driven by the " dual-carbon" strategy, the demand for lightweighting in the aerospace and automotive sectors is becoming increasingly urgent. 6061 aluminum alloy has emerged as a critical structural material due to its excellent specific strength and corrosion resistance. However, traditional molten-state additive 6061 aluminum alloy repair technologies face bottlenecks such as hot cracking susceptibility and elemental burn-off. In contrast, Cold Spray solid-state additive repair technology, which utilizes low-temperature, high-speed particle deposition, can circumvent molten-state defects and exhibits significant advantages in aluminum alloy repair. This study establishes a tripartite technological system for Cold Spray repair of 6061 aluminum alloy, termed " Multi-scale Enhancement-Process Synergy-Dynamic Post-processing" . It reveals the competitive mechanism of temperature-pressure on densification/reinforcement phase integrity and establishes a medium-temperature, high-pressure process window. A dual-scale micro-nano synergistic strengthening model is created to achieve the co-optimization of hardness and wear resistance. By integrating theoretical modeling with COMSOL finite element simulation, the critical size ofnano-SiC particles required for effective embedding onto the aluminum matrix surface was determined. Increasing the gas pressure (to 5 MPa) significantly enhanced deposition efficiency and densification, resulting in a 13.98% increase in hardness and an 11.4% improvement in fracture toughness of the repair layer compared to the single-scale micron-reinforced layer. The dispersion of nanoparticles also contributed to a reduced wear rate. Optimizing the nano/micron particle mass ratio to 5:15 yielded the best overall performance of the repair layer, achieving a peak hardness of 100.64 Hv and a 41.92% reduction in wear rate compared to the coating reinforced solely with micron-SiC particles.

Keywords: 6061 aluminum alloy; cold spraying; surface repair; micro-nano strengthening

Effect of Heat Treatment Temperature on Microstructure and Properties of Micro/Nano TiN Reinforced Cold Sprayed Aluminum Coatings

Liu Jiaying[1,2], Gong Chunzhi[1], Tian Xiubo[1]

[1] State Key Laboratory of Precision Welding & Joining of Materials and Structures, Institute of Technology, Harbin, China

[2] Zhengzhou Research Institute, Harbin Institute of Technology, Zhengzhou, China

E-mail address: 15843796199@163.com (Liu Jiaying)

Abstract: Cold spraying is an effective method for producing aluminum-based protective coatings. However, the hardness and wear resistance of pure aluminum coatings often require improvement. This work systematically investigates the synergistic effect of micron-scale (25 μm) and nano-scale (500 nm) titanium nitride (TiN) particles on cold-sprayed pure aluminum (Al) coatings. It primarily focuses on the influence and underlying mechanisms of subsequent heat treatment at various temperatures (150 ℃-350 ℃) on the coating microstructure and mechanical properties. Dual-scale TiN/Al composite powder was prepared by ball milling, and coatings were deposited using optimized cold spray parameters. The cold sprayed coatings were then heat-treated at a series of temperatures (150 ℃, 250 ℃, 350 ℃) for 2 hours. The microstructure, hardness and wear resistance of the coating before and after heat treatment were characterized by SEM、EDS、XRD, Vickers hardness tester and rotary friction and wear test system. Results demonstrate that cold spraying successfully produces dense micro/nano dual-scale TiN/Al composite coatings. The incorporation of dual-scale TiN significantly enhances coating hardness to 38.7 ± 2.03 $HV_{0.5}$. This paper presents the mechanical property test results and compares the effects of heat treatment at different temperatures.

Keywords: cold-spray; heat treatment; micro/nano TiN/Al composite coating

Effect of CNT on Microstructure and Properties of Plasma Sprayed Al_2O_3/TiO_2-CeO_2 Composite Coating

Xu Zhifu
Shandong Jiao Tong University, China
E-mail address: 2933290553@qq.com (Xu Zhifu)

Abstract: Four CNT-Al_2O_3/TiO_2-CeO_2 composite coatings with varying carbon nanotube (CNT) contents were prepared using plasma spraying technology, and the effects of CNT content on the wear and corrosion resistance of the coatings were investigated. Nanoscale CNT-Al_2O_3/TiO_2-CeO_2 powders with nanostructures were produced using mechanical ball milling and spray drying techniques. These powders were then used to prepare CNT-Al_2O_3/TiO_2-CeO_2 composite coatings on the surface of 45 steel substrates through plasma spraying. The effects of different CNT mass fractions on powder feeding, the microstructure of the coatings, and their wear and corrosion resistance were studied using a scanning electron microscope, Raman spectrometer, X-ray diffraction instrument, microhardness tester, tribometer, electrochemical workstation, and salt spray tester. The mechanisms behind the wear and corrosion resistance of the coatings were analyzed. The results showed that the addition of CNT did not significantly affect the sphericity, particle size, or flowability of the powder. The average particle size of the powders with different CNT contents was approximately 55.44 μm, with a flow rate of about 11.43 g/min, and the surfaces were smooth. The four coatings prepared using plasma spraying technology mainly consisted of α-Al_2O_3 and γ-Al_2O_3 phases. As the CNT content increased, the internal structure of the coatings became more compact and uniform, reducing grain boundary defects and refining the grains. The distribution of O, Al, Ti, Ce, and C in the coating is more uniform. Studies on the hardness and wear resistance of the coating show that the hardness initially increases and then decreases, while the wear resistance improves. When the mass fraction of CNT is 3 wt. %, the average microhardness of the coating reaches a peak of 899 $HV_{0.2}$, with the lowest average friction coefficient of 0.36 and the least grinding loss of only 1.5 mg. The wear resistance is 9.3 times higher than that of the substrate and 2.4 times higher than that of the coating without CNT. The modified coating is primarily affected by adhesive wear and tribo-particle wear. Electrochemical tests on the coating reveal that, compared to the coating without CNT, when the mass fraction of CNT is 3 wt. %, the coating exhibits optimal corrosion resistance. In a 3.5 wt. % NaCl solution, the corrosion potential increases by 73.5 mV, and the self-corrosion current decreases by 76.14%. In a 6 wt. % Na_2SO_4 solution, the corrosion potential increases by 81.6 mV, and the self-corrosion current decreases by 73.66%. In a 5% HCl solution, the corrosion potential increases by 217.3 mV, and the self-corrosion current decreases by 84.67%. Neutral salt spray tests show that when the mass fraction of CNT is 3 wt. %, the coating has the smallest weight gain of 9.76 mg/cm^2, with a significant reduction in surface pores and cracks, and no obvious corrosion products or holes, demonstrating excellent corrosion resistance.

Keywords: atmospheric plasma spraying; Al_2O_3/TiO_2-CeO_2; CNT; wear resistance; corrosion resistance

Promoting $Yb_2Si_2O_7$ Environmental Barrier Coatings for Service at 1425 ℃

Chen Lin[1], Li Shuwen[1], Li Changjiu, Yang Guanjun*

National Key Laboratory for Mechanical Behavior of Materials, School of Materials Science and Engineering, Xi'an Jiaotong University, Xi'an 710049, China

Abstract: Si-based environmental barrier coatings (EBCs) are indispensable for SiC composites. However, the service temperature of next-generation composites exceeds 1400 ℃, leading to melting of the Si bond coat. To mitigate this issue, $Yb_2Si_2O_7$ (YbDS) and mullite/YbDS coatings without Si bond coat were proposed. The results indicate that YbDS coatings can survive for 150 h at 1425 ℃, while mullite/YbDS coatings experienced significant spallation during this period. Both SiO_2 phase transition and interfacial holes cause spallation of the YbDS and mullite/YbDS coatings. This research provides valuable insights for designing next-generation EBCs.

Design and Property Study on High-Entropy Cerium Oxide-Based Thermal Barrier Coatings

Xue Yun

Lanzhou Institute of Chemical Physics, Chinese Academy of Sciences, China

E-mail address: yxue20@licp.cas.cn (Xue Yun)

Abstract: The rapid development of high-performance aero-engines and gas turbines urgently requires thermal barrier coatings (TBCs) with lower thermal conductivity, higher coefficient of thermal expansion, higher temperature resistance, corrosion resistance, and thermal shock resistance. The emergence of high-entropy ceramics has created more possibilities for the design of ceramic top-layer materials in TBCs. Therefore, based on the research of the traditional $A_2B_2O_7$-type rare earth cerium oxide (such as $La_2Ce_2O_7$) thermal barrier materials, the A-site component was designed and regulated by high entropy, leading to the generation of high-entropy rare earth cerium oxide ceramics for TBCs. The compositional relationship between high-entropy configuration and thermophysical properties of the materials was studied. Besides, the corrosion resistance was investigated in depth. The mechanism of high-entropy ceramics' special effects on improving the corrosion resistance of TBCs was clarified, and the corrosion mechanism of high-entropy ceramics was revealed. On this basis, different-structured high-entropy TBCs were fabricated by atmospheric plasma spraying, and the thermal shock resistance of the coatings was systematically evaluated. The research results showed that the reasonable design of the composition and microstructure of high-entropy TBCs enabled them to exhibit better thermal shock resistance, thereby establishing the enhancement mechanism of the high-entropy configuration with the structure of the coating. The accumulation of previous research work provides an important reference for the development and design of advanced high-entropy TBCs.

Keywords: atmospheric plasma spraying (APS); thermal barrier coatings (TBCs); high-entropy ceramics; thermal protection performance

5d Electronic Structure Micro-Mechanism of Room Temperature Plastic Deformation of Ceramic Materials

Yang Li

Xidian University, China

E-mail address: 2678708344@qq.com (Yang Li)

Abstract: Advanced ceramics are critical for extreme-environment applications due to their unmatched thermal stability, corrosion resistance, and mechanical strength. However, reconciling high fracture toughness with strength remains a fundamental challenge, particularly for thermal barrier coating (TBC) materials. Here, we report a breakthrough in the 5d electronic structure of ferroelastic YTaO4-based TBCs through self-organized that activates coordinated ferroelastic switching. In situ micromechanical testing reveals that this domain switching enables unprecedented compressive plasticity (16% strain at RT) and fracture stress (~3.5 GPa), with the yield of the platform as high as 3.19%, which is remarkable for advanced ceramics. Multiscale characterization coupled with first-principles calculations demonstrates that nanotwin boundaries serve as preferential nucleation sites for transformation dislocations, and collective domain reorientation accommodates shear strains via low-energy (010) [001] slip systems. Finite element simulations further confirm that engineered domain patterns induce ceramic toughness characteristics. This nanotwin-regulated plasticity mechanism of 5d electronic structure material establishes a new paradigm for damage-tolerant ceramic coatings in next-generation turbine engines.

Keywords: thermal barrier coating; 5d electronic structure; nanotwin domains; in-situ TEM; strengthening-toughening mechanisms

Rare-Earth Tantalates for Next-Generation Thermal Barrier Coatings

Feng Jing
Kunming University of Science and Technology, China
E-mail address: 20240222@kust.edu.cn (Feng Jing)

Abstract: Ceramic thermal barrier coatings (TBCs) have attracted significant research attention owing to their utility in the thermally insulating alloy components of gas turbines and aircraft engines that operate at high temperatures. Most TBCs comprise yttria-stabilized zirconia (YSZ); however, YSZ no longer meets the demands of modern TBC applications due to its low working temperature and high thermal conductivity. It is therefore imperative to develop a ferroelastic ceramic to replace YSZ in TBC applications. Ferroelastic rare-earth tantalates ($RETaO_4$) possess many desirable properties, such as ferroelastic toughening, low thermal conductivity, high thermal expansion coefficients, and excellent comprehensive mechanical properties, and thus, they are promising next-generation TBCs, which are expected to operate at ultra-high temperatures (≥1600 ℃). This review summarizes the thermophysical properties, CaO-MgO-$AlO_{1.5}$-SiO_2 (CMAS) corrosion resistance, coatings, and shortcomings of three types of tantalate ceramics ($RETaO_4$, RE_3TaO_7, and $RETa_3O_9$) and outlines the direction of future work in this field.

Keywords: rare-earth tantalates; thermal barrier coatings (TBCs); thermal properties

Research Development of Liquid Plasma Spray

Ma Wen, Bai Yu, Gao Yuanming, Yin Yichuan

Science and Engineering, Inner Mongolia University of Technology, Hohhot, China

E-mail address: 15843796199@163.com (Ma Wen)

Abstract: Liquid plasma spray (LPS) technologies, including suspension plasma spray (SPS) and solution precursor plasma spray (SPPS), have gained significant attention for their ability to deposit advanced nanostructured and finely structured coatings. SPS utilizes micrometer-sized or sub-micrometer particles suspended in a liquid carrier, which are injected into the plasma jet, enabling the formation of coatings with enhanced mechanical, thermal, and functional properties. Recent international research has focused on optimizing suspension formulations, plasma torch parameters, and substrate cooling to improve coating homogeneity, porosity control, and adhesion. Notable advancements include the development of YSZ thermal barrier coatings (TBCs) with superior thermal cycling resistance and Y2O3 plasma-resistant etching coating for semiconductor industry applications. SPPS, on the other hand, involves the direct injection of chemical precursors into the plasma, leading to in-situ synthesis and deposition of coatings. This method allows precise control over composition and microstructure, facilitating the fabrication of complex oxides coatings, such as rare-earth doped ZrO2 and SrZrO3, as well as high-entropy rare-earth zirconates TBCs for aerospace industry applications. International studies have explored precursor chemistry, plasma-particle interactions, and post-treatment effects to enhance coating crystallinity and functionality. Challenges remain in achieving uniform deposition, different microstructures and scaling up the process, but SPPS shows promise for next-generation functional coatings. Overall, both SPS and SPPS represent innovative approaches in plasma spray technology, with ongoing research aimed at expanding their industrial applicability in aerospace, energy, and biomedical fields.

Keywords: suspension plasma spray; solution precursor plasma spray; thermal barrier coatings; high-entropy ceramic coatings

Silicon-Tailored Hard Al-N-Si Coatings with Structural Chromaticity and Blue-Light Regulation

Li Fengji

[1] Center for Advanced Thin Films and Devices, School of Materials and Energy, Southwest University, 2 Tiansheng Road, Beibei, Chongqing 400715, China

E-mail address: lee0818@ swu. edu. cn (Li Fengji)

Abstract: The development of advanced optical coatings exhibiting both exceptional mechanical durability and tunable chromatic properties has emerged as a critical requirement for next-generation photonic applications. This investigation presents the synthesis of Al-N-Si thin films withdifferent Si concentrations deposited on Si wafers and glass substrates through reactive magnetron co-sputtering technology. Comprehensive material characterization was performed utilizing field emission scanning electron microscopy, X-ray photoelectron spectroscopy, grazing-incidence X-ray diffraction, high-resolution transmission electron microscopy, chromatic coordinate analysis, UV-Vis spectroscopy, variable-angle spectroscopic ellipsometry, and nanoindentation techniques. Key findings reveal that the stoichiometric $Al_{0.92}N$ coating develops columnar grain architecture extending through the film thickness. Si incorporation at 6. 8 at. % induces microstructural densification and enhances oxidation resistance while maintaining a hexagonal wurtzite crystallographic structure with predominant (101) orientation. At 10. 7 at. % Si concentration, grain refinement to 11. 0 nm occurs with preferential (100) orientation development. Optical analysis demonstrates maintained zero extinction coefficients coupled with refractive index reduction to 2. 08-1. 99 across visible wavelengths. Mechanical testing reveals maximum hardness of 27. 1 ± 2. 3 GPa, while optical transmission in the hazardous blue spectrum (400-450 nm) decreases to 68. 2-79. 1%. Notably, the coatings on the Si substrate demonstrate structural coloration spanning visible spectrum (red: 300 nm, green: 241 nm, cyan: 208 nm, blue: 200 nm, purple: 186 nm) through interference modulation. A mechanistic model elucidating Si-mediated structural evolution was established, with particular emphasis on solid solution strengthening mechanism, thin-film interference phenomena, and blue light attenuation principle. This systematic investigation provides fundamental insights into structure-property relationships for multifunctional optical coatings, advancing the development of intelligent surfaces with combined mechanical robustness and controllable optical responses.

Keywords: aluminum nitride; reflectance; transmittance; structural color; hardness; blue light attenuation

Improving Hardness & Toughness of Ceramic Coatings and Their Dislocation-Related Deformation Mechanisms

Dr. Shenghao Zhou

Institute ofNew Materials, Guangdong Academy of Sciences, Guangzhou, China

Abstract: Renowned for their high hardness, wear-resistance and corrosion-resistance, ceramics are ideal candidate materials for protective layers. But their applications are severely restrained by their brittle nature. Hardening-toughening has been an everlasting topic in this field. However, the hardness and/or toughness mechanisms related to microstructure response during deformation remain less understood. Firstly, we performed systematic investigations on the deformation mechanisms on hard-yet-tough TiMoN coatings guided by VEC (valence electron concentration) -CEC (core electron count) solid solution theory. Transmission electron microscopic (TEM) studies operated in two-beam bright field mode proved that the nano-crystals of as-deposited TiMoN coatings can store high-density dislocations due to reduced stress concentration at dislocation core as well as strong ion bombardment. The numbers of nanoscale dislocations increase by about 50 % after deformation, endowing strain strengthening effective and plastic deformation as extra path for dissipating strain energy. The results verify dislocation-mediated plastic process is responsible for the high hardness and toughness of TiMoN coatings. Next, encouraged by the results, we successfully fabricated coating with superhardness (48 GPa) yet good toughness with the benchmark TiMoN composition via a high-throughput experiment. The superhard yet tough TiMoN coatings have high Mo content (74.73 at%) and large N deficit (N content: 18.29 at%). The super-hardness is attributed to the dense structure with extremely small grain size of about 24.9 nm. More importantly, the nanocrystals are severely distorted, containing numerous dislocations, stacking faults, 9R phase and nanotwins. These defects interact with each other, not only impede dislocation motion improving hardness, but also provide paths for energy dissipation, enhancing deformability of the coating. Consequently, the superhard coatings are concurrently characterized by high H/E value ~0.1 and metal-like pile-up flow at the edge of cube corner indent. Microstructure observation also unveils that grain size was reduced from 24.9 nm down below 7 nm, and crystalline morphology change from columnar grains to equiaxial grains, another proof for dislocation activity. Our works show there are plenty of rooms to tune dislocation-related mechanical properties of ceramics coatings, showing new potential approaches towards designing hard-yet-tough protective coatings.

Keywords: TiMoN coatings; hardness; toughness; dislocations; microstructure

High Performance Single-Crystalline Diamond Nanowire Solar-Blind Photodetector

Yang Bing, Lu Jiaqi, Jiang Xin, Sun Dongming
Shenyang National Laboratory for Materials Science, Institute of Metal Research (IMR),
Chinese Academy of Sciences (CAS), Shenyang, China
E-mail address: byang@ imr. ac. cn (Yang Bing)

Abstract: Diamond solar blind photo-detectors have received considerable attentions during the past decades owing to its excellent physical properties such as wide band gap (5.47 eV), high thermal conductivity and radiation resistance. In order to optimize the photodetector performance, much effort has been made on the device geometries such as metal-semiconductor-metal structure with ohmic or Schottky contact and p-n junctions for diamond. Inheriting the intrinsic physical and chemical properties of bulk ultrawide bandgap semiconductors, one-dimensional (1D) nanostructure devices exhibit improved photoelectric performance due to efficient enhancement in light absorption (large surface-to volume ratio and comparable Debye length with their small size) and good confinement and transport of photogenerated carriers. In our work, 1D single crystalline diamond nanoneedles or nanowires were prepared by thermal oxidation of [001] orientated micro/nano crystalline diamond films. With selective etching of nanocrystalline diamond fraction, the microcrystalline diamond grains were retained, forming vertical 1D nanostructure arrays. These diamond nanowires were crystallized in single grains. Single nanowire photodetectors were fabricated with transferring diamond nanowires on SiO_2 substrate and depositing Ti/Au electrode. The 1D single-crystalline diamond nanowires exhibited a responsivity of 5.29×10^3 A/W at 222 nm and a UV/visible rejection ratio of 35. The photoelectric performance was further optimized by nanowire regrowth and device design.

Keywords: diamond

High Pressure Preparation of NovelSuperhard Carbon Materials

Dong Jiajun

State Key Laboratory of High Pressure andSuperhard Materials, College of Physics, Jilin University, Changchun 130012, China

E-mail address: dongjj@ jlu. edu. cn (Dong Jiajun)

Abstract: High pressure, as an extreme condition, can effectively shorten interatomic distances, alter atomic interactions, and even modify bonding configurations. This enables the creation of novel materials inaccessible under ambient conditions and serves as a critical pathway for synthesizing artificial diamonds. Unlike the conventional high-pressure synthesis of cubic diamond, the structural transformations of carbon materials under high pressure exhibit rich tunability, leading to diverse novel carbon structures with exceptional properties. This report presents recent research advances by our team in the high-pressure synthesis of new diamond forms, including: 1) Discovery of a new mechanism for hexagonal diamond formation *via* post-graphite phases, achieving the synthesis of near-pure bulk hexagonal diamond; 2) Proposal of a novel strategy using pressure-induced " amorphous carbon clusters" to construct sp^3-bonded amorphous carbon, resulting in the synthesis of nearly fully sp^3-bonded amorphous superhard carbon monoliths; 3) Utilization of laser-heated diamond anvil cell (DAC) techniques to synthesize novel carbon materials such as porous diamond aerogel films and diamond-graphite hybrid composite structures. These results provide new insights into the graphite-diamond transformation mechanism and will be instrumental in guiding the further synthesis of novel carbon materials under high pressure.

Keywords: high temperature and high pressure; carbon materials; superhard; diamond

Effect of Interface Regulation on the Thermal Conductivity of Diamond-Copper Composites

Yi Tao, Li Chengming, Wei Junjun, Liu Jinlong, Chen Liangxian, Zhang Jianjun

University of Science and Technology Beijing, Beijing, China

E-mail address: yt292797@163.com (Yi Tao)

E-mail address: weijj@ustb.edu.cn (Li Chengming)

Abstract: Diamond/copper composites have attracted much attention in the field of thermal management of electronic packaging due to their potential ultra-high thermal conductivity, but the extremely poor wettability and interfacial bonding between diamond and copper severely limit their thermal properties. In this study, a prealloying strategy was proposed to optimize the interface structure and improve the thermal conductivity by pre-preparing copper-chromium alloy powders (Cu-Cr) with different chromium contents, mixing them with diamond powder, and using Spark plasma sintering (SPS) technology. The effects of chromium content (0.5-1. wt%) and sintering temperature (850-1000 ℃) on the microstructure and thermophysical properties of composites were systematically studied. The results show that chromium alloying effectively promotes the interfacial reaction, and the intermediate phase of chromium carbides (such as Cr_3C_2/Cr_7C_3) is formed on the diamond surface. Chromium content significantly affects thermal conductivity, reaching a peak of 661 W/(m · K), at this time, scanning electron microscopy (SEM) observation showed that the interface was the most tightly integrated, and the intermediate phase was appropriate and evenly distributed. Further optimization of the sintering temperature showed that the thermal conductivity of the composite materials obtained by sintering at 950 ℃ was up to 731W/(m · K), at which the interface reacts avoids diamond damage. This study confirms that the prealloy Cu-Cr powder combined with SPS process (optimal parameter: 0.9 wt% Cr, 950 ℃) is an effective way to significantly enhance the interface bonding of diamond/copper composites and obtain high thermal conductivity.

Keywords: diamond-copper composites; Interfacial bonding; chromium alloying; thermal conductivity

Properties of Low Nitrogen Content AlCrSiN Alternating Coatings: Experimental and First-Principles Studies

Liu Jiyuan[a,b], Hu Shubing[a,b,*], Huang Yixuan[a,b], Chen Yefei[a,b], Yu Jinke[a,b], Zhang Shuo[a,b], Guo Yuhang[a,b], Li Hongya[a,b]

[a] School of Materials Science and Engineering, Huazhong University of Science and Technology, Wuhan 430074, China

[b] State Key Laboratory of Material Processing and Die & Mould Technology, Wuhan 430074, China

E-mail address: hushubing@ hust. edu. cn (Hu Shubing)

Abstract: The extreme thermomechanical conditions inherent to hot forging processes impose rigorous demands on mold coating performance. While industrial applications predominantly employAlTiN and AlCrN coatings with 50 at. % nitrogen content, research addressing low-nitrogen alternatives remains notably scarce. This study bridges this critical knowledge gap through the development of AlCrSiN/AlCrN alternating multilayer coatings deposited on 3Cr2W8V die steel using a hybrid multi-arc ion plating (AIP) and high-power impulse magnetron sputtering (HiPIMS) approach. We systematically investigated the influence of nitrogen content on coating architecture, mechanical properties, tribological behavior, and high-temperature oxidation resistance. The low-nitrogen AlCrSiN_ Alt coating exhibited a unique CrN/Cr_2N/AlN/amorphous multiphase structure, demonstrating exceptional comprehensive performance: superior adhesion strength (77.9 N) and an ultra-low friction coefficient (0.78), attributable to the formation of lubricious Al/Si oxide transfer films during sliding. In stark contrast, the high-nitrogen AlCrSiN_ Sync coating achieved significantly higher hardness (38.2 GPa) via a single-phase cubic (Cr, Al) N structure dominated by Al solid-solution strengthening. However, this high-hardness coating suffered catastrophic cracking under thermal load due to deleterious volumetric expansion (exceeding 20%) during high-temperature decomposition of the cubic phase into hexagonal AlN. Microstructural analysis revealed that the AlCrSiN_ Alt coating formed a thermally stable Cr-Al-Si crystalline/amorphous composite oxide layer at 1000 ℃, significantly outperforming the monolithic $(Al, Cr)_2O_3$ oxide layer formed on the AlCrSiN_ Sync counterpart. The pronounced volume expansion during oxidation generated substantial internal stresses within AlCrSiN_ Sync, leading to pervasive microcrack nucleation and accelerated coating failure. Furthermore, in-situ precipitation of nano-dispersed $SiCr_3$ particles within the AlCrSiN_ Alt coating provided additional strengthening, enhancing its high-temperature mechanical resilience. First-principles calculations provided atomic-scale insights into the fundamental oxidation mechanisms. The AlN (100) surface demonstrated the lowest oxygen molecule adsorption energy (−1.21 eV), indicating superior oxidation resistance. This passivation behavior stems from strong N-O covalent bonding stabilizing the surface upon oxygen exposure. Silicon incorporation strategically reduced the density of stable adsorption sites on the cubic (Cr, Al) N surface by 33% and lowered the adsorption energy to −1.89 eV. Electronic structure analysis confirmed that robust covalent Si-O and Si-N bonds effectively suppressed metal atom diffusion, with enhanced electron localization observed at Si-O interfaces. This research conclusively establishes that the multiphase architecture of the low-nitrogen AlCrSiN_ Alt coating enables superior stress accommodation during thermal cycling, while its composite oxide layer provides exceptional barrier protection. Conversely, the high-hardness AlCrSiN_ Sync coating's single-phase structure sacrifices thermal stability, as its cubic-to-hexagonal phase transformation induces irreversible damage. The strategic engineering of low-nitrogen AlCrSiN coatings thus achieves an optimal balance between outstanding tribological performance (low friction/wear) and exceptional high-temperature stability, offering significant potential for next-generation hot-forging die applications demanding prolonged service life under extreme conditions.

Keywords: AlCrSiN coating; low nitrogen content; friction; oxidation; first-principles calculation

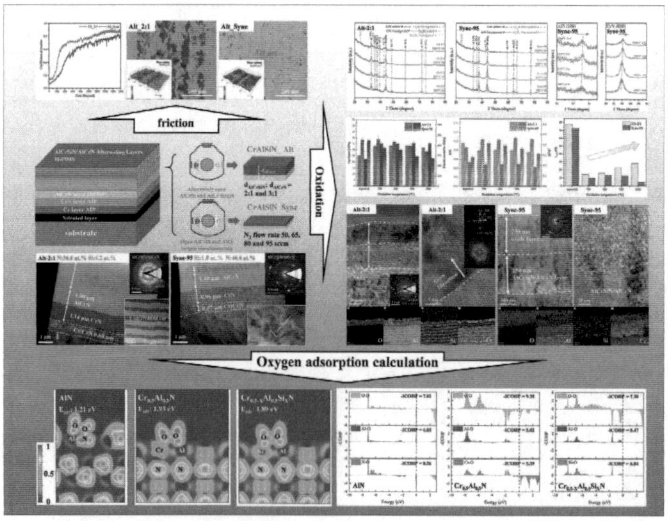

Toughening Design and Molten Aluminum Erosion Resistance of CrAlN-Based Hard Coatings

Wang Zhenyu

State Key Laboratory of Advanced Marine Materials, Zhejiang Key Laboratory of Extreme-environmental Material Surfaces and Interfaces, Ningbo Institute of Materials Technology and Engineering, Chinese Academy of Sciences, Ningbo 315201, China

Abstract: Integrated high-pressure die casting of aluminum alloys is one of the key pathways for achieving the dual-carbon goals in the automotive industry. However, die surfaces frequently suffer from cracking, erosion, melt damage, and adhesive wear. Applying high-performance coatings via physical vapor deposition (PVD) has proven to be an effective strategy for extending die service life. Among them, CrAlN coatings are widely used for die protection due to their good adhesion to tool steel and excellent oxidation resistance. Nevertheless, their application in Al alloy die-casting conditions remains limited by insufficient toughness and poor resistance to molten metal erosion. In this study, CrAlYN coatings with varying Y contents were successfully fabricated using cathodic arc ion plating via a high-throughput approach. Detailed structural characterization revealed that Y atoms are substitutionally incorporated into the CrAlN lattice, and increasing Y content leads to progressive grain refinement. The CrAlYN-1.2 coating (Y: 1.2 at.%) achieves a high hardness of 36.6 ± 2.5 GPa while exhibiting a fracture toughness (K_{IC}) of 9.65 MPa · $m^{1/2}$—more than double that of the undoped CrAlN coating. The enhanced performance is attributed to Y-induced weakening of Cr – N bonds and the formation of lubricious Cr_2O_3 phases at elevated temperatures, which promote the formation of adherent oxide layers within the wear tracks. As a result, the CrAlYN-1.2 coating demonstrates superior wear resistance at 700 ℃. Molten aluminum immersion tests further reveal that the CrAlN coating exhibits aluminum adhesion and localized pitting after only 30 min exposure. In contrast, the CrAlYN-1.2 coating shows no signs of melt damage or aluminum adhesion even after 6 h immersion, highlighting its excellent resistance to molten Al erosion.

Keywords: hard coatings; CrAlN; aluminum erosion; toughness; PVD

Advances in Heteroepitaxial Nucleation and Growth of Single-Crystalline Diamond

Bing Dai, Zhu Jiaqi, Zhu Hongda

Harbin Institute of Technology

E-mail address: daibinghit@ vip. 126. com/daib@ hit. edu. cn (Bing Dai)

Abstract: Diamond, renowned for its exceptional physicochemical properties, finds applications in a myriad of advanced technological domains. Single-crystal diamond, particularly large-sized, high-quality, electronic-grade single-crystal diamond, has garnered significant attention due to its structural integrity, high purity, and low defect density. Hailed as the " ultimate semiconductor" and the " Mount Everest" of electronic materials, single-crystal diamond exhibits substantial application potential in power devices, detectors, quantum computing, and beyond, offering robust support for the development of emerging fields such as aerospace, 5G/6G technology, and artificial intelligence. The mass production of large-sized, high-quality diamond wafers is indispensable, thereby necessitating the advancement of heteroepitaxial synthesis techniques for single-crystal diamond. Currently, single-crystal diamonds grownheteroepitaxially on iridium (Ir) substrates boast the largest dimensions and superior growth quality. Consequently, several research institutions and teams have embarked on studies aimed at the heteroepitaxial fabrication of large-sized single-crystal diamonds. Notably, Schrec et al. at the University of Augsburg have successfully produced single-crystal diamonds up to 92 mm in diameter on Ir (100) substrates using heteroepitaxial methods. Domestic universities have also initiated preliminary experimental research on the heteroepitaxial synthesis of large-sized single-crystal diamonds. The crux of heteroepitaxial single-crystal diamond technology lies in substrate selection and the nucleation and growth processes of diamond. This presentation introduces substrates of various structures suitable for diamond epitaxy and elucidates the nucleation and growth processes of diamond on iridium (Ir) substrates, with a particular emphasis on the mechanisms of bias-enhanced nucleation (BEN) and epitaxial lateral overgrowth (ELO), as well as substrate patterning nucleation and growth techniques. It also highlights the current limitations in research and offers a prospective outlook on the theoretical and experimental directions in diamond heteroepitaxy.

Fabrication and Properties of Hard Coatings by a Hybrid PVD Method

Wang Q. M.[1], Xu Y. X.[1], Wu Z. T.[1], Li L. S.[2]

[1] Guangdong University of Technology, 510006 Guangzhou, P. R. China

[2] Guangdong Huasheng Nanotechnology Co. Ltd., 523000 Dongguan, P. R. China

Abstract: By integratingcathodic arc evaporation (CAE) with magnetron sputtering (MS) or high-power impulse magnetron sputtering (HiPIMS), hard coatings with diverse multicomponent compositions, superior mechanical properties, modulated residual stresses, and enhanced high-temperature properties can be fabricated. Among these coatings, some have been specifically designed to achieve outstanding cutting performance in high-speed cutting applications. In this presentation, several nitride and oxide hard coatings, such as AlTiN, TiAlN/TiSiN, AlCrN/Cu, and AlCrO, were deposited using a hybrid PVD coating system. The microstructures, mechanical properties, and cutting performance of these coatings will be discussed.

Keywords: hybrid coating method; cathodic arc evaporation (CAE); magnetron sputtering (MS); hard coatings; microstructure, mechanical properties

Structural Regulation and Tribological Properties of Molybdenum Disulfide Films Prepared by HiPIMS

Zhang Bin[1,2], Gao Haiyang[1,2]

1. State Key Laboratory of Solid Lubrication, Lanzhou Institute of Chemical Physics, Chinese Academy of Sciences, Lanzhou 730000, Gansu, China

2. School of Materials Science and Optoelectronic Technology, University of Chinese Academy of Sciences, Beijing 101408, China

E-mail address: bzhang@ licp. cas. cn (Zhang Bin)

Abstract: The high ionization efficiency characteristic of HiPIMS technology enables deposited films to exhibit superior adhesion and densification. This advantage can address the low hardness and poor adhesion issues of traditional magnetron-sputtered MoS_2 films while offering a means for microstructural regulation. By optimizing the MoS_2 film interlayer process, regulating film structure through HiPIMS plasma, and modifying films via doping, this study explores performance evolution to enhance tribological properties and load-bearing capacity. A series of MoS_2 films were prepared by adjusting HiPIMS pulse frequencies and compared with mid-frequency magnetron-sputtered films. The presence of an amorphous phase between nanocrystalline MoS_2 grains improved mechanical properties, achieving a film hardness of 4. 6 GPa. Under experimental conditions, HiPIMS films maintained stable friction under a high load of 30 N with excellent wear resistance. Additionally, the films achieved a long friction life exceeding one million cycles under a 10 N load in high vacuum. In subsequent doping studies, trace noble metal doping significantly increased the S/Mo atomic ratio and enhanced crystallization of MoS_2 in the films. HiPIMS ensured uniform dispersion of dopants, enabling performance improvements with minimal additives. During friction, the ordered structure at the interface reduced the friction coefficient and wear rate. A wear rate of 3.3×10^{-8} $mm^3/N \cdot m$ was achieved under a 10 N load, with a reciprocating friction life exceeding 300, 000 cycles, demonstrating significantly enhanced tribological performance.

Keywords: HiPIMS; molybdenum disulfide (MoS_2); mechanical properties; tribology; noble metal doping

Ultrahard and Ultralow-Friction: Synergistic TiC/Soft-Elastomer Interfaces with Record-Low Friction

Li Weiye[1,2], Chen Li[2], Zhang Bin[1]

[1] Lanzhou Institute of Chemical Physics, Chinese Academy of Sciences, Lanzhou, China

[2] School of Petrochemical Technology, Lanzhou University of Technology, Lanzhou, China

15998074180@ 163. com (Li Weiye)

bzhang@ licp. cas. cn (Zhang Bin)

Abstract: In aerospace applications, rubber is an indispensable sealing material, effectively preventing contaminants from entering housing structures and lubricant leakage. However, severe wear and friction induced by sliding motion remain the primary causes of seal failure, significantly compromising operational safety and service life. This study aims to address the technical challenge of achieving long-term performance in engineering-scale rubber seals. It innovatively employs filtered cathodic arc deposition to deposit ultra-hard Cr/Ti metal transition layer TiC films (with an ultra-hard of 35 GPa) on fluoroelastomer (FKM) substrates, demonstrating exceptional friction longevity exceeding 2, 000, 000 cycles at 1, 000 rpm. The results reveal that the Cr/Ti transition layer with arc current ratios of 160/120 A (Cr/Ti) achieves the optimal performance: a minimized coefficient of friction (0. 305) and superior mechanical properties. These findings provide critical material selection guidelines for high-performance FKM seals in tribologically demanding environments. The study not only advances the fundamental understanding of the effects of metal-layer composition on coating performance but also proposes a novel surface modification strategy for FKM.

Keywords: Metal Transition Layer; TiC Films; Ultra-hard; Fluoroelastomer; Mechanical Properties; Friction longevity

The Effect of Ultra-Thick Diamond Microstructure on High-Quality Polishing

An Kang[1,2], Liu Fengbin[1], Liu Jinlong[2], Wei Junjun[2], Chen Liangxian[2], Li Chengming[2]

[1] School of Mechanical and Materials Engineering, North China University of Technology, Shijingshan, Beijing, 100144

[2] Institute for Advanced Materials and Technology, University of Science and Technology Beijing, Haidian, Beijing, 100083

Abstract: In recent years, with the advancement of deposition technology, the research and application of ultra-thick CVD diamond have gradually increased due to its excellent properties. During the CVD diamond deposition process, the orientation gradually changes with the increase in thickness. This study adopts a method of step-by-step grinding thinning and batch testing of multiple sets of diamond with different thicknesses to conduct comparative analysis. It statistically analyzes the evolution of the microstructure during the increase in thickness. The impact of this process is modeled and analyzed to reveal the reasons for the changes.

Keywords: ultra-thick diamond; crystal orientation; microstructure; polishing

Self-Adaptive and Self-Healing Tribological Behaviors of Magnetron Sputtered WS (C, N) Nanocomposite Coatings

Cao Huatang

Huazhong University of Science and Technology, Wuhan, China

caoht@ hust. edu. cn

Abstract: This report introduces a breakthrough in self-lubricating sputtered WSC/WSN tribocoatings, engineered for demanding applications across industries requiring adaptive durability and high performance. WSC/WSN coatings demonstrate exceptional tribological behavior across a range of extreme conditions, including constantly discrete high temperatures, ramping temperature either in heating or cooling and wide temperature cycling from room temperature to 500 ℃. The WSC/WSN coatings exhibit an extremely low coefficient of friction (CoF = 0. 02) up to 400 ℃, with high thermal stability and superior triboperformance. Moreover, the coatings possess favorable tribo-reversibility under 400 ℃ ↔ room temperature cycles. Transmission electron microscopy analysis verified the self-lubricating, tribologically reversible, and ultralow lubrication mechanisms of the WSC/WSN coatings. Although under high-temperature tribosliding, the WS_2 layer still dynamically forms a self-organized, layered interface structure that continuously adapts to sliding conditions, ultimately enabling sustained superlubricity and tribological reversibility. Oxidation during high-temperature tribosliding actually has only a minor degrading effect on friction provided that the coatings retain sufficient sulfur to predominantly form WS_2 lubricant agents. This study provided novel insights into the development of advanced tribocoatings exhibiting adaptive ultra-lubrication in wide variable temperature conditions.

Some Issues on Materials Science in PVD Film Growth

Huang Feng

Vidali Industry (Shenzhen) Co., Ltd, Shenzhen, China

E-mail address: feng.huang@vtlnk.com (Huang Feng)

Abstract: Knowledge of underlying phase equilibrium thermodynamics is critical to the efficient processing or synthesis of candidate materials; this is best illustrated by the classic Fe-C equilibrium phase diagram in heat treatments of steels. Similar phase diagrams, however, have found limited usage in hard coatings prepared by far-from-equilibrium processes such as physical vapor deposition (PVD). This presentation aims to introduce the concept of metastable phase diagrams and, on this basis, to re-interpret experimental observations on some popular nitrides, carbides, and borides. Primary attention will be devoted to multi-element Ti-containing nitrides to highlight the effect of equilibrium thermodynamics on the formation of metastable phases and microstructural evolution. We demonstrate that qualitative topological arguments, rather than quantitative thermodynamic calculations, seem adequate in many cases to provide important guidelines for accelerated process development.

Design of Macroscale Solid Superlubricating Coatings Towards Engineering

Ji Li

State Key Laboratory of Solid Lubrication, Lanzhou Institute of Chemical Physics, Chinese Academy of Sciences, Lanzhou, China

E-mail address: jili@ licp. cas. cn (Ji Li)

Abstract: Superlubricity is a frontier subject in tribology research. From a theoretical perspective, it refers to the phenomenon where the frictional force between objects in relative motion approaches zero; From an engineering perspective, it refers to the reduction of friction coefficient by 1-2 orders of magnitude compared to traditional lubricants, reaching the level of 0. 001 or below. The breakthrough of superlubricity technology is expected to promote the revolutionary progress of industrial technological civilization, but there are still many challenges in obtaining macroscale superlubricity performance. Based on the three superlubricity principles of hydrogen passivation of amorphous carbon films, incommensurate structrual superlubricity and molecular rolling, this report provides some innovative research ideas such as the designs of medium-range ordered structure, heterostructure and self-curling rolling lubrication, and the relevant research progresses are introduced.

Covalent Graphene-Diamond Heterostructure with Ultrahigh Mechanical Robustness

Shen Bin

Shanghai Jiao Tong University, Shanghai, China

E-mail address: binshen@sjtu.edu.cn (Shen Bin)

Abstract: This presentation focuses on the fabrication, properties, and applications of graphene-diamond covalent heterostructures. Our research team successfully synthesized high-quality few-layer graphene sheets via liquid metal gallium-catalyzed CVD and achieved covalent bonding with diamond substrates. Systematic studies revealed the growth mechanism: graphene sheets epitaxially grow along the diamond (111) crystal plane, forming a stable hexagonal C6 covalent bonding configuration. Furthermore, through nano-scratch testing and molecular dynamics simulations, we investigated the interfacial bonding strength and mechanical behavior, revealing an interfacial tensile strength of 150 GPa—exceeding the intrinsic strength of graphene itself. The heterostructure exhibits exceptional wear resistance and low friction coefficients at both macro- and micro-scales. Leveraging this high-performance heterostructure, we developed a fabrication process for graphene-diamond covalent heterostructure coatings and applied them to cutting tools and abrasives. Experiments demonstrate that the coating significantly reduces cutting forces and tool wear, enhances polycrystalline diamond polishing efficiency by over 150%, and substantially improves surface finish quality. This research pioneers new directions for high-performance cutting tool coatings and abrasive technologies, expands the hybrid-dimensional heterostructure technology framework, and holds promise for advancing efficient precision machining in aerospace and third/fourth-generation semiconductors.

Exploring New Diamonds: High-Pressure Synthesis and Property Characterization

Yao Mingguang

Jilin University, Changchun, China

E-mail address: yaomg@jlu.edu.cn (Yao Mingguang)

Abstract: The inherent challenges of diamond include brittleness, pronounced anisotropy, susceptibility to cleavage along crystalline planes, and bandgap engineering difficulties, etc. This presentation will introduce our group's synthesis of high-quality transparent sp^3 amorphous carbon bulk materials and hexagonal diamond. Utilizing our independently developed large-volume press ultrahigh-pressure technology, these novel ultrahard materials demonstrate breakthrough properties. The amorphous carbon achieves unprecedented performance among non-crystalline materials: a Vickers hardness of 102 GPa, Young's modulus of 1182 GPa (rivaling diamond), thermal conductivity of ~30 W/mK, and tunable optical bandgap (1.8-2.7 eV). This material further exhibits superior formability and isotropy compared to diamond, coupled with exceptional optoelectronic properties, revealing substantial application value.

Research Hotspot and Industry Upgrading of Diamond

Zhao Xinghao

Journal Manager of Sinomach Diamond (Henan) Co., Ltd., Zhengzhou, China

E-mail address: zxh@jgszz.cn (Zhao Xinghao)

Abstract: The current hotspot and the future trend of diamond research is summarized from the perspective of bibliometrics, in combination of surveys from academic journals and industrial associations. Country-disaggregated statistics indicate that the diamond research in China has achieved rapid growth, significantly narrowing the gap with global frontiers. The groups are further analyzed, especially the top-ranked teams, across major research countries, breaking down by individual research groups. Their recent publications, as well as the contacts, are revealed. Finally, the report ends with the prospect for future industry-academia-research collaboration integrating China's industrial history in diamond production.

Research on Gradient Energy-Regulated Surface Modification Technology for High-End Equipment Components

Jin Jie
Associate Researcher, Tsinghua University
E-mail address: jj-king@ mail. tsinghua. edu. cn (Jin Jie)

Abstract: This presentation introduces an advanced surface modification technology based on gradient energy regulation, aimed at enhancing the performance and longevity of critical components in high-end equipment. The technology combines low-energy diffusion and high-energy ion implantation to create gradient-modified layers with superior hardness, wear resistance, and corrosion resistance. Key innovations include multi-level energy regulation systems, heterointerface matching, and composite ion implantation techniques, and developed multifunctional coatings, including DLC and WS_2/Ag films, tailored for extreme conditions like high temperatures and ultra-low temperatures. Applications span aerospace, defense, and precision manufacturing, such as helicopter transmission splines, satellite bearings, and rocket engine turbopumps. Results demonstrate significant improvements: rolling contact fatigue life increased by over 100%, friction coefficients reduced by 50%, and wear resistance enhanced by tenfold. Future directions include AI-driven intelligent manufacturing and atomic-level precision. This technology addresses " bottleneck" challenges in high-end equipment, offering transformative solutions for reliability and performance in demanding environments.

Keywords: tribology; gradient energy regulation; surface modification technology; ion implantation; extreme conditions

High-Throughput Preparation for Coatings with Ultra-High Strength Yet Toughness and Their Efficient Characterization Technology and Equipment of Fracture Toughness

Zhang Shihong

Key Laboratory of Green Fabrication and Surface Technology of Advanced Metal Materials, Ministry of Education, Anhui University of Technology, Maanshan 243002, PR China

E-mail address: shzhang@ ahut. edu. cn (Zhang Shihong)

Abstract: The development of physical vapor deposition (PVD) coating technology and the preparation of high-strength and tough hard coatings on alloy substrates are of great significance for improving the service life of alloy tools/key components in harsh environments such as high-load wear, wide- range temperature, alternating stress, impact fatigue and erosion wear. Focusing on typical applications in high precision manufacturing such as processing tools, aerospace devices, weapons/energy equipment, etc. , the design criteria and synergistic strengthening mechanisms of multi-component, multi-layer, and multi-scale interfaces for PVD hard coatings are clarified through the high-throughput computing and experimental methods. The growth mechanism of self-assembled coatings regulated by multi-field PVD plasma is studied, and the high-throughput deposition equipment is designed and manufactured. The regulation and optimization technologies including of coating composition/structure screening, interface structure, mechanical properties and service life are established. The efficient characterization technology and equipment for sub-micron/micron thickness coating fracture toughness are developed. The coating design-preparation-characterization platform and database are established to achieve performance evaluation, demonstration application and technology standardization.

Keywords: PVD coatings, Hard yet tough coatings, High-throughput preparation, Fracture toughness, Standardization.

Ion Beam Preparation and Characterization of Thick Amorphous Carbon Films

Chen Lin, Shen Yongqing, Wu Shuai, Liao Bin, Zhang Xu *

Faculty of Arts and Sciences, Beijing Normal University, China

E-mail address: linchen@bnu.edu.cn (Chen Lin)

E-mail address: zhangxu@bnu.edu.cn (Zhang Xu)

Abstract: Tetrahedral amorphous carbon (ta-C) films, exhibiting diamond-like hardness approaching 80 GPa and properties, show promising potential for extreme applications such as high-load wear and high-speed sand erosion. However, their thickness and broader implementation are constrained by high internal stress. The fabrication of high-performance thick ta-C films on steel substrates remains a significant challenge globally. In this study, 50-μm-thick ta-C films with a hardness of 7500 HV and 71% sp^3 bond content were successfully deposited on stainless steel substrates using filtered cathodic vacuum arc deposition (FCVAD) under low ion energy conditions. The structure and properties of thick a-C films with low internal stress and high mechanical properties prepared by high-voltage pulse method were studied. Additionally, the performance of these thick coatings in harsh environments was systematically investigated.

Keywords: filtered cathodic arc; amorphous carbon; thick film; ion energy

HPHT Synthesis of High-pure and Co-doped Single Crystal Diamonds

Ma Hong'an*, Wang Zhiwen, Jia Xiaopeng

State Key Laboratory of High Pressure and Superhard Materials, Jilin University, China

E-mail address: maha@jlu.edu.cn (Ma Hong'an)

E-mail address: 1505000191@qq.com (Wang Zhiwen)

Abstract: Diamond is an ultra-performing material featuring a unique combination of exceptional properties, including the highest known hardness, unsurpassed thermal conductivity, the broadest optical transmission window, high charge carrier mobility, and superior breakdown voltage. These attributes render it indispensable for industrial, technological, defense, and biomedical applications. As the core technology for producing high-quality diamond, the high-pressure high-temperature (HPHT) method offers distinct advantages in synthesizing both high-pure and doped diamond crystals. This work successfully synthesized high-pure diamond and co-doped diamond single crystals via HPHT. High-pure diamond serves as a critical material for high-end optical windows and quantum devices, while doped diamond provides essential components for high-temperature semiconductors and power electronics. By optimizing the growth system, innovating synthesis chamber design, and leveraging doping element synergism, we achieved controllable fabrication of high-crystallinity diamonds tailored to specific elemental doping requirements. These results establish an experimental and theoretical foundation for advancing HPHT synthesis processes toward practical applications of high-purity and doped diamonds.

Keywords: HPHT; high-pure diamond; co-doped diamond; defect; high-pressure annealing

Diamond-Graphene Nanohybrid Composite Films: Preparation, Characterization and Application

Zhang Wen, Liu Huiqiang, Wang Bing, Xiong Ying*

State Key Laboratory of Environmental-friendly Energy Materials, Southwest University of Science & Technology, Mianyang 621010, P. R. China

E-mail address: xiongying@ swust. edu. cn (Xiong Ying)

Abstract: In this work, a novel microwave plasma chemical vapor deposition based on liquid organic molecules as carbon and nitrogen sources was employed to synthesize diamond-graphene nanphybrid composite films, which could be grown on several different substrate such as Si, carbon cloth, and silicon carbide fibers. In these films, the nanohybrid nanosheets, consisted of diamond as core and graphene as sheet, would be interlaced together and vertically arranged to form three-dimensional porous structure, exhibiting large specific surface area, well electrical and thermal conductivity characteristics, and abundant surface termination states. Based on these features, these diamond-graphene nanohybrid composite films have been used as field emission emitter to generate high-quality electron beams, as supercapacitive electrodes to storage electrical energy, as SERS substrates to detect explosive molecules with high sensitivity and high selectivity, and as thermal managements to dissipate the heat from high-power electrical devices.

Keywords: diamond-graphene nanohybrid films; microwave plasma chemical vapor deposition; electron emission; thermal managements; SERS substrates

Achieving High-Adhesion Coatings on the Inner Walls of Slender Tubes throughPlasma Energy Modulation

Hu Tianshi[a], Geng Huiyuan[a], Tian Xiubo[a]

[a] State Key Laboratory of Precision Welding and Joining of Materials and Structures, Harbin Institute of Technology, Harbin150001, China

Abstract: Uniformly depositing high-quality physical vapor deposition (PVD) coatings on the inner walls of slender tubular components remains a formidable challenge in the industry. One of the most effective approaches is cylindrical target magnetron sputtering (CMS). However, as the inner diameter of the tube decreases, the conventional cathode-anode-bias discharge configuration becomes impractical, requiring the cylindrical target to serve as the cathode and the tube as the anode. The absence of bias voltage and the reduced target-substrate distance pose significant obstacles to plasma energy control. On one hand, the lack of bias results in uncontrollable particle energy; on the other, the short plasma transport distance leads to minimal collisions between sputtered particles and gas molecules, with the particle energy following a Thompson distribution. This allows a small fraction of high-energy particles to cause severe irradiation damage. By integrating bipolar high-power impulse magnetron sputtering (Bipolar-HiPIMS) with CMS, we modulated the plasma energy inside the tube by adjusting the working gas pressure (0.75 Pa, 1.5 Pa, 2.2 Pa, and 2.9 Pa) while applying a 50 V positive pulse voltage during bipolar pulsing. This method was employed to deposit Cr coatings on the inner walls of 20# steel tubes with a 40 mm diameter and 120 mm length. Analysis of discharge waveforms and plasma spectral characteristics revealed that as the gas pressure increased from 0.75 Pa to 2.9 Pa, the negative pulse voltage required for discharge decreased from 570 V to 516 V, while the electron temperature rose from 0.284 eV to 0.314 eV, attributed to enhanced electron capture efficiency. Microstructural and mechanical analyses of the coatings indicated a pronounced (110) preferred orientation across all samples. As gas pressure increased, enhanced particle collisions facilitated plasma thermal equilibrium, reducing coating hardness and brittleness while improving plasticity and toughness. However, excessively high gas pressure led to reduced coating density. At 2.2 Pa, the coating exhibited the smallest grain size of 8.03 nm and the highest adhesion strength of 75 N.

Keywords: inner tubes; Bipolar-HiPIMS; CMS; plasma energy modulation; cr coating

Control of Film Composition During Reactive Magnetron Sputtering of Composite Targets

Dmitriy A. Golósov[1], Geng Huiyuan[2], Tian Xiubo[2], Sergey M. Zavadski[1], Sergey N. Melnikov[1]

[1] Belarusian State University of Informatics and Radioelectronics, Plasma Processing Research center (Minsk, Republic of Belarus)

[2] Harbin Institute of Technology, School of Materials Science and Engineering (Harbin, PRC)

Abstract: Currently, thin films are widely used in optics, micro- and optoelectronics, and as wear-resistant and protective-decorative thin films. It should be noted that simple materials and binary compounds (oxides, nitrides, carbides) almost completely exceeded their potential. Significant progress in thin film technology can be achieved only if multicomponent materials are used. Multicomponent materials are very diverse in composition, with complex oxides being of particular interest in thin film technology. These materials include, for example, ferrites, cobaltates, cuprates, niobates, and other compounds. Currently, several methods are used for the deposition of multicomponent thin films by magnetron sputtering. These methods include reactive sputtering of alloy targets, multi-target sputtering, RF magnetron sputtering of multicomponent oxide or nitride targets and sputtering of composite (mosaic) targets. Each method has its own advantages and disadvantages. Among them, the sputtering of composite targets method offers significant benefits. The method makes it possible to deposit multicomponent films with any number and content of elements using a single magnetron. It is also suitable for sputtering dissimilar materials with low mutual solubility or large differences in melting temperatures. The method is easy to scale up. One of the main disadvantages of the method is the difficulty of defining the size and number of inserts needed to obtain the required concentration and uniform distribution of elements in the deposited film. Therefore, this study investigated the magnetron discharge voltage, deposition rate and composition of complex oxide films during reactive magnetron sputtering of two-component (Ti-Al, Ta-Al, Zr-Hf, Ti-Zr) composite targets in Ar/O_2 gas mixture. It was found that the magnetron discharge voltage and film deposition rate during sputtering of composite targets and changing the reactive gas concentration are determined by the effective sputtering yield and the secondary electron emission coefficient of the target material, which depend on the area occupied by metals on the target, the degree of their oxide coverage, sputtering yield and secondary electron emission of these metals and their oxides. Both with joint and separate gas supply to the chamber and a change in Γ_{02}, the relative content of metals in the film and the film deposition rate distribution profile change. The change in the metal content in the film is associated with the processes of oxide film formation on the surface of the target parts and the difference in the oxidation rates of the metals. The change in the distribution profile of the deposition rate occurs as a result of the generation of a directed flow of negatively charged ions of sputtered particles in the magnetron discharge region. Optical emission spectroscopy (OES) showed that there is a linearcorrelation between the ratio of the metal content in the film and the ratio of the intensity of the reference lines in the optical radiation of atoms of these metals. This makes it possible to effectively use the OES method to predict and control the metal content in films during reactive magnetron sputtering of composite targets. Based on the obtained experimental data, a model of magnetron sputtering of a two-componentcomposite target was developed. This model takes into account the distribution of the ion current density on the target, sputtering yields and secondary electron emission coefficient of metals and their oxides, and the rate of the chemical reaction of oxide formation. The model was experimentally tested during sputtering of Ti-Al composite targets in Ar and Ar/O_2 working gases. It was established that the developed model allows predicting the relative concentration of metals in the deposited films with an error of less than 5% during composite target sputtering in an Ar atmosphere and with an error of less than 10% during reactive sputtering in an Ar/O_2 gas mixture.

Keywords: reactive magnetron sputtering, composite target, multicomponent thin films, elemental composition, optical emission spectroscopy, reactive sputtering model.

Research on the Electromagnetic Shielding Performance of Al Coating on Carbon Fiber Resin Based Composite Materials Using HiPIMS

Xu Jianping*, Lei Jiangtao

School of materials science and engineering, Heilongjiang Institute of Technology, China

E-mail address: CLxjp@126.com (Xu Jianping)

E-mail address: 1648902120@qq.com (Lei Jiangtao)

Abstract: To augment the electromagnetic shielding properties of carbon fiber reinforced resin composites (CFRRCs), high-power impulse magnetron sputtering (HiPIMS) technology was utilized to apply an aluminum (Al) coating onto the surface of the CFRRCs. This research endeavor focused on examining the sputtering behavior of heterogeneously conductive composites under conditions of high bias voltage, evaluating the discharge characteristics of HiPIMS at varying pulse frequencies, and conducting a thorough analysis of the morphology and microstructure of the deposited Al coating. The electromagnetic shielding effectiveness (EMSE) of the resin-based composites was assessed using a vector network analyzer. The results obtained demonstrated that as the HiPIMS pulse frequency escalated, the target voltage, target current, and peak power decreased progressively during the stable discharge phase. Specifically, at a pulse frequency of 50 Hz, the target current attained a peak value of 71 amperes (A). The incorporation of an electrical insulating ring during the bias self-glow discharge significantly mitigated plasma sputtering damage to the CFRRCs, thereby facilitating the deposition of transition metal coatings at elevated bias voltages. The Al coating exhibited a state of high compressive stress, which enhanced the adhesion strength between the coating and the substrate. When subjected to high bias voltage with the electrical insulating ring in place, the Al coating surface exhibited a smooth and uniform appearance, characterized by fine particles. Transmission electron microscopy (TEM) analysis confirmed a strong bond between the Al coating and the CFRRC surface, with the Al coating displaying a non-columnar crystalline morphology. Notably, the application of the Al coating resulted in an increase in the EMSE of the CFRRCs by up to 34.5%.

Keywords: carbon fibre; composites; bias voltage; HiPIMS; electromagnetic shielding; Al coating

" Metal-like" Behavior of CrN Ceramics

Liu Liangliang, Wu Zhongzhen*

Peking University Shenzhen Graduate School, Shenzhen, China

E-mail address: liull620@ 163. com (Liu Liangliang)

E-mail address: wuzz@ pku. edu. cn (Wu Zhongzhen)

Abstract: Ceramic coatings that simultaneously exhibit high hardness and significant plastic deformability are highly desirable due to their enhanced service stability under impact loading. However, achieving a balance between hardness and deformability remains a major challenge for ceramic coatings. Introducing nanotwins into ceramic coatings holds promise for simultaneously enhancing both hardness and plasticity. Nevertheless, the inherently high stacking fault energy in ceramic materials renders twin formation extremely difficult. In this study, we address this challenge by tuning the ion energy during deposition, aiming to overcome the nucleation barrier of twins in CrN ceramics. The results demonstrate that ion energy modulation enables the formation of high-density nanotwins in CrN coatings. These nanotwinned CrN ceramics exhibit room-temperature plastic deformation behavior analogous to that of metals. Furthermore, investigation into the deformation mechanisms reveals that the nanotwinned CrN undergoes detwinning processes and displays a grain refinement strengthening inflection point, similar to those observed in metallic materials.

Keywords: nano-twins; CrN; plasticity; metal-like

The Establishment of Superlubricity in Engineering Field for H-DLC Composite Ininert and Active Multi-Environments

Abstract: The application of hydrogenated diamond-like carbon (H-DLC) with superlubricity can significantly decrease friction and largely save energy dissipation in mechanical systems. Thus, the realization of the near-frictionless state at the engineering scale in both inert and active atmospheres for H-DLC film is of interest. However, the film can achieve the state only under inert and vacuum conditions, while it fails in active atmospheres, such as oxygen and moist air, which seriously hinders wide application of the superlow friction. In previous studies, we have successfully solved this challenge by combining H-DLC film with MoS_2 flakes. Nevertheless, the life of the superlubricity achieved by the MoS_2/H-DLC composite in an inert gas was extremely short, which was attributed to the failure of MoS_2-rich transfer films. To establish the near-frictionless state in various atmospheres, graphene oxide with numerous oxygen-containing groups was used to prepare a GO/MoS_2/H-DLC triple composite in this study. The tribological performances were investigated systematacially. The results demonstrated that the triple composite could establish superlubricity in inert gases and ultralow friction in active atmosphere by forming robust transfer films at sliding interfaces. The structures at frictional interfaces including wear tracks and scars were thoroughly investigated. This work provides a valuable approach for the H-DLC composite to achieve an extremely low friction in various atmospheres and can guide the application of superlow friction in engineering fields.

Ultralow Friction of MoS_2/H-DLC Composite at Elevated Temperature

Abstract: The H-DLC films are most ideal solid lubrication materials to decrease friction and wear, which attract the attention from the number of the researchers. The superlubricty established by the H-DLC film failed in oxygen-containing and moist air. By adding the 2D material MoS_2 to the H-DLC films, the MoS_2/H-DLC composites could achieve superlubricity in the oxygen-containing and moist air, which has an excellent friction performance. The friction performance of the composite under harsh environment is unclear, such as high temperature and pressure, severely limited practical application of the MoS_2/H-DLC composite. Aiming to the problem, the friction property of the MoS_2/H-DLC composite under elevated temperature is studies. The ultralow friction with friction coefficient from 0.02 to 0.062 is achieved by the MoS_2/H-DLC composite at the high temperature. The friction performance of the composite was tested under a load of 7 N in the open air with relative humidity of approximately 26%. It is also found that the composite achieves ultralow friction from the room temperature to 573 K, while the composite is worn-out at the temperature of 473 K, and when the temperature rises to 573 K, the friction coefficient significantly increases. The study is expected to provide some suggestions for the application of the H-DLC composite under high temperature.

Anomalous Target Poisoning and Film Composition Evolution in the Erosion Process of High-Power Reactive Magnetron Sputtering

Cui Suihan*, Wu Zhongzhen*

Peking University Shenzhen Graduate School, Shenzhen, China

E-mail address: cuish@pku.edu.cn (Cui Suihan)

E-mail address: wuzz@pku.edu.cn (Wu Zhongzhen)

Abstract: During magnetron sputtering erosion, the change in the morphology and magnetic field of the target surface affects the plasma discharge. Especially in high power and reactive conditions, the plasma evolution during erosion determines the discharge and the deposition. In this work, a global model with iterative modification of the erosion profile is established, by which the reactive sputtering of an Al target in Ar/N_2 is simulated to study the plasma dynamics in the erosion process at different power densities. With increasing discharge power density, the electron density and electron temperature increase significantly to enhance plasma ionization. Consequently, N_2 adsorption is replaced by N deposition and N-containing ion sub-plantation. In the erosion process of high-power discharge, N_2 adsorption is further weakened, and the proportion of N-containing ion sub-plantation increases further from 25% to 37%, becoming the main reason for target poisoning. AlN coverage rate of target surface rises all the time, leading to a severe target poisoning. However, the discharge enhancement promotes the sputtering, and offsets the inhibition of target poisoning on sputtering, resulting in increased densities of the Al element components. Therefore, the total flux of Al element output from the discharge region increases, resulting in an increase in the Al / N ratio of the film, which is completely opposite to the variation of the poisoning state.

Keywords: high-power reactive magnetron sputtering; cathode erosion; target poisoning; film composition

Plasma Dynamics in Bipolar-Pulse HiPIMS

Han Mingyue[1], Luo yang[1], Li Liuhe[1]

[1] Department of Material Processing and Control Engineering, School of Mechanical Engineering and Automation, Beihang University, Beijing, 100191, People's Republic of China

Abstract: Investigation of the plasma decay mechanism from energy and density temporal/spatial evolutions together is important and urgent in the novel bipolar-pulse high power impulse magnetron sputtering (BP-HiPIMS) discharge for adjusting the deposited ion energy flux. In this work, temporal and spatial characteristics of the electron energy distribution function (EEDF) have been systematically investigated using a time-resolved (250 ns) Langmuir probe to obtain the plasma decay process clearly. The plasma decay has a typical characteristic of three-step during the positive pulse. At the initial period of positive pulse (Step 1), the existence of hot electrons and potential gradient force can accelerate the plasma density decay even up to ~10^{15} $m^{-3}/\mu s$. As the electron energy are cooled to several eVs, the plasma decay is dominated by the density gradient pressure with an ion sound velocity (0.5-0.7 kms^{-1}), which can increase the downstream electron density (Step 2). As the redistribution of plasma density in the whole discharge domain, the electron density decay is exponential (Step 3) and the maximum decay rate is near the target due to the higher density gradient, while the maximum electron temperature is away from the target. Along the density gradient diffusion, the relation between electron density and temperature in BP-HiPIMS satisfies the well-known Boltzmann relation $n_e = n_0 \exp(eV_p/kT_e)$. In addition, the EEDF characteristics in the BP-HiPIMS operated with an auxiliary anode and solenoid coil have also been investigated in this work, where an increase in electron density and plasma diffusion mobility has been observed after applying the anode or solenoid coil. These temporal and spatial EEDFs allow us to understand the complex plasma physics in the emerged BP-HiPIMS discharge clearly, especially with the view of high-energy and low-energy electron loss and balance.

Simulation of Focusing Control for High-Density Ion Beams Using a "Micro-Lens Arc Array—Deep-Hole—Substrate" Focusing System

Tao Zhiqun, Cui Suihan*, Wu Zhongzhen*

School of Advanced Materials, Peking University Shenzhen Graduate School, Shenzhen 518055, China

E-mail address: cuish@pku.edu.cn (Cui Suihan)

E-mail address: wuzz@pku.edu.cn (Wu Zhongzhen)

Abstract: Ion beams enable direct material printing without external heat sources, demonstrating significant potential for applications in electronic circuit direct writing and microstructure fabrication. However, high-density ion beams generally suffer from divergence effects and electron interference, posing challenges for high-precision beam focusing. In this study, the divergent ion beam is considered as a combination of numerous ion microbeams in which the ion directions are parallel. A two-stage focusing scheme is proposed, incorporating a micro-lens arc array, a deep-hole electrode, and a substrate electrode, and its focusing performance is systematically investigated through simulation. The results show that for an ion beam with a density of 1×10^{19} m^{-3} and an average electron temperature of 3 eV, the overlapping plasma sheaths in the micro-lens enable complete electron filtering while maintaining maximal ion transmission (>77.4%) with -100 V potential, 0.29 mm inner diameter and 0.05 mm spacing. Optimal focusing occurs at 14.3 m^{-1} arc array curvature, 20 mm electrode diameter, 105 mm electrode spacing, and 1400 V potential difference, achieving an ion beam compressed from 140 mm to 1.50 mm, with the central density of 4.70×10^{20} m^{-3} and the ion utilization of 65.3%. With deep-hole electrode of 1.5 mm diameter × 2.0 mm depth, 0.5 mm electrode spacing, and 4500 V potential difference, the balanced effects of central ionic space-charge repulsion and peripheral electric field convergence enable uniform output of the ion beam. The beam diameter can be further reduced to 0.78 mm with the average ion density of 2.88×10^{20} (density fluctuation amplitude: ±5.8%, variation coefficient: 4.8%) and the ion utilization of 57.3%.

Keywords: high-density ion beams; high-precision focusing; micro-lens; arc array; deep-hole electrode

Effect of Substrate Bias on the Microstructure, Mechanical and Tribological Properties of Cr/WC/C Coatings

Wu Xiaopan, Ou Yixiang

Joint Laboratory of Advanced Energy Materials and Intelligent Equipment, Beijing Academy of Science and Technology, Beijing 100000, China

E-mail address: ouyx16@ tsinghua. org. cn (Ou Yixiang)

Abstract: WC/C coatings synergistically integrate the mechanical robustness of the WC phase with the lubricity and corrosion resistance of the amorphous carbon (a-C) phase, where the top a-C layer, typically composed of diamond-like carbon (DLC) or graphitic carbon, serves as a physical barrier against corrosive ions and facilitates the formation of a self-lubricating transfer film under sliding contact. This work employed a hybrid deposition technique combining High-Power Impulse Magnetron Sputtering (HiPIMS) and Pulsed DC Magnetron Sputtering (PDCMS) to deposit Cr/WC/C multilayer coatings onto AISI austenitic stainless steel (SS316L) substrates. Deposition parameters for the Cr interlayer remained constant, while the substrate bias applied to the WC/C composite coating was systematically varied from −30 V to −200 V (−30, −60, −90, −120, −150, −200 V). Advanced characterization techniques—including scanning electron microscopy (SEM), X-ray diffraction (XRD), Raman spectroscopy, nanoindentation, Vickers hardness testing, scratch testing, and tribological wear testing—were utilized to systematically investigate the influence of substrate bias on the coating microstructure, mechanical properties, and tribological performance. Key findings reveal stable coating thicknesses (4.5-4.8 μm) for all samples, comprising Cr adhesion layer, WC transition layer and top C layer. The phase composition of all coatings contains Cr and β-WC_{1-x} phases. Increasing substrate bias significantly reduced microstrain from 0.35 % at −30 V to 0.11 % at −200 V, with the (111) plane grain size reaching a minimum of approximately 3 μm at −120 V. Coatings deposited at −120 V substrate bias demonstrated optimal overall performance, achieving the highest hardness (16.63 GPa) and effective Young's modulus (210.09 GPa), maximum H/E* (0.079) and H^3/E^{*2} (0.104) ratios, highest elastic recovery (47.8%), greatest energy dissipation capacity (32.63 nJ · μm^{-3}), and strongest coating-substrate adhesion (32 N). While all coatings exhibited a consistently low friction coefficient (0.06-0.07) under dry sliding conditions (4 N, 3 Hz), the −200 V coating showed the highest wear rate (8.44×10^{-6} $mm^3 \cdot N^{-1} \cdot m^{-1}$), significantly exceeding the substantially lower rates ((0.18-0.40) $\times 10^{-6}$ $mm^3 \cdot N^{-1} \cdot m^{-1}$) of the others. These results highlight that a substrate bias of -120 V is critical for optimizing the comprehensive properties of Cr/WC/C coatings. The optimized coating developed herein offers an innovative solution for enhancing the surface protection reliability and service lifetime of SS316L components in demanding environments such as marine drilling equipment and biomedical implants, demonstrating significant academic merit and promising engineering application potential.

Keywords: SS316L substrate; Cr/WC/C coating; substrate bias; adhesion; wear rate

Breaking the Hardness-Toughness Dilemma in Protective Coatings: Controllable Preparation and Mechanism Exploration of Superhard Yet Tough Coatings

Wang Haoqi, Associate Researcher

[1] Radiation Technology Institute, Beijing Academy of Science and Technology, Beijing 100875, China

[2] Advanced Energy Material and Intelligent Equipment Joint Laboratory, Beijing Academy of Science and Technology, Beijing 100875, China

E-mail address: 243492428@ qq. com (Wang Haoqi)

Abstract: Superhard coatings have been widely applied for surface protection of components. However, due to the intrinsic trade-off between hardness and toughness in surface engineering, most conventional superhard coatings achieve their exceptional hardness at the expense of compromised toughness. Low-toughness hard coatings are prone to brittle fracture during service life, leading to premature failure via microcracking propagation. This study addresses this challenge by developing superhard and tough surface coatings through high-intensity pulsed magnetron sputtering (HIPIMS) . An innovative approach combining 3D layer-by-layer scanning reconstruction with high-resolution transmission electron microscopy (HRTEM) characterization enables multiscale visualization of indentation-induced microcrack networks in the coatings. By systematically analyzing geometric features, size distributions, and spatial structural relationships of microcracks across nanoscale to micrometer scales, new insights into the toughening mechanisms of superhard coatings are revealed. This research provides a novel pathway for advancing the development of high-performance protective coatings with balanced hardness and toughness.

Keywords: Superhard yet tough coatings; HiPIMS; Visualization of cracks

Research on Surface Modification of Hydrogen Injector for Hydrogen Fuel Cell

Jiang Qili[1], Feng Yi[1], Wu Xiaopan[1], Zhang Yue[2], Ou Yixiang[1]

[1] Beijing Academy of Science and Technology, Beijing, China

[2] Nanjing Tech University, Nanjing, China

Abstract: The impact of the injector valve core and valve seat at higher frequencies could seriously affect the lifespan of the hydrogen injector in hydrogen fuel cell. The magnetron sputtering technology was employed to modify the working surface of a hydrogen injector. Its nanocoating is composed of chromium tungsten carbide with tungsten carbide as the bonding layer, and tetrahedral amorphous carbon as the surface layer, which aim to improve the performance and lifespan of hydrogen injectors. X-ray diffractometer was employed to analyze the phase structure. Raman spectrometer was employed to characterize the molecular structure. Nano indentation tester was employed to investigate the hardness. The micrometer scratch tester was employed to test the adhesion of the coating. These results showed that compared to the surface without coating preparation, the modified surface structure was smooth and dense with higher hardness. The scratch morphology is smooth without peeling failure. The coating abrasion shows hardly furrows. The wear mechanism is a composite mechanism of abrasive wear and oxidative wear. All data indicate that after surface modification of the prepared coating while heat treatment is not important. Directly coating untreated samples can help improve preparation efficiency. The working surface after coating preparation shows higher hardness and toughness, thereby improving the coating's ability to resist crack initiation and propagation. The hydrogen injector with new nanocoating exhibits excellent wear resistance and hydrogen resistance, which could postpone the failure of the hydrogen injector.

Keywords: magnetron sputtering; hydrogen injector; nanocoating; ta-C layer

Power Supplies for High power impulse magnetron sputtering and recent applications

Tian Xiubo, Geng Huiyuan, Ou Kejun

State Key Laboratory of Precision Welding and Joining of Materials and Structures, Harbin Institute of Technology

New-Films Technology Co. Ltd

Abstract: Recently, various waveforms of high-power impulse magnetron sputtering power supplies have been studied. The use of small perturbations in plasma can modulate the waveform of discharge current, effectively optimizing the structure and performance of the film layer. This paper presents several pulse discharge waveforms for metal target, magnetron discharge with full oxygen mode, and high-energy pulse discharge using oxide targets. At the same time, multiple matching waveforms for synchronous pulse bias are also provided. Finally, some applications using high-power impulse magnetron discharge were presented, including wear resistance, corrosion resistance , coating for TGV or biomaterials, etc.

Study on Deposition and Performance Modification of ta-C Films by High Power Impulse Magnetron Sputtering Technology

LI Jianzhong

Northeastern University, Shenyang City, 110819

Abstract: ta-C film is a diamond-like carbon filmby carbon elements. It has high hardness, excellent wear resistance, low friction coefficient, excellent chemical inertness, and good optical properties. In this paper, an attempt was made to change the output waveform of the power supply by applying reverse positive pulses after negative pulse voltage, in order to control the output energy of the power supply and optimize relevant process parameters to solve the problem of low carbon atom ionization rate by High power impulse magnetron sputtering technology (HiPIMS). Firstly, the effects of arc enhanced glow discharge etching, HiPIMS metal discharge etching, and transition layer pre-treatment methods on the adhesion of ta-C films were compared. The results showed that the adhesion of ta-C films prepared on the surface of high-speed steel after different pre-treatments was higher than that of ta-C films prepared on the surface of WC alloys. In addition, the energy of ta-C film deposition process can be changed by adjusting the substrate bias voltage and HiPIMS power frequency. The research results indicate that as the bias voltage increases, the sp^3 bond content in the ta-C shows a trend of first decreasing and then increasing. The high-power impulse magnetron sputtering technology using reverse positive pulse not only improves the deposition efficiency and layer performance of ta-C film through a target free acetylene gas source, but also regulates the problem of large particle contamination in the magnetron sputtering deposition process. Lastly, cylindrical silicon/titanium targets were designed to modify the performance of ta-C films. Although the properties of the ta-C film treated by Ti doping process has also improved, the overall performance is slightly lower than that of silicon doping.

Keywords: High Power Impulse Magnetron Sputtering (HiPIMS); positive reversal; Tetrahedral Amorphous Carbon film (ta-C); elements doping

Format Requirements for Extended Abstract to the 30th World Congress of International Federation for Heat Treatment and Surface Engineering

Liu Xingguang*, Wang Yuqi, Wu Lishuo, Liu Liting, Ran Songlin, Zheng Jun*

Key Laboratory of Green Fabrication and Surface Technology of Advanced Metal Materials,

Ministry of Education, Anhui University of Technology, Ma'anshan 243002, China

E-mail address: sdwfcllxg@ icloud. com (Liu Xingguang)

E-mail address: Jzheng@ ahut. edu. cn (Zheng Jun)

Abstract: This studyaims to improve the solid-lubrication property of a-C: H films by accelerating the formation and enhancing the durability of tribofilms. Results showed that introducing a thin Cu or Ti cover layer with proper thickness would significantly enhance the tribological performance of a-C: H films at low sliding speed. After introducing a Cu cover layer, the CoF of a-C: H was reduced to 0.004 (superlubricating) at the optimal Cu-layer thickness of ~32 nm. By adding a ~40 nm thick Ti layer could also significantly improve a-C: H's tribological performance, i. e. reducing the CoF from 0.04 ±0.016 to 0.01 ±0.003, and the wear rates from 7.33×10^{-7} $mm^3/(N \cdot m)$ to 1.01×10^{-7} $mm^3/(N \cdot m)$, respectively. These dramatic improvements can be attributed to the formation of a dense and compact carbonous transfer layer consisted of nanocrystalline Cu/Cu_2O or TiO_2/Ti_2O_3 and an amorphous-carbon matrix. Enhancement on the nanocomposite transfer layer is probably related to the formation of Cu-O-C bridging bonds for the Cu cover layer, or O-Ti-C and Ti-O-C bridging bonds for the Ti cover layer. A double-reinforcement mechanism on the tribofilm named of "physical pinning" and "chemical bonding", after introducing a proper metallic cover layer, was proposed.

Keywords: tribological behaviour; a-C: H film; transfer-layer enhancement; ti cover layer; cu cover layer

HiPIMS-Enabled Hydrogen-Free Carbon-Based Nanocomposite Coatings for High-Performance PEMFC Bipolar Plates

Ou Yixiang

Professor/Research Fellow Director, Joint Laboratory of Advanced Energy Materials and Intelligent Equipment

Beijing Academy of Science and Technology, Beijing 100000, China

E-mail address: ouyx16@ tsinghua. org. cn (Ou Yixiang)

Abstract: Amorphous carbon (α-C) coatings play a critical role in enhancing the electrical conductivity, corrosionresistance, and chemical stability of stainless steel bipolar plates in proton exchange membrane fuel cells (PEMFCs). However, traditional deposition methods, such as magnetron sputtering and chemical vapor deposition, suffer from limitations including suboptimal sp^2/sp^3 ratios, high residual stress, and insufficient mechanical durability, restricting their application. This study develops Ti-doped α-C coatings using a combined high power impulse magnetron sputtering (HiPIMS) and pulsed DC magnetron sputtering (PDCMS) approach to overcome these challenges. By varying Ti target power (0-1 kW), the microstructure, electrical properties, corrosion resistance, and mechanical performance of the coatings were systematically investigated. Ti doping increased the sp^2/sp^3 ratio and promoted low-resistivity carbide phase formation, reducing the interfacial contact resistance from 5. 8 to 2. 6 mΩ cm2 (0. 6 kW). At an optimal Ti target power of 0. 6 kW, the coatings achieved excellent corrosionresistance, with a corrosion current density of 0. 12 μA cm 2. Additionally, mechanical properties were significantly enhanced, with hardness reaching 13. 64 GPa and elastic recovery improving to 43 %. These findings demonstrate that Ti-doped α-C coatings not only address the limitations of traditional α-C deposition techniques but also offer a cost-effective, high-performance solution for stainless steel bipolar plates, advancing the commercialization of PEMFCs.

Graphene-Diamond Heterostructures: Toward All-Carbon Devices with Enhanced Electrical Properties

Yuan Xiaolu[1,2], Zhao Zichen[2], Liu Jinlong[2], Wei Junjun[2], Chen Liangxian[2], Li Chengming[2*], Wang Wenrui[1*]

[1] School of Mechanical Engineering, University of Science and Technology Beijing, China

[2] Institute for Advanced Materials and Technology, University of Science and Technology Beijing, China

E-mail address: chengmli@ mater. ustb. edu. cn (Li Chengming)

E-mail address: gmbitwrw@ ustb. edu. cn (Wang Wenrui)

Abstract: Graphene-diamond (G-D) heterostructures have been attracting considerable attention due to its unique features for all carbon sp3-sp2 electronic applications. Whereby the electrical properties of diamond surface can be purposely tailored and significantly altered through transformed graphene layers. In this work, G-D heterostructures were prepared by nickel (Ni) -catalyzed high-temperature rapid annealing, and were analyzed by Raman, Hall effect measurement and Transmission electron microscopy (TEM). The results indicated that the difference in surface conductivity of GOD composite structure is mainly related to the number of transformed graphene layers, which is mainly affected by annealing temperature, annealing time and the thickness of nickel film. Notably, linear ohmic contact characteristics are established between Pd/Ti/Au contacts and the in situ G-D heterostructure without requiring additional postannealing treatment. An unprecedentedly low specific contact resistivity of $1.4 \times 10^{-7}\ \Omega\ cm^2$ has been achieved. This study paves new pathways for advancements in graphene-based devices or all-carbon electronics.

Keywords: diamond; graphene; heterostructures; electrical properties

The 30th World Congress of International Federation for Heat Treatment and Surface Engineering
August, 19-22, 2025, Suzhou, China

Molten Salt Synthesis of Metastable Perovskite Nickelates Electronic Phase Transition Semiconductors Within MPa-High Oxygen Pressures

Chen Jikun

University of Science and Technology Beijing, China

E-mail address: jikunchen@ ustb. edu. cn (Chen Jikun)

Abstract: The multiple magneto-/electrical quantum transitions discovered with d-band correlated metastable perovskite nickelate (RENiO3) is promising for correlated electronic applications in the post-Moore Era. Nevertheless, their material synthesis requires a GPa-high pressure, as provided by cubic press, to stabilized their highly distorted perovskite structure (in particular for the ones exhibiting the heavy rare-earth composition), but this largely limits their synthetic amount as required by discrete device applications. Herein, we synthesized the powder of RENiO3 in grams/batch with ~1000 times lower pressure and ~300 ℃ lower temperature in comparison to the previous tens of milligram/batch results, assisted by their eutectic precipitation and heterogeneous growth within alkali-metal halide molten salt at MPa oxygen pressures. This provides solid base for further exploring the new functionalities of RENiO3 under extreme conditions, such as high pressures. The in situ characterizations of their transportation properties under high pressures within a diamond anvil cell reveal a distinguishing pressure predominated bad metal transport within the nonequilibrium state of RENiO3 showing high-pressure sensitivity up to 10 GPa, and the temperature dependences in electrical transportations are effectively frozen, shedding a light on their potential applications in pressure sensors at GPa-ranges.

Keywords: electronic phase transition; metastable materials

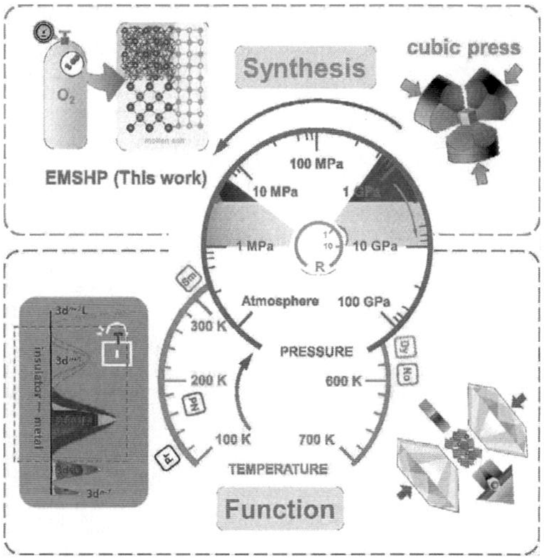

Tailoring Atomic Interfaces in One-Dimensional van der Waals Heterostructures for Semiconductor Applications

Kang Lixing

CAS; Suzhou Institute of Nano-tech and Nano-bionics, China

E-mail address: lxkang2013@ sinano. ac. cn (Kang Lixing)

Abstract: Owing to their pronounced nanoscale interfacial coupling effects, one-dimensional (1D) van der Waals (vdW) heterostructures exhibit highly distinctive electronic and magnetic behaviors. Among them, carbon nanotube (CNT) -derived 1D atomic crystal heterojunctions represent a novel class of vdW systems, where atomic-level interface engineering becomes possible. CNTs provide an ideal nanoconfinement platform due to their exceptional chemical stability, precisely defined inner cavities on the sub-nanometer to few-nanometer scale, and cylindrical surface topologies, enabling fine control over nanoscale heterointerface formation, phase stabilization, and interfacial orbital interactions. Building on this principle, we have developed a general confinement-guided synthesis strategy to fabricate diverse 1D atomic chains and heterostructures within single-walled CNTs. The encapsulation not only drives the formation of metastable or previously inaccessible 1D phases but also induces strong interfacial charge redistribution between the guest atomic chains and the host CNTs. Using low-voltage aberration-corrected transmission electron microscopy (AC-TEM) and advanced spectroscopic analyses, we achieve atomic-resolution structural characterization and probe the nature of charge transfer and orbital hybridization at the heterointerface. Magnetic and transport measurements reveal that these 1D confined heterostructures exhibit emergent magnetic phases—such as spin-glass-like states, superparamagnetism, and enhanced ferromagnetic ordering—which are markedly distinct from their bulk counterparts. These findings underscore the critical role of dimensional reduction and interfacial confinement in tuning quantum functionalities and offer new opportunities for interface-driven applications in nanoelectronics, spintronics, and surface engineering.

Keywords: semiconductor; CNTs (carbon nanotubes); nanostructures

On-site Formation of Graphene Lattices on Top of a Diamond Surface with Partial Covalent Bonding to the Substrate

Zou Rui[1], Liang Genhao[1], Chen Yinghao[1], Chen Xingyu[1], Zhao Lishan[1*]

College of Advanced Interdisciplinary Studies, National University of Defense Technology, Changsha City, Hunan, China

E-mail address: lishanzhao@ nudt. edu. cn (Zhao Lishan)

Abstract: Graphene and diamond, both as allotropes of carbon, have attracted a lot of attention in scientific research due to their excellent properties. Graphene has ultra-high carrier mobility in the plane owing to the Dirac cone-shaped energy band. But on account of the weak inter-layer force of two-dimensional materials, the out-of-plane thermal conductivity is poor, and the zero band gap halts to a large extend its application in the semiconductor industry . Although diamond has simultaneously a ultra-large band gap, high thermal conductivity and carrier mobility, its carrier concentration is very low in undoped state, and it is easy to achieve shallow level dopants in neither p-type nor n-type doping. Therefore, due to the disadvantages of their own characteristics, the industrialization value of graphene and diamond is greatly discounted. Although the graphene-diamond structure formed by transferring graphene to the diamond surface can combine the advantages of graphene and diamond, it is still unable to be directly applied because the structure is unstable due to the weak bonding force between them. Here we have developed a method where graphene lattices are formed via on-site phase transition of carbon atoms at the top of the (100) surface of a CVD grown single crystal diamond sample, and the top layer of granphse is tightly bonded to the diamond substrate with covalent bonding. Results of Raman spectroscopy shows clearly signatures of of the diamond substrate and mono- or few layer graphene. XPS results reveal that the prepared diamond-graphene heterojunction has covalent bonds in between, which greatly enhanced the bonding strength of them. At the same time, we also conduct research on its heat conduction and electrical properties. This research provide a way of combining the high carrier mobility of graphene with the high thermal conductivity (~2000 w/m · K) of diamond, providing a solution for the preparation of high-power RF devices working at high temperature (such as 5g/6g communication devices) or high-frequency transistors, and to solve the heat dissipation bottleneck of traditional semiconductors, helping to break through the performance boundary of single material. It has the potential to change in the fields of next-generation electronic devices, quantum technology, sensing and energy, and provides a new research dimension for the interface science of low dimensional materials and bulk materials.

Keywords: graphene; diamond; covalent bonds; interface; thermal conductivity

The Effect of Annealing Treatment On the Electrical Properties of Boron-doped Single Crystal Diamond

Li C., Li J., Zhao Z. C., Wang Y. Q., Tu J. P., Chen L. X., Wei J. J., Liu J. L.*, Li C. M.

Institute for Advanced Materials and Technology, University of Science and Technology Beijing, Beijing 100083, China

E-mail address: liujinlong@ustb.edu.cn (Liu J. L.)

Abstract: When using microwave plasma chemical vapour deposition to prepare boron doped single crystal diamond, the introduction of trimethyl borate leads to a more complex plasma composition, resulting in the existence of more impurities in the prepared samples, which affects the stability of electrical properties. The annealing treatment can eliminate the defects and impurities to a certain extent to make its properties stable. This study used microwave plasma chemical vapour deposition equipment to prepare a boron doping concentration of $10^{20}cm^{-3}$ single crystal diamond, its conductivity type is not stable, in less than 25μA excitation current presents n-type conductivity, in more than 2 mA excitation current presents p-type conductivity, the excitation current in the two in between the conductivity type will appear irregular changes. After thermal annealing at 800 ℃ for 30 minutes, the resistivity was slightly increased from 0.9 mΩ · cm to 11 mΩ · cm, and stable p-type conductivity was obtained at different excitation currents. This study shows that thermal annealing contributes greatly to the elimination of impurities and defects in boron-doped single-crystal diamond, and this result proves that the annealing treatment has a great role in the preparation of boron-doped single-crystal diamond power devices with stable and excellent electrical properties.

Keywords: boron doping; doping concentration; anneal; resistivity

Controlled Growth and Atomic-level Flattening and Polishing of High-purity Diamonds

Li J., Li C., Zhao Z. C., Li K., Wang Y. Q., Chen L. X., Wei J. J., Liu J. L. *, Li C. M.

Institute for Advanced Materials and Technology, University of Science and Technology Beijing, Beijing 100083, China

E-mail address: liujinlong@ustb.edu.cn (Liu J. L.)

Abstract: Diamond, as a wide-band semiconductor material, suffers from roughness scattering due to surface and body defects that severely limit device performance. In this study, high-purity homogeneous epitaxial single-crystal diamond films with impurity concentration < 5 ppb were grown using MPCVD. a combination of mechanical polishing and chemical mechanical polishing (CMP) was used to achieve atomic-level surface planarisation on high-purity (100) single-crystal diamond (SCD). By optimising the CMP parameters (load of 2500 g, rotational speed of 55 r/min, and flow rate of the polishing solution of 0.5 mL/min), the surface roughness was reduced to less than Ra = 0.1 nm in the area of 5×5 μm^2 and the subsurface damages were completely removed by oxidation after 48 h of polishing. Atomic force microscopy (AFM) observation showed that the scratches were gradually eliminated during the polishing process and an atomic-level step structure was formed。 According to the principle of silicon and silicon germanium heterojunction: Si and SiGe lattices are different, the forbidden band width is adjusted to enhance the carrier mobility. The preliminary exploration of SiC epitaxy on high-quality flat diamond surface using Molecular Beam Epitaxy (MBE) has laid the foundation for SiC/diamond heterojunction in high-temperature electronic devices.

Keywords: atomic-level planarisation; heterojunction; roughness; single-crystal diamond

Epitaxial Growth of High-Quality AlGaN/GaN Heterostructures on Diamond Substrates with High Interfacial Thermal Conductance

Yang Xuelin[1], Yang Hongcai[1], Chen Beile[1], Song Bai[2], Shen Bo[1]

[1] School of Physics, Peking University, Beijing 100871, P. R. China

[2] Department of Energy and Resources Engineering, Peking University, Beijing 100871, China

E-mail address: xlyang@pku.edu.cn (Yang Xuelin)

Abstract: The direct epitaxial growth of GaN on diamond substrates offers a fundamental solution for thermal management in high-power-density GaN electronic devices. However, the chemical inertness of diamond surfaces, with their robust C-C bonds, has persistently challenged high-quality III-nitride nucleation. Here, we demonstrate the successful growth of high-quality III-nitrides on diamond (111) by employing an interface engineering. This approach simultaneously achieves complete removal of the amorphous surface layer and breaking of C-C bonds, enabling high-quality AlN nucleation with well-aligned in-plane and out-of-plane orientations. We observe an ultra-high interfacial thermal conductance ($\geqslant 600$ MW$m^{-2}K^{-1}$) at the interface between AlN and diamond. By combining atomically-resolved electron energy-loss spectroscopy with machine-learning molecular dynamics, the physics behind the high conductance was discussed. On top of this high-quality AlN nucleation, low-dislocation-density GaN layers and high mobility AlGaN/GaN heterostructures have been fabricated. The FWHM values of the XRD rocking curves for the GaN (002) and (102) planes are 651 and 788 arcsec, respectively. The AlGaN/GaN heterostructures grown on diamond substrates exhibit a record room-temperature electron mobility of 1640 $cm^2/(V \cdot s)$. This work demonstrates that GaN-on-diamond is a high-performance platform for next-generation power electronics, where superior thermal management and electron transport are simultaneously realized.

Keywords: GaN on diamond

Research Progress on High-quality 8-inch Silicon Carbide Epitaxial Technology

Fang Yulong
Hebei Semiconductor Research Institute, China
E-mail address: yvloong@163.com (Fang Yulong)

Abstract: Silicon carbide (SiC) has demonstrated significant potential in high-temperature, high-voltage, high-frequency, and high-power electronic applications due to its wide bandgap, high saturation electron drift velocity, and excellent thermal conductivity. In recent years, driven by market demand in sectors such as new energy vehicles and photovoltaics, SiC technology has entered a golden era of rapid development. Currently, the mainstream SiC materials and devices in the market are still predominantly based on 6-inch wafers. Accelerating the research and development of 8-inch SiC materials and device technologies is crucial for further reducing material and wafer processing costs, as well as promoting the widespread adoption and advancement of the entire SiC industry chain. Based on the domestic 8-inchSiC CVD reactor, this study breaks through several key technologies of 8-inch silicon carbide epitaxy. By adopting a chlorine-based process route involving TCS and HCl, a high-speed epitaxial growth rate exceeding 60 μm/h was realized, with thickness uniformity of less than 1% and doping uniformity below 2%. The epitaxial material exhibited a surface killer defect density of less than 0.2/cm^2 and a basal plane dislocation (BPD) density below 0.1/cm^2. The performance of the 8-inch epitaxial wafers met or even surpassed that of their 6-inch counterparts, fully satisfying the requirements for 8-inch SiC device fabrication. Additionally, an 8-inch ammonia-based N-type doping process was developed, enabling continuous epitaxial growth across 15 wafers with an inter-wafer doping concentration variation of less than 2%. This demonstrated excellent batch consistency and stability, addressing the long-standing challenge of doping concentration instability between production batches in SiC epitaxy. These advancements lay a solid foundation for the mass production and automated manufacturing of 8-inch SiC epitaxial materials.

Keywords: SiC; epitaxial wafer; uniformity; defect

Reliability Mechanism of RF Diamond Semiconductor Devices

Xu Yuehang

University of Electronic Science and Technology of China

E-mail address: Yuehangxu@ uestc. edu. cn (Xu Yuehang)

Abstract: Diamond semiconductor is characterized by high breakdown electric field, high carrier saturation velocity, and high thermal conductivity, and is hailed as the "ultimate semiconductor". At present, hydrogen-terminated diamond devices show good performance in cutoff frequency, output current and output power, and demonstrate good application potential in the fields of microwave power electronics. However, hydrogen-terminated diamond devices still face severe reliability challenges. This report will present the progress of our team in the reliability mechanism and model of RF diamond semiconductor devices.

Keywords: reliability mechanism; RF; diamond device; semidconductor device

Interface and Near Interface Trap Analysis for Al_2O_3/Diamond MOS with High-Temperature Annealing

Zhang Xufang

North China University of Technology, China

E-mail address: zhangxufang@ ncut. edu. cn (Zhang Xufang)

Abstract: We focused on inversion-type diamond MOSFETs with normally-off properties with realizing diamond n- and p-doping, OH-diamond termination, oxide/diamond interface and near interface trap characterization. We developed novel OH-terminated diamond surface and focus on the key point-diamond/Al_2O_3 interface, which is the key to the channel mobility of diamond MOSFETs. Specifically, we employed the high-low C-V method, simultaneous C-V method, and conductance method by considering the surface potential fluctuation to analyze the properties of interface traps. In terms of the wide bandgap semiconductors, the near interface traps near the band edge of the semiconductor are also harmful, which enable to capture carriers and degrade the channel mobility and device stability. Therefore, we focused on the frequency-dependence of capacitance and conductance in the accumulation condition and established an equivalent circuit model to estimate the possible distribution of near interface traps close to the valence band edge of diamond. The comprehensive analysis of interface and near interface traps in diamond MOS structures would be important to find out effective approaches to improve the interface quality and device performance.

Keywords: near interface

Research Progress on Silicon-Terminated Diamond Field-Effect Transistors

Fu Yu

XidianUniversity, China

E-mail address: 741993138@ qq. com (Fu Yu)

Abstract: In recent years, silicon-terminated diamond field-effect transistors have become a new research hotspot in diamond electronic devices due to their good dielectric adhesion and compatibility with CMOS processes. This report reviews the research progress of silicon-terminated diamond devices and introduces some of the latest research advances of Xidian University in this type of devices. In response to the lack of depletion-mode materials and high voltage characteristics, as well as the complex process of silicon-terminated diamond devices, Xidian University has developed high-voltage enhancement-mode hydrogen/silicon compositely terminated diamond devices, high conductivity depletion-mode hydrogenated silicon terminated diamond materials and devices, and proposed a new low resistance boron-silicon co-doped diamond ohmic contact technology.

Keywords: diamond; semiconductor device

Research on Terminal Structures of Wide Bandgap Nitride Power Semiconductor Devices

Xiao Ming

Xidian University, China

E-mail address: xm427@126.com (Xiao Ming)

Abstract: We focus on the terminal structure design and performance verification of nitride-based power semiconductor devices, represented by gallium nitride and extending to ultra-wide bandgap materials. Terminal structures play a pivotal role in determining breakdown voltage, electric field distribution, and overall device reliability. Targeting both lateral and vertical device architectures, we investigate high electron mobility transistors, vertical PN diodes, SBD, and junction field-effect transistors, and propose high-voltage, high-reliability terminal designs that effectively suppress electric field crowding, enhance breakdown capability, and improve avalanche robustness. Among these, the PN junction serves as a critical element in terminal engineering. Furthermore, we explore the potential of ultra-wide bandgap materials such as aluminum nitride, leveraging their extreme physical properties to enable terminal designs suitable for high-voltage and high-temperature applications. Our research aims to deliver scalable, high-voltage terminal structure solutions for next-generation power electronic devices with superior efficiency and reliability.

Keywords: power device; HEMT; wide bandgap; III-nitride semiconductors; diode

Diamond with Preferred Orientation, From Thermology to Electronics

Liu J. *, Chan S., Zhu X, Zhao Z., Lai Y., Chen L., Wei J., Kawarada H., Li C.

[1] University of Science and Technology Beijing

[2] Waseda University

E-mail address: liujinlong@ ustb. edu. cn (Liu J.)

Abstract: Dark defects tend to be generated during the growth process, significantly affecting the performance of the diamond. We find that a preferred orientation along the (220) planes is conducive to the densification growth of the samples, thereby enhancing their optical and thermal properties. Furthermore, the dark defects formation mechanism was proposed, that featuring numerous contact twins that alleviate competition from penetration twins and original grains during growth. (111) oriented diamond film exhibits numerous penetration twins, and holes form between them, showing the obvious dark defects. By suppressing dark features, 5inch diamond film with (220) preferred orientation was obtained. And it was used to be the semiconductor with H-termination and the MISFET devices were fabricated. Both the normally on and normally off devices were found on the diamond substrate. Among the devices, the highest current density of 2000 μm mA mm^{-1} was reported.

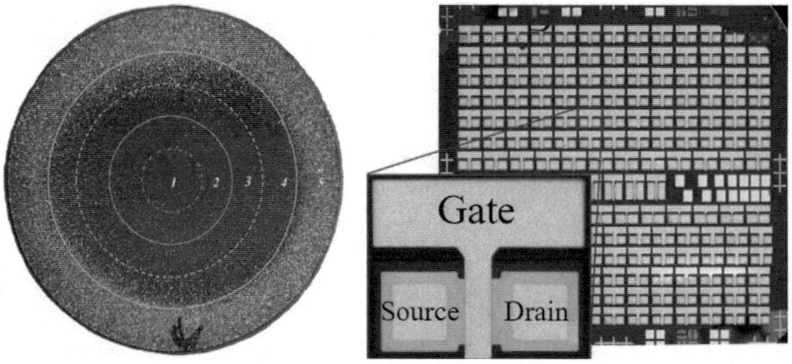

Fig. 1 5 inch diamond film with (220) preferred orientation and devices

Keywords: optical properties; preferred orientation; current density; MISFET devices

Two-Dimensional Semiconductor Materials and Devices for Advanced IC Manufacturing

Zhang Zheng[2]

Academy for Advanced Interdisciplinary Science & Technology, University of Science and Technology Beijing

E-mail address: zhangzheng@ ustb. edu. cn (Zhang Zheng)

Abstract: On the new track for 1nm advanced process non-silicon materials, two-dimensional transition metal dichalcogenides (TMDC) materials have emerged as one of the most promising non-silicon material systems due to their stable structure surpassing the size limit and semiconductor performance comparable to bulk silicon. The United States, the European Union, South Korea, and other countries have introduced national strategies to support research into non-silicon materials. In 2021, Samsung, IMEC, Intel, TSMC, and others successively unveiled technology roadmaps featuring two-dimensional TMDC materials for the next generation of 1 nm advanced process chips. The International Semiconductor Alliance Technology Roadmap also includes 2D materials as part of the 1. 5 nm process technology roadmap for commercial chips by 2028. Focusing on the application of 2D semiconductor materials in future advanced process integrated circuits, our team has conducted research on wafer-scale high-quality 2D semiconductor material preparation, design and fabrication of high-performance 2D van der Waals heterojunction electronic devices, and integration and interconnection of 2D semiconductor devices compatible with silicon-based processes. We achieved the single-crystal growth of 2-inch 2D semiconductor materials, obtained a two-dimensional semiconductor device approaching the Mott-Schottky theoretical contact performance limit, and developed a pico-watt-level pseudo-CMOS logic gate circuit based on molybdenum disulfide transistors.

Keywords: semiconductor material

Growth Mechanism of Photoresist-Assisted Patterned Vertical Aligned Carbon Nanotubes for Infrared Image Generation

Yuan Xiaolu[1,2], Huang Yidan[2], Ren Feitong[2], Hao Sijia[2], Feng Xurui[2], Wei Junjun[2], Chen Liangxian[2], Liu Jinlong[2], Li Chengming[2*], Wang Wenrui[1*]

[1] School of Mechanical Engineering, University of Science and Technology Beijing, No. 30, Xueyuan Road, Haidian District, Beijing, 100083, China

[2] Institute for Advanced Materials and Technology, University of Science and Technology Beijing, No. 30, Xueyuan Road, Haidian District, Beijing, 100083, China

E-mail address: chengmli@ mater. ustb. edu. cn (Li Chengming)

E-mail address: gmbitwrw@ ustb. edu. cn (Wang Wenrui)

Abstract: This study focuses on the growth mechanism of photoresist (PR) -assisted patterned vertical aligned carbon nanotubes (VACNTs). Controlled growth of patterned VACNTs was achieved on silicon-based silicon dioxide (Si/SiO_2) substrates using standard photolithography combined with magnetron sputtering and plasma-enhanced chemical vapor deposition (PECVD) techniques. The experiments revealed that at a low temperature of 650 ℃, the denatured PR formed wrinkled structures that enveloped the iron catalyst particles, suppressing their surface activity and preventing CNT growth. When the temperature was raised to 750 ℃, the PR decomposed, releasing metallic iron catalyst while simultaneously providing a carbon source, thus promoting the directional growth of carbon nanotubes. This mechanism provides a theoretical basis for suppressing thermal crosstalk in infrared imaging technology and simplifying pixel fabrication processes.

Keywords: VACNTs; patterned growth; photoresist; thermal crosstalk

2D Nanoelectromechanical Resonators with Thermal Treatment and Their Surface Effect

Yang Rui

Shanghai Jiao Tong University, China

E-mail address: rui. yang@ sjtu. edu. cn (Yang Rui)

Abstract: Ultra-thin 2D materials such as MoS_2 have excellent mechanical properties, such as ultra-low weight, high Young's modulus, and high strain limit, and only require picowatt (10^{-12} W) range of power to maintain strong and stable resonance, making them ideal for 2D NEMS resonators. However, the quality (Q) factor, electrical signal transmission characteristics, measurement and tuning methods of 2D NEMS resonators have not been fully studied and optimized, thus it is difficult to fully leverage the advantages of these NEMS resonators. In addition, while mechanical resonators have been used in sensors and radio-frequency (RF) signal processing previously, their memory and computing mechanisms are still not clear. To integrate 2D devices, the large-scale growth and transfer of 2D materials is important, but the fabrication of large-scale suspended 2D NEMS arrays is still challenging. To solve the problems, we study the thermal treatment and surface effects on damping and Q factor. In NEMS resonators, a larger Q indicates less energy dissipation, which leads to stronger oscillations at resonance under the same external excitation, and sharper resonant peaks with better frequency selectivity. We have achieved strain tuning of Q in 2D MoS_2 NEMS resonators with different boundary conditions, established a model about how tensile strain improves the Q, and experimentally measured strain tuning of Q by up to $\Delta Q/Q = 448\%$. This provides a new way to develop ultra-low-power, low-loss, and tunable devices for on-chip information processing. In addition, we discover the surface nonideality effect on Q factor, and optimize the Q factor in 2D NEMS resonators. Moreover, we fabricate large-scale arrays of 2D NEMS resonators based on thermal treatment, and develop NEMS memories based on thermal hysteresis effect. The results are important for on-chip sensing and computing based on 2D NEMS resonators.

Keywords: 2D; NEMS; resonator

Surface Modification and Applications in Electronic Devices of Diamond

Ren Zeyang[1,2], Zhang Jinfeng[1,2], Fu Yu[1], Su Kai[1], Hao Yue[1], Zhang Jincheng[1]

[1] The National Key Laboratory of Wide Bandgap Semiconductor Devices and Integrated Technology, Xidian University, Xi'an 710071, China

[2] Xidian-Wuhu Research Institute, Wuhu 241002, China

E-mail address: zeyangren@ xidian. edu. cn (Ren Zeyang)

Abstract: Diamond is a quintessential ultra-wide bandgap semiconductor, renowned for its ultra-wide bandgap, high carrier mobility, and exceptional thermal conductivity. These properties endow it with tremendous potential for applications in high-temperature, high-frequency, and high-power electronic devices. However, the development of diamond-based electronic devices has been hindered by the persistent challenge of low doping activation efficiency. Surface modification of diamond with elements such as nitrogen, hydrogen, oxygen, or silicon can alter its electron affinity, thereby tuning its surface electrical properties to meet the requirements for electronic device applications. This report outlines the methodologies for achieving n-type and p-type surface modifications of diamond and highlights recent advancements in electronic devices based on surface conductivity. Furthermore, the prospects and advantages of diamond surface conductive devices are discussed.

Normally-off Hydrogen-Terminated Diamond Field-Effect Transistor and Its Logic Inverter

LIANG Yuesong[1,2], WANG Wei[1,2*], WANG Hongxing[1,2]

[1] School of Electronic Science and Engineering, Xi'an Jiaotong University, Xi'an 710049, China

[2] Key Laboratory of Physical Electronics and Devices, Ministry of Education, Xi'an 710049, China

E-mail address: wei_ wang2014@ xjtu. edu. cn (WANG Wei)

Abstract: : Diamond, considered as the fourth-generation wide band gap semiconductor material, has excellent properties such as large band gap, high breakdown voltage, high mobility, and high thermal conductivity. It has broad application prospects in the fields of high frequency, high power, and high temperature. The diamond normally-off field effect transistor (FET) has the advantages of safety and energy saving. This article proposes the implementation of an enhancement-mode hydrogen terminated diamond FET using a low work function gate. The reason for the normally-off operation is that electrons flow into the hydrogen terminal diamond surface from the low work function material, compensating and depleting the two-dimensional hole gas (2DHG), thereby pinching the channel. The advantage of this technology is that it does not damage the characteristics of 2DHG such as carrier concentration and mobility, avoids the potential interface contamination problem of multi-layer gate deposition, has a simple manufacturing process, and is compatible with industrialized semiconductor processes, which will promote the development of enhancement single crystal diamond FETs. The directly coupled hydrogen-terminated diamond FET logic (DCHDFL) circuit is fabricated. The E-mode and D-mode FETs are assigned as driver and load devices of the DCHDFL circuit to achieve inversion characteristics. The E-mode FET showcases high I_{DSmax} of 53. 3 mA/mm, V_{TH} of −0. 8 V, low SS of 98 mV/dec and on/off ratio of 109, which enable input/output logic level matching with a low drive/load ratio of 1. 0. The peak gain of circuit increases from 12. 57 to 36. 3 V/V with V_{DD} ranging from −5 V to −25 V, which is the highest gain achieved of diamond inverters, due to the high on/off ratio and low SS of E-mode FET. This circuit exhibits proper functions up to 200 ℃, demonstrating a good thermal stability. These results indicate the great potential and possibilities for diamond smart power integrated circuit application.

Keywords: diamond; hnormally-off; inverter; transistor

Proceedings of IFHTSE 2025

The 30th World Congress of International Federation for Heat Treatment and Surface Engineering
August, 19-22, 2025, Suzhou, China

SiC-Diamond Composite: Surface Morphology, Interface Structure and Stress

Peng Yan
ShandongUniversity, China
E-mail address: pengyan@ sdu. edu. cn (Peng Yan)

Abstract: Diamond, with its exceptional thermal conductivity, serves as a critical heat dissipation material. Its use enables higher device and system power densities while enhancing stability and reliability during high-load operation. Demand is rapidly increasing for diamond-based thermal management solutions capable of performing in extreme and complex environments, such as those found in aerospace and defense military applications. To address these challenges, we have reported a new strategy: the direct deposition of polycrystalline diamond films onto 4H-SiC substrates using microwave plasma chemical vapor deposition (MPCVD). The synthesized diamond-SiC composite substrates were comprehensively characterized, including analysis of surface morphology, interface structure and crystal orientation. The composite substrate exhibits 1. 62 times higher thermal conductivity than a standard 4H-SiC substrate of equivalent thickness. Furthermore, temperature-dependent Raman spectroscopy was employed to track interfacial stress within the diamond-SiC composite as a function of temperature and film thickness, enabling accurate GPa-level stress quantification. The observed stress-temperature relationship exhibited nonlinear behavior, directly linked to the diamond film's quality and thickness. This study offers quantitative insights into the thermal stress mechanisms within SiC-diamond composite and establishes a design framework for optimizing these structures for heat sink applications.

Keywords: Silicon carbide; diamond; interface structure; stress quantification

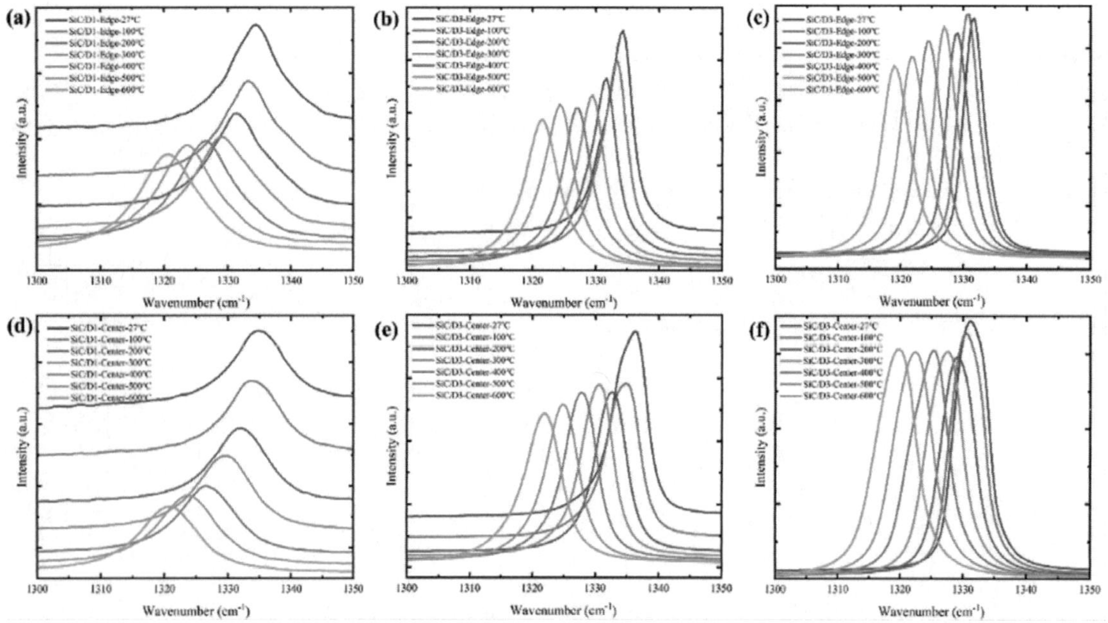

Fig. 1 Temperature-dependent Raman spectra at the center and edge of SiC / D1, SiC / D2 and SiC / D3

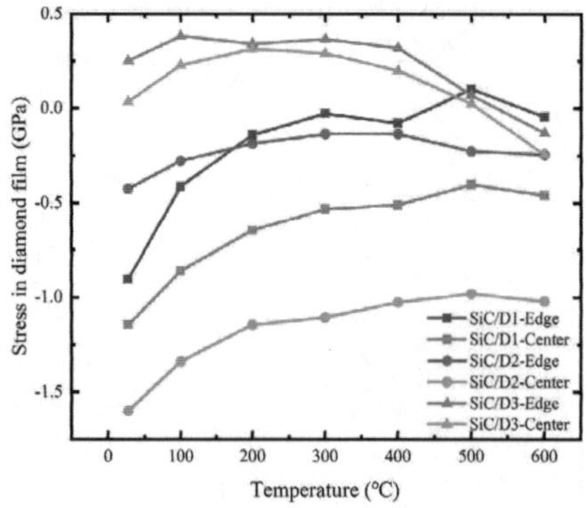

Fig. 2 Comparison of stress measured at the center and edge of the diamond film for samples as a function of temperature

MPCVD Diamond Boron Doping and Semiconductor Applications

Liu Benjian[a,b,*], Viktor Ralchenko[a,b], Andrey Bolshakov[a,b], Zhu Jiaqi[a,b], Mikhail Shestakov[c], Tomas Samuely[d]

[a] National Key Laboratory of Science and Technology on Advanced Composites in Special Environments
Harbin Institute of Technology Harbin 150080 China

[b] Zhengzhou Research Institute Harbin Institute of Technology Zhengzhou 450008 China

[c] Moscow Institute of Physics and Technology, Russia

[d] P. J. Šafárik University, 04001 Košice, Slovakia

E-mail address: liubenjian@hit.edu.cn (Liu Benjian)

E-mail address: mikle.shestakov.science@gmail.com (Mikhail Shestakov)

Abstract: Diamond is a wide bandgap semiconductor material with a bandgap width of up to 5.5 eV. It has high breakdown field strength, high thermal conductivity, electron mobility, hole mobility, and radiation resistance, as well as high and low temperature resistance, making it suitable for semiconductor devices in extreme environments. Diamond is usually non-conductive, but in order to achieve its semiconductor applications, we need to regulate its electrical properties. We have successfully controlled the p-type semiconductor properties of diamond MPCVD growth by introducing a solid-state boron source. By controlling the atmosphere, the doping concentration of boron impurities can be effectively regulated. Based on p-type diamond doping technology, we have developed multiple diamond devices, including nuclear cells, field-effect transistor devices, and Schottky diode devices, and verified their performance. We have obtained isotope battery devices with open circuit voltage up to 2V using heterojunction technology. A low subthreshold swing field-effect transistor device with a voltage below 60mV/dec was achieved using a negative capacitance oxide gate dielectric, and its NOT gate circuit was developed.

Keywords: MPCVD; diamond; boron doping; nuclear battery; MOSFETs; schottky diode

High-Temperature Remote Plasma Pretreatment for Interface State Suppression and Performance Enhancement in GaN Power Devices

Deng Kexin, Wang Xinhua, Huang Sen
Institute of Microelectronics, Chinese Academy of Sciences
E-mail address: wangxinhua@ime.ac.cn (Wang Xinhua)

Abstract: Interface states have consistently posed a core challenge in the research of III-N material systems, limiting the scaling and practical application of gallium nitride (GaN) -based power devices. During the transition from the off-state to the on-state, electrons trapped by deep-level interface states under high gate-drain voltage cannot be released promptly, leading to severe current collapse. Under high gate bias in the on-state, charging occurs at interface states beneath the gate. The dynamic trapping/emission processes at these gate-adjacent interface states during switching cause threshold voltage drift. Under gate overdrive conditions, electrons can be further injected into dielectric defects, resulting in more pronounced threshold voltage drift. Unlike traditional silicon (Si) semiconductors, high-quality oxide layers are difficult to prepare on GaN via high-temperature oxidation. Therefore, it is essential to develop effective methods to remove the low-quality native oxides from the GaN surface and deposit suitable dielectric layers. The native oxide layer on GaN is a mixed system containing various oxidation states. First-principles calculations indicate that Ga2O is the primary source of deep-level interface states. To address this, we developed a high-temperature plasma surface modification technique operating at 500℃. This technique efficiently processes the native oxides, particularly Ga2O, successfully restoring the step-flow morphology on the GaN surface. Both theoretical calculations and constant-capacitance deep-level transient spectroscopy (CCDLTS) measurements demonstrate that after removing Ga2O from the GaN surface, interface states located deeper than E_C-E > 0.4 eV within the bandgap are significantly suppressed. Applying this high-temperature plasma surface modification technique for GaN surface treatment prior to passivation dielectric deposition substantially improved the device's current collapse characteristics. In drain lag mode, the current collapse rate of field-plate-free devices under pulsed testing decreased from 82% to 33% (static drain bias of 100 V).

Keywords: GaN; interface state; power devices

Diamond (111) Facet Growth and Self-Aligned NV Centers for Quantum Precision Sensing

Zhang Sen[a,b*], Zhu Jiaqi[a,b], Mikhail Shestakov, Tomas Samuely

[a] National Key Laboratory of Science and Technology on Advanced Composites in Special Environments

Harbin Institute of Technology Harbin 150080 China

[b] Zhengzhou Research Institute Harbin Institute of Technology Zhengzhou 450008 China

[c] Moscow Institute of Physics and Technology, Russia

[d] P. J. Šafárik University, 04001 Košice, Slovakia

E-mail address: szhang@hit.edu.cn (Zhang Sen)

E-mail address: mikle.shestakov.science@gmail.com (Mikhail Shestakov)

Abstract: The development of robust and highly sensitive quantum sensors is paramount for advancements in diverse fields, from biomedical imaging to fundamental physics. Among the promising candidates, nitrogen-vacancy (NV) centers in diamond stand out due to their exceptional spin coherence properties even at room temperature. However, achieving optimal sensor performance critically depends on the precise control and alignment of these quantum defects within the diamond lattice. This work addresses this challenge by presenting a novel approach that leverages controlled diamond (111) facet growth to achieve the self-alignment of NV centers. By carefully tailoring the growth conditions, we demonstrate a method that inherently promotes the preferential incorporation and crystallographic alignment of NV centers along desired orientations. This technique significantly enhances the ensemble's spin coherence and optimizes the NV centers' sensitivity, overcoming limitations associated with randomly oriented defects. The successful realization of self-aligned NV centers through this targeted (111) facet growth paves the way for the development of high-performance diamond-based quantum sensors with unprecedented precision in detecting magnetic fields, electric fields, temperature, and strain, ultimately advancing the frontier of quantum precision sensing.

Keywords: MPCVD; diamond; NV center; (111) growth

Leveraging Large Language Models for Multimodal Data Integration in Heat Treatment Research: MHTR-Agent Development and Applications

Wang Xuefei[1], Luo Chunyang[1], Wang Haojie[2], Wang Zhaodong[1]

[1] State Key Laboratory of Digital Steel, Northeastern University, Shenyang, 110819, China

[2] School of Mechanical Engineering, Shenyang University of Technology, Shenyang 110870, China

E-mail address: is_ wxf@ 163. com (Wang Xuefei)

E-mail address: luochunyang2020@ 163. com (Luo Chunyang)

Abstract: The rapid development of large language models (LLMs) is transforming artificial intelligence applications across various fields, including materials science. This paper introduces a novel multimodal heat treatment research agent (MHTR-Agent) that leverages LLMs to enhance data integration and model analysis in the heat treatment domain. By employing OpenAI's API and natural language processing techniques, MHTR-Agent efficiently extracts key features from unstructured data, revealing trends and insights in the " composition-microstructure-property" relationships. Utilizing carburized steel as a case study, the agent integrates multimodal data-comprising metallographic images, process parameters, and material compositions—to accurately predict hardness and wear performance. The model's mean absolute error (MAE) was reduced to 2. 88, demonstrating superior precision over traditional methods. Additionally, the RAG-based intelligent assistant, MHT-Advisor, significantly optimizes process parameters, mitigating hallucinations typical of generic LLMs. This research not only enhances data analysis and model construction efficiency but also sets a new paradigm for AI-assisted scientific discovery and complex process development in materials science. Future work will focus on expanding the framework's adaptability and updating LLMs with incremental learning techniques for broader industrial applicability.

Keywords: large language models; multimodal learning; heat treatment; performance prediction

Numerical Simulation and Experimental Study of Thermal Compression Deformation Inhomogeneity of an Eutectoid Pearlite Steel

Liu Haonan, Fu Wantang

State Key Laboratory of Metastable Materials Science and Technology, College of Materials Science and Engineering,

Yanshan University, 438W. Hebei St., Qinhuangdao, Hebei Province, 066004, P. R. China

E-mail address: wtfu@ysu.edu.cn (Fu Wantang)

Abstract: It is one of the most commonly used methods to study the evolution law of microstructure and properties of metal materials in the process of rolling, forging, deformation and heat treatment by using hot working simulation machine. The closeness between the physical simulation results and the actual deformation conditions (or the deformation conditions set by the test) will directly affect the reference significance of the physical simulation. In the process of hot compression deformation of cylindrical specimens, due to the friction between the indenter and both ends of the specimen, the specimens tend to appear waist-like after deformation, that is, uneven deformation occurs. Especially when the deformation is large, this situation is more prominent in the longitudinal section of the sample. The precipitation kinetics and deformation recrystallization process of many materials are significantly affected by the differences in thermal deformation conditions, resulting in different microstructure and properties of the corresponding regions. Therefore, it is of great significance to reversion the original meaning of physical simulation to deeply study the deformation inhomogeneity and its rule on the longitudinal section of the compressed specimen, and clearly give the region which is basically consistent with the set deformation conditions. The microstructure and properties of samples at different positions after hot compression deformation are often very different, which is mainly related to the differences in deformation conditions or the non-uniformity of deformation. Although these studies also involve numerical simulation of the deformation process, the relationship between the actual strain variable and the test set point has not been clearly given and the effect of strain rate has not been taken into account. In this paper, the finite element analysis of the strain field distribution on the longitudinal section of the specimen after hot compression is combined with the microstructural characteristics of the corresponding region to determine the area where the actual deformation conditions on the specimen are consistent with the set conditions of the test, which provides a theoretical basis for further improving the reliability of the material physical simulation. In this paper, the thermal compression deformation process of a eutoid pearlitic steel at 700℃ from 0.1 to 10 s^{-1} is numerically simulated. The distribution of the isoeffect variation and strain rate in the longitudinal profile of the specimen after deformation is analyzed. The area where the actual deformation conditions are in good agreement with the experimental conditions is determined and the corresponding microstructure changes are verified. The results show that the equivalent strain and strain rate of the core of the specimen are about 40% larger than the set value, while the face value of the end is 1/5 ~ 1/4 of the set value. The actual strain and strain rate in the region of 2/5-3/5 from the specimen end face to the center are basically consistent with the deformation conditions set by the test, and the consistency is the best at 1/2.

Keywords: thermal compression; deformation Inhomogeneity; numerical simulation; microstructure

Study on Isothermal Precipitation Kinetics of Nitride in a High Nitrogen Austenitic Stainless Steel

Liu Haonan, Fu Wantang

State Key Laboratory of Metastable Materials Science and Technology, College of Materials Science and Engineering, Yanshan University, 438W. Hebei St., Qinhuangdao, Hebei Province, 066004, P. R. China

E-mail address: wtfu@ysu.edu.cn (Fu Wantang)

Abstract: Steel with nitrogen content of more than 0.4wt % is known as high-nitrogen steel, which is a new material with excellent performance and broad application prospects, among which high-nitrogen chromium-manganese austenitic steel (such as 18Mn18Cr0.5N steel) is currently one of the materials with the best match of strength and toughness in all metal materials in the world. During the " Seventh Five-Year Plan" period, 18Mn18Cr0.5N steel was introduced to manufacture components such as generator rotor protection rings. However, in the actual production of the protective ring, this kind of steel in the high-temperature deformation (such as forging, hot bending, etc.) and the cooling process after deformation, if the carbide or nitride are precipitated from the austenite, its performance will be greatly affected. However, the study on the kinetics, process and type of precipitation of carbide (or nitride) in these steels is not so clear up to now. Therefore, the isothermal precipitation of nitride in 18Mn18Cr0.52N steel was studied in this work to provide a theoretical basis for the formulation of hot deformation and direct treatment process after hot deformation. In present work, the isothermal precipitation of nitride from austenite in 18Mn18Cr0.5N steel at 800 ~ 950 ℃ has been studied by means of metallography and X-ray diffraction (XRD). The results show that for 18Mn18Cr0.5N steel, there is no nitride precipitation above 950 ℃, the nose tip temperature of Cr2N isothermal precipitation is about 860 ℃, and the critical cooling rate without the precipitation is about 0.75 ℃/s. The quantitative relationship between Cr2N precipitation (f/%) and austenite lattice constant ($a\gamma$/nm) is as: $-9.2312 \times 10^4 + 5.13683 \times 10^5 a_\gamma - 7.14503 \times 10^5 a_\gamma^2$.

Keywords: high nitrogen austenitic steel; isothermal precipitation kinetics; nitride

Research and Application Progress of Digital Twin Technology in Heat Treatment

Shang Yujiao, Jiang Chao, Li Xianjun, Zhang Wenliang, Luo Ping, Dong Chengzhi, Zhang Minghao

China Academy of Machinery Beijing Research Institute of Mechanical & Electrical Technology Co., Ltd., China

E-mail address: 382888241@qq.com (Shang Yujiao)

Abstract: With the continuous advancement of intelligent manufacturing, Industry 4.0 and the concept of lean production, as well as the popularization of technologies such as big data and artificial intelligence (AI), the manufacturing industry is exhibiting a highly digitalized and automated trend, with the production process and mode gradually becoming intelligent and integrated. Traditional heat treatment processes and equipment have become inadequate to meet the changes and upgrades of the industry in the new era. Therefore, digital twin technology has emerged as a critical breakthrough in transitioning heat treatment from empirical-driven approaches to model-driven methodologies. This paper comprehensively examines the research and application progress of digital twin in heat treatment from the aspects of technology and equipment, and it identifies current challenges and proposes future development directions based on existing limitations.

Keywords: digital twin; heat treatment; digitalization; intelligent manufacturing

Utilization of PINN Supporting Heat Treatment Processes

Imre Felde[1], Shi Wei[2]

[1] Óbuda University, Budapest, Hungary

[2] Tsinghua University, Peking, P. R. China

Abstract: Physics-informed neural networks (PINNs) have become increasingly popular in several engineering domains due to their efficiency in addressing real-world problems characterized by noisy data and occasionally incomplete physics information. In the realm of PINNs, automatic differentiation plays a pivotal role in evaluating differential operators without succumbing to discretization errors. Furthermore, a multi-task learning approach is defined to concurrently accommodate observed data fitting and adherence to the fundamental governing principles of physics. In this context, wepresent the application of PINNs to heat treatment operations, with a specific focus on handling realistic conditions that are challenging for traditional computational methods. The estimation of unknown thermal boundary conditions on surfaces is facilitated by predicting them through sparse temperature measurements. We illustrate the applicability of PINNs through several industrial applications related to quenching, underscoring the effectiveness of neural networks in tackling complex heat transfer problems encountered in industrial settings.

Quality Fluctuations Simulation Method by Using Multi Agent System in Large Scale Load Setting of Heat Treatment

Tsuyoshi Sugimoto

National Institute of Technology, Asahikawa College, Mechanical Systems Engineering Department,

Shunko-dai 2Jo 2chome, 1-6, Asahikawar, Hokkaido, 071-8142, Japanl

E-mail address: t_sugi@ asahikawa-nct. ac. jp (Tsuyoshi Sugimoto)

Abstract: As the trend toward CASE in automobiles, ships and aircraft field, there is a demand for ultra-long life materials, high precision shapes, and rapid response to new design shapes in power transmission parts. On the other hand, the number of materials and heat treatment engineers is decreasing due to the declining birthrate, and the retirement of experienced engineers is making it difficult to manufacture products that rely on intuition and skill. To address this issue, research and development of model-based development method (MBD) and process integration (PI) for heat treatment is being promoted every day in the materials and process fields, and the same is true for the heat treatment process, which is the basis of the strength of steel parts. In MBD, heat treatment conditions are determined by heat treatment simulation, so simulations with a level of quality that can replace experiments are required. We confirmed the controlling factors of quality through heat treatment simulation analysis and found that cooling conditions, materials, and effects from previous processing processes are important factors. In this study, we clarified methods for evaluating the influence of each of these factors, investigated methods for comprehensively evaluating these sources of variation, and established a quality prediction method by applying the stochastic finite element method and conducting simulations using a multi-agent method that uses a quality genome, in which each part is treated as an agent with independent will.

Keywords: quality fluctuations; quenching; vapor film; material composition; variation

Robust Additive Manufacturable Ni Superalloys Designed by the Integrated Optimization of Local Elemental Segregation and Cracking susceptibility criteria

Yu Hao[1,*], Li Jinguo[2], Sybrand van der Zwaag[3], Xu Wei[1]

[1] State Key Laboratory of Rolling and Automation, Northeastern University, Shenyang 110819, China

[2] Institute of Metal Research, Chinese Academy of Sciences, Shenyang 110016, China

[3] Faculty of Aerospace Engineering, Delft University of Technology, 2629 HS, Delft, the Netherlands

E-mail address: yuhao@ral.neu.edu.cn (Yu Hao)

Abstract: To achieve an effective design of additively manufacturable Ni superalloys with decent service performance, a hybrid computational design model has been developed, where the strategy to tailor local elemental segregations was integrated within a scheme of minimizing the cracking susceptibility. More specifically, the phase boundary of primary NbC/γ matrix was introduced into the design routine to tune the spatial distribution of critical solutes at an atomic scale, thereby inhibiting the formation of borides and segregation-induced cracking. Based on the output of the design, new grades of Ni superalloy have been developed with excellent additive manufacturability, as confirmed by the robustness of printing parameters in fabricating low-defect-density samples. The capability of the phase boundaries to evenly distribute boron atoms was validated experimentally, and the cracking induced by uncontrolled boron segregation at grain boundaries was effectively prevented. The newly designed alloys showed good tensile properties and decent oxidation resistance at different service temperatures, which are comparable to those of conventionally produced superalloys. The finding that phase boundaries can be employed to prevent undesirable clustering of boron atoms can be extended to manipulate the distributions of other critical elements, which provides a new path for designing novel Ni superalloys with balanced printability and mechanical properties.

Keywords: additive manufacturability; boron segregation; computational design; phase boundaries; ni superalloys

Kinetics of Phase Transformation Revealed by a Distribution of Rate Constants

Zhang Yanxiang, Yan Mufu*

Harbin Institute of Technology, Harbin, Heilongjiang, China

E-mail address: hitzhang@ hit. edu. cn (Yan Mufu)

Abstract: The precipitation kinetics for carbides and/or nitrides during themochemical treatments has often been described by the so-called Johnson-Mehl-Avrami-Kolmogorov (JMAK) equation. The JMAK equation also provides the foundation for the construction of TTT diagram and CCT diagram. However, as the JMAK equation usually assume a constant rate coefficient for the specific transformation procecss, limited information for the evolution of rate coefficients can be resolved. Here we propose a concept of Distribution of rate constants (DRC), and a method to obtain the DRC from experimental measurements is proposed. Several examples including carburizing and nitriding are given for resolving the precipitation kinetics for carbides and/or nitrides and/or pores. We show that a constant rate constant (e. g. nucleation rate) is not rational for these processes.

Keywords: phase Transformation; TTT diagram; JMAK

Alloy Design Guided by Physical Metallurgical Knowledge Graph

Wang Chenchong

Northeastern University, China

E-mail address: wangchenchong@ral.neu.edu.cn (Wang Chenchong)

Abstract: Traditional alloy design typically relies on trial-and-error methods, which are time-consuming, costly, and inefficient. Whilst physical metallurgical (PM) models offer some predictive capabilities, their reliability is limited by errors accumulating across space scales. In recent years, artificial intelligence (AI) technologies have been increasingly applied to materials design, enhancing the efficiency of performance prediction. However, conventional AI methods usually depend on large-scale experimental data and lack a deep understanding of PM principles. As a result, they struggle to effectively address the complex coupling among composition, processing, and performance, and exhibit limitations in generalization and physical interpretability. To overcome these challenges, this study proposes a novel framework that integrates physical metallurgical knowledge graphs (PMKGs) with graph neural networks (GNNs) to efficiently predict the ultimate tensile strength (UTS) and total elongation (TEL) of quenching and partitioning steels. using genetic algorithms for dual-objective optimization. Compared to traditional AI models, this approach significantly improves prediction accuracy and robustness, especially in TEL prediction, where the R^2 value increases by approximately 15%. Moreover, the method enables an effective balance between UTS and TEL, leading to the design of alloys with superior overall performance. The designed alloys, with a composition of approximately 0.3 wt.% C, 3 wt.% Mn, 1.2 wt.% Si, and minor amounts of Cr and Al, achieve a UTS exceeding 1500 MPa and TEL near 20%, aligning with PM principles and validating the rationality and feasibility of this method. This study offers new insights into applying AI in complex multi-objective alloy design, highlighting the potential of integrating expert knowledge with GNNs.

Keywords: alloy design; knowledge graphs; physical metallurgy; dual-objective optimization; graph neural networks

Precipitation and Mechanical Property Prediction of Mg Alloys during Aging Treatment

Luo Qun

School of Materials Science and Engineering, Shanghai University, Shanghai 200444, China

Abstract: Aging treatment is a crucial process for strengthening magnesium alloys. However, the nanoscale dimensions and metastable characteristics of precipitates formed during aging pose challenges in predicting both the precipitation process and mechanical properties. As a key system for high-strength magnesium alloys, Mg-Gd alloys exhibit superior comprehensive performance. In this work, the evolution of β′ phase in Mg-Gd alloys was systematically characterized using high-angle annular dark-field scanning transmission electron microscopy (HAADF-STEM). By integrating the metastable phase boundaries and precipitation kinetics of β′, a thermodynamic description of β′ was established. This approach overcomes the limitations of traditional methods, which are inapplicable to metastable precipitates. Precipitation kinetics simulations were employed to elucidate the nucleation, growth, and coarsening behaviors of β′, clarifying the dynamic mechanisms underlying age strengthening. A quantitative relationship was established among the precipitation kinetics of β′ phase in Mg-Gd alloys, alloy composition, aging temperature, and time. Furthermore, by coupling strengthening models, a quantitative correlation linking "alloy composition-aging process-precipitates-mechanical properties" was developed. The predicted temperature-time-hardness (TTH) relationship provides intuitive theoretical guidance for designing aging processes in Mg-Gd-based alloys.

Keywords: precipitation; aging treatment; Mg-Gd alloys

A Metallographic Image Translation Method from Optical to Scanning Electron Microscopy for GCr15 Bearing Steel

Sun Yixiao[1], Wang Jiangang[2], Lian Xiaolong[2], Zhang Wenyu[1], Wang Mingzhou[3], Liu Xindong[3], Ju Dongying[3]*

[1] University of Science and Technology Liaoning, 114051, School of Electronic and Information Engineering, Anshan, Liaoning, China

[2] Hebei University of Science and Technology, 050018, College of Material Science and Engineering, Shijiazhuang, Hebei, China

[3] Zhejiang XCC Group Co., Ltd., 312500, Xinchang, Zhejiang, China

E-mail address: robinsunyixiao@163.com (Sun Yixiao)

E-mail address: dyju.sitec@gmail.com (Ju Dongying)

Abstract: Microstructural analysis is essential for controlling the performance and quality evaluating of steel materials. During heat treatment, metallographic imaging and interpretation are critical for monitoring microstructural evolution, optimizing process parameters, and supporting material grading. The resolution and clarity of metallographic images directly affect the accuracy of microstructure identification and quantitative evaluation. Optical microscopy (OM) is widely employed in industrial settings for observing the microstructure of heat-treated steel due to its low cost, high accessibility, and ease of use. However, OM images often suffer from limited resolution, low contrast, and poor hierarchical representation, hindering the clear identification of key microstructural features such as grain boundaries, carbides, and retained austenite. This compromises the reliability of quality evaluation after heat treatment. Incontrast, scanning electron microscopy (SEM) provides higher spatial resolution and superior structural fidelity, offering more detailed and realistic images of microstructures. Nevertheless, SEM imaging requires expensive equipment, complex sample preparation, and long processing time, impeding its practicality for high-throughput industrial inspection. To address this limitation, A novel deep learning-based image translation method was proposed for reconstructing SEM-style high-resolution images from low-quality optical inputs. This enables more accurate analysis of steel microstructures without relying on direct SEM imaging. In this study, GCr15 high-carbon chromium bearing steel is used as the research object. A paired metallographic image dataset was constructed, consisting of 240 spatially aligned image pairs with optical images acquired at 500 × magnification and corresponding SEM images at 1000 × and 2000 × magnifications. Based on this dataset, A deep learning-based image-to-image translation method is proposed, using a conditional generative adversarial network (cGAN). The generator is designed with an encoder-decoder architecture to learn the mapping from optical metallographic images to their SEM-style counterparts. The discriminator, jointly trained, guides the generator to produce outputs closely resembling real SEM images. To preserve key microstructural details, A structural similarity (SSIM) loss is incorporated into the training objective. This enhances the network's ability to retain grain boundaries, phase interfaces, and morphological features. Experimental results show that the proposed method outperforms traditional enhancement and translation approaches in terms of structural accuracy, texture fidelity, and visual coherence. Stable and consistent outputs are also maintained in areas with complex or blurry structures. This study presents an intelligent and cost-effective approach for enhancing metallographic images of heat-treated steel. The proposed method supports intelligent metallographic analysis, automated grading, defect detection, and data-driven materials research, contributing to the advancement of digital and intelligent manufacturing in the steel industry.

Keywords: optical metallography; scanning electron microscopy; image-to-image translation; microstructure analysis; deep learning

Development of a Large Language Model-Based Manufacturing Execution System for Heat Treatment

Sun Yixiao[1], Liu Chao[2], Xu Rongbin[2], Yu Menghong[2], Ju Dongying[2*]

[1] University of Science and Technology Liaoning, 114051, School of Electronic and Information Engineering, Anshan, Liaoning, China

[2] Zhejiang XCC Group Co., Ltd., 312500, Xinchang, Zhejiang, China

E-mail address: robinsunyixiao@163.com (Sun Yixiao)

E-mail address: dyju.sitec@gmail.com (Ju Dongying)

Abstract: Metal heat treatment is an essential process across several critical industries, including mechanical manufacturing, automotive, aerospace, and energy sectors. The technique directly affects the microstructure and service performance of materials, serving as a foundation for ensuring the reliability of high-end equipment and the precision of advanced manufacturing. In heat treatment production management, it is necessary to analyze a large volume of multi-source heterogeneous data, including material compositions, specifications, and standards embedded in order and product information; parameterized process paths and equipment runtime data; as well as multi-dimensional quality indicators such as metallographic images, hardness values, and deformation measurements. These data are highly fragmented, semantically complex, and dynamically evolving. Traditional Manufacturing Execution Systems (MES) typically provide basic capabilities such as data collection, archiving, and visualization, they lack semantic understanding and integrated reasoning capabilities. As a result, decision-making remains heavily dependent on human expertise,, posing significant limitations for intelligent workshop management. To address these limitations, this study proposes and implements a heat treatment MES enhanced by a large language model. The system is powered by a locally deployed Qwen3.0-30B model and constructs an intelligent agent with task understanding, semantic extraction, and autonomous reasoning capabilities. Utilizing the Model Context Protocol (MCP), the language model can securely access modular MES data through a standardized interface, accessing order information, work orders, processing routes, equipment logs, inspection records, and quality assessments. The system enables natural language-based production data querying. Furthermore, unlike traditional manual query-based approaches, the model autonomously poses questions, aggregates, and invokes data based on real-time production context, achieving self-driven production management and optimization. The developed system supports various intelligent functionalities, including natural language querying of production data; automatic identification and warning of order delays and delivery conflicts; proactive detection of abnormal quality trends and cause inference; and risk prediction based on multi-source fusion of equipment operating status. In practical deployment at a high-precision bearing manufacturing enterprise, the system achieved a 90% success rate in natural language query tasks and an average recall of 70% in abnormal scenario identification. These outcomes illustrate the system's substantial contribution to improving human-computer interaction efficiency and minimizing production risks. This research represents the first to integrate large language models, characterized by advanced semantic understanding and reasoning capabilities, into heat treatment MES environments. By addressing traditional system limitations, which primarily depend on manual input, this study proposes a unified architecture and deployment framework for intelligent manufacturing management. Ultimately, these findings provide critical decision-making support for heat treatment processes, surface engineering, and essential component manufacturing, marking a significant advancement from mere digital transparency toward enhanced semantic understanding and intelligent operational interventions.

Keywords: heat treatment manufacturing execution system; large language model; intelligent manufacture; multi-source heterogeneous data

An Approach to Microstructure Recognition of Spheroidizing Annealed for GCr15 Steel Based on Semi-Supervised Learning

Liu Lichuan[1], Zhang Yujun[1*], Li Xusheng[2], Sun Yixiao[1], Zhang Kunlong[2], Ju Dongying[1*]

[1] University of Science and Technology Liaoning, 114051, School of Computer and Software Engineering, Anshan, Liaoning, China

[2] Zhejiang XCC Group Co., Ltd., 312500, Xinchang, Zhejiang, China

E-mail address: 1997zyj@163.com (Zhang Yujun)

E-mail address: dyju.sitec@gmail.com (Ju Dongying)

Abstract: This study proposes a metallographic image recognition method based on semi-supervised learning, to identify the microstructure of GCr15 steel after spheroidizing annealing. GCr15 steel is a typical high-carbon chromium bearing steel, widely used in the manufacture of high-precision, high-load bearings due to its excellent hardness and wear resistance. Among the various heat treatment processes, spheroidizing annealing plays a crucial role. It not only significantly improves the machinability of steel but also facilitates the formation of a uniform and refined initial microstructure. This lays a stable and reliable foundation for subsequent quenching and tempering processes. Therefore, accurate identification of the microstructure after spheroidizing annealing is of great significance for ensuring product quality and optimizing process parameters. In practical industrial production, metallographic image analysis serves as a critical tool for microstructure evaluation and quality control. Traditionally, this task has relied on manual inspection by experienced metallographers. However, such methods are time-consuming, labor-intensive, and prone to subjective bias, leading to inconsistency and low efficiency. With the growing demand for intelligent manufacturing and quality traceability, there is an urgent need for efficient, reliable, and highly automated image analysis tools. In recent years, artificial intelligence, particularly deep learning, has shown great potential in image classification, segmentation, and pattern recognition, making automated microstructure recognition based on computer vision a research hotspot. Nevertheless, although supervised learning methods have achieved remarkable results in image recognition tasks, their reliance on large quantities of high-quality labeled data limits their widespread application in industrial scenarios. To address this bottleneck, this study introduces a semi-supervised learning strategy that leverages both limitedlabeled samples and large amounts of unlabeled data. By integrating techniques such as pseudo-label generation and consistency regularization, the method effectively exploits the underlying structural information in unlabeled data, thereby improving model generalization and reducing dependence on manual annotation. During training, this study systematically compares multiple mainstream semi-supervised learning algorithms and evaluates their performance on a dataset of spheroidized microstructures. Considering the characteristics of metallographic images, such as blurred boundaries and class imbalance, the network architecture, loss functions, and data augmentation strategies are optimized accordingly. In addition, detailed ablation experiments are conducted to validate the individual contribution of each module and assess the model's performance and generalization ability from multiple perspectives. To further enhance model performance, the study incorporates advanced consistency regularization methods to improve robustness against data perturbations, and effectively utilizes unlabeled data via high-confidence pseudo-labels. The network architecture employed in this study is capable of learning both coarse-and fine-grained features, thereby enabling more accurate localization of microstructural boundaries. The model is trained and validated on high-resolution metallographic images to ensure the authenticity and representativeness of industrial samples. This method not only outperforms several existing baseline models, but also shows good transferability across different steel grades, providing a robust and scalable solution for practical metallurgical inspection. Experimental results show that the proposed method outperforms traditional fully supervised models across multiple evaluation metrics, particularly under low-label conditions, demonstrating superior robustness, generalization, and scalability. Moreover, the method significantly reduces the need for manual annotation, offering promising industrial application potential. In summary, this study demonstrates the feasibility and effectiveness of applying semi-supervised learning to metallographic image recognition, offering a promising direction for intelligent microstructure analysis in the field of materials science.

Keywords: GCr15 steel; spheroidizing annealing; metallographic image; semi-supervised learning; microstructure recognition; deep learning

Multiscale Design for High-Performer Metallic Alloys with Synergistic Improvement of Strength and Ductility

Du Jinglian
Northwestern Polytechnical University
E-mail address: dujl666@126.com (Du Jinglian)

Abstract: Following the classical nucleation and growth theory in combination with atomistic simulations, in this work, the underlying mechanism behind synergistic improvements of strength and ductility for high-performer alloys is investigated from the thermo-kinetic synergistic perspective. Bridged by the key thermo-kinetic parameter of phase transition and deformation involved in materials processing, quantitative models for the correlation between microstructure and strength-ductility were established, based on which potential physical origins for the exclusive phenomenon of strength and ductility were clarified. Accordingly, a multiscale thermo-kinetic design scheme for synergistic improvement of strength and ductility was proposed by coupling the driving force and energy barrier for defect structural formation. This scheme was successfully confirmed in the heat-treatable aluminum alloys. Our investigation provides innovative perspective for synergistically improving the strength and ductility of metallic alloys, and thus offer theoretical guidance for the quantitative connections among alloys composition, processing parameters, microstructural parameters and mechanical performances of metallic materials.

Keywords: metallic alloys; strength-ductility; thermo-kinetic synergy

Simulation of Microstructure and Texture Evolutions During Cold Rolling and Annealing of Ferritic Stainless Steel

Zhang Chi, Zhang Liwen, Song Kangjie

School of Materials Science and Engineering, Dalian University of Technology, Dalian 116024, Liaoning, China

E-mail address: zhangchi@ dlut. edu. cn (Zhang Chi)

Abstract: The formability of ferritic stainless steel is strongly influenced by its grain size distribution and crystallographic texture. The microstructure and texture evolutions during cold rolling and annealing exhibit complex behaviors along with the initial microstructure, texture and processing parameters. In this work, an integrated crystal plasticity (CP) and cellular automaton (CA) model was developed to investigate the microstructure and texture evolutions of ferritic stainless steel during cold rolling and subsequent annealing. To address convergence issues under high deformation, a remeshing method compatible with CP was introduced to transfer inhomogeneous microstructural information to the undeformed configuration. The cold rolled microstructure, subgrain structure, texture and dislocation density was simulated using the CP model. Then the deformed polycrystalline structure was used as the initial microstructure for recrystallization simulation using the CA model. The CA framework incorporated nucleation criteria based on critical dislocation density and grain boundary migration dynamics governed by disorientation. The microstructure and texture evolutions of interested grains with typical initial orientations were tracked by the developed method. Furthermore, the effects of initial orientation, grain morphology and interaction between neighbor grains on the microstructure and texture evolutions were analyzed.

Artificial Intelligence Technology Driving Intelligent Development of Heat Treatment: Research Progress and Applications

Zhu Jia, Su Miao, Zhang Tian, Chen Yi, Jia Yunke

China Academy of Machinery Beijing Research Institute of Mechanical & Electrical Technology Ltd., Beijing 100083, China

E-mail address: jsrclzj@vip.sina.com (Zhu Jia)

E-mail address: jsrclsm@vip.sina.com (Su Miao)

Abstract: As a core process for material performance control, heat treatment is undergoing intelligent transformation driven by artificial intelligence (AI) technology. This paper reviews the AI technology developments and applications in heat treatment, covering on key scenarios such as material design, properties prediction, process optimization, defects detection and identification, equipment intelligence and production managements. Forcing on the cases of AI system driven by multimodal large models, AI technology can play an important role in the development of heat treatment technology, such as multi-phase and multi-scale synergistic regulation of microstructure and properties of materials, full-flow process control of heat treatment distortion, high efficiency-high precision-high reliability green heat treatment, online dynamic control of heat treatment quality and energy efficiency optimization scheduling of heat treatment production, etc. Finally, the key challenges and future directions to achieve deep integration of multimodal large models and green manufacturing are discussed.

Keywords: artificial intelligence; multimodal large models; heat treatment; green manufacturing.

Simulation of Microstructure Evolution for SA508-3 Steel Hot Compression Bonding Joints During Aging Treatment

Yang Xingyun*, Tong Daming, Gu Jianfeng

School of Materials Science and Engineering, Shanghai Jiao Tong University, China

E-mail address: 16340824150548@ sjtu. edu. cn (Yang Xingyun)

E-mail address: daming_ tong@ sjtu. edu. cn (Tong Daming)

Abstract: A mesoscale cellular automata (CA) model was developed to simulate the microstructural evolution during aging treatment for SA508-3 steel hot compression bonding. The main mechanisms for the deformed joints undergoing aging treatment, including static recovery (SRV), static recrystallization (SRX), meta-dynamic recrystallization (m-DRX) and grain coarsening are all integrated in the modelling framework. The grain morphology, recrystallization fraction and interfacial bonding rate are obtained during aging treatment. The effect of aging temperature and pre-strain are investigated by the proposed CA model, showing that the aging temperature is significant, while the pre-strain is less on the recrystallization behaviour during aging treatment. The simulated results demonstrate a good consistency with the experimental data in terms of the average grain size, indicating that the developed model can provide an insight into how the complicated recrystallization processes during aging treatment affect the overall bonding process and enable quantitative prediction of the microstructural evolution and interfacial bonding rate for SA508-3 steel hot compression bonding joints during aging treatment.

Keywords: cellular automaton; meta-dynamic recrystallization; mesoscale modeling; SA508-3 steel; hot compression bonding

Prediction and Application of Steel Material Parameters Based on Data-Driven

Xiu Yongshuai[1], Deng Xiaohu[1*], Sun Yixiao[2], Tian Ruqiang[3], Li Xusheng[3], Ju Dongying[3]

[1] Tianjin University of Technology and Education, 300222, School of Mechanical Engineering, Tianjin, China

[2] University of Science and Technology Liaoning, 114051, School of Electronic and Information Engineering, Anshan, Liaoning, China

[3] Zhejiang XCC Group Co., Ltd., 312500, Xinchang, Zhejiang, China

E-mail address: 15506435784@163.com (Xiu Yongshuai)

E-mail address: dengxh@tute.edu.cn (Deng Xiaohu)

Abstract: The accurate acquisition of material parameters is a fundamental prerequisite for building finite element analysis (FEA) models of steel heat treatment. The reliability of simulation results and their applicability in engineering contexts are directly influenced by the precision of these parameters. However, traditional methods of obtaining such data face significant challenges. Experimental measurements are time-consuming, costly, and often unable to comprehensively cover the full range of chemical compositions. On the other hand, empirical models based on simplified assumptions generally suffer from limited generalization ability, making them unsuitable for real-time or high-throughput industrial applications. Therefore, given the challenges posed by the complexity and variability of industrial heat treatment scenarios, it is imperative to develop parameter prediction models that offer both high accuracy and rapid responsiveness. This model can provide real-time and reliable input data to support finite element simulations and improve the overall efficiency and reliability of the heat treatment process. In the present work, we propose a model integrating data-driven algorithms to enhance both the accuracy and efficiency of key steel material parameter prediction. The goal is to improve the prediction accuracy and efficiency of key steel properties. Ahigh-quality and structured training dataset is constructed by combining simulation data from computational materials software with experimental measurements. This dataset contains thousands of samples. It systematically covers the elemental composition, typical microstructural states, and temperature range of steel materials. It possesses statistical significance and robustness to noise, providing a reliable foundation for model generalization. Input features include 18 elemental compositions (such as Fe, C, Si, Mn, Cr), temperature, and three typical microstructures (austenite, bainite, and martensite). Microstructures are encoded using binary indicators (0/1). The prediction targets are three key performance indicators: thermal conductivity, specific heat capacity, and yield strength. To assess predictive performance, five machine learning models were evaluated, including Random Forest, Gradient Boosting, XGBoost, Lasso Regression, and Ridge Regression. Among them, the XGBoost model demonstrated superior performance across all evaluation metrics. After hyperparameter optimization, the XGBoost model achieved a coefficient of determination (R^2) exceeding 0.99 and an experimental validation error of less than 0.5%. In addition, we also developed a Steel Material Data Management System (S-MDMS). The system integrates multidimensional information such as material composition, microstructure distribution, thermophysical parameters, and phase transformation behavior. It supports full lifecycle functions, including data collection, validation, query, visualization, and model feedback. This significantly enhances data quality, traceability, and reusability. It also provides a solid platform for collaborative research and continuous model improvement. Furthermore, a three-stage modeling workflow is established. The stages include data management, feature extraction, and predictive feedback. In the feature extraction stage, statistical analysis and principal component analysis are used to reduce dimensionality. In the predictive feedback stage, model outputs are returned to the database. This guides further data collection and experimental validation. It also improves the scientific rigor and closed-loop nature of the prediction process. In summary, we establish an intelligent prediction system that integrates data management, parameter prediction, and dynamic updating. High-accuracy prediction of steel material parameters is achieved using an optimized XGBoost model. Supported by a database system that continuously stores historical data and prediction results, the system provides real-time and reliable data support for the intelligent optimization of heat treatment processes and finite element simulations.

Keywords: thermo physical parameter; heat treatment; machine learning; forecasting model; database construction

Numerical Simulation ofthe Influence of Cold Bulging Process on Quenching Residual Stresses in 7050 Aluminum Alloy Ring

Wang Pengyu*, Li Hao, Wei Shi*

Department of Mechanical Engineering, Tsinghua University, China

E-mail address: wpd23@ mails. tsinghua. edu. cn (Wang Pengyu)

E-mail address: shiw@ tsinghua. edu. cn (Wei Shi)

Abstract: In this work, a cold bulging process for reducing quenching residual stresses of large aluminum alloy rings was numerically calculated. During the quenching process, the large temperature gradient in an aluminum alloy ring causes uneven distribution of thermal expansion, and results in quenching residual stresses when it is cooled to room temperature. The subsequent cold bulging process generates plastic deformation in the ring, and reduces magnitudes of quenching residual stresses in the ring. The thermal expansion coefficient of 7050 aluminum alloy in a wide temperature range was obtained by thermal expansion experiment, the high temperature density of 7050 aluminum alloy was obtained by using the density measured at the room temperature and the thermal expansion coefficient. The specific heat and thermal diffusion coefficient of 7050 aluminum alloy were measured by a laser flash experiment, and the thermal conductivity was obtained based on the density dependent on temperatures. Flow stress curves of 7050 aluminum alloy were obtained by hot compression tests on a Gleeble thermal test machine, and the elastoplastic constitutive model was established by combining the elastic parameters published in literatures. Finally, the finite element analysis model of 7050 aluminum alloy for predicting stress evolutions during ring quenching and the cold bulging process was established. The temperature and stress changes in the ring during quenching were calculated by using ABAQUS software, and the influence of cold bulging processes on residual stresses of the ring after quenching was explored. The results indicate that residual compressive stresses (both radial and circumferential) are present on the upper and lower end surfaces of the ring after quenching, while residual tensile stresses exist in the core of the ring. The cold bulging process can effectively reduce the quenching residual stresses in the ring. Four cold bulging processes with 0. 1%, 0. 5%, 1% and 3% bulging amount were simulated respectively, the process with 3% bulging amount reduces quenching residual stresses effectively.

Keywords: 7050 aluminum alloy ring; quenching; cold bulging; numerical simulation

Numerical Simulation-Based Prediction of Heat Treatment Cracking and Process Optimization

Abstract: A finite element numerical simulation incorporating a damage mechanics model was employed to predict the quench cracking behavior of V-notched blocks made from two different tool steels, Cr12MoV and H13, subjected to identical heat treatment processes. Experimental validation demonstrated the effectiveness of the simulation approach and confirmed that the underlying cracking mechanisms differed significantly between the two materials. Finally, through numerical simulation-based optimization, the heat treatment process was improved, successfully preventing quench cracking in V-notched blocks of both materials.

Keywords: tool steel; heat treatment; quench cracking; oyane damage model; finite element method; simulation

Numerical Simulation and Experiment of Vacuum Carburizing and Quenching of G13Cr4Ni4Mo4V Bearing Steel

Tao Yuan[1*], Xia Tao[1], Fu Zhongyuan[2], Cai Chenhuan[3], Zhu Jingchuan[1], Zhou Lina[2], Liu Yong[1]

[1] School of Material Scienceand Engineering, Harbin Institute of Technology, China

[2] AECC Harbin Bearing Co. Ltd, Heilongjiang, China

[3] Faculty of Business, Economics and Informatics, University of Zurich, Switzerland

E-mail address: 17335376878@163. com (Tao Yuan)

E-mail address: 19112354078@163. com (Xia Tao)

Abstract: This study presents a predictive model developed to simulate the carbon concentration distribution and hardness evolution in G13Cr4Ni4Mo4V aerospace bearing steel during vacuum carburizing and subsequent quenching. The model is constructed based on a saturation value adjustment method, which enables dynamic switching between the strong carburizing and diffusion stages. It integrates thermodynamic calculations derived from the CALPHAD approach with diffusion kinetics governed by Fick's second law, while also incorporating phase transformation and hardness estimation submodels. The model was implemented using the Python programming language and is capable of automatically generating pulse time schedules corresponding to different total carburizing durations. This feature allows for the quantitative prediction of carbon concentration gradients and mechanical property distributions under a variety of heat treatment conditions. To validate the model, a series of vacuum carburizing and quenching experiments were conducted. Samples were subjected to heat treatment following model-generated pulse schedules, and the resulting carbon concentration profiles were measured using electron probe microanalysis (EPMA). Hardness distributions were also determined via Vickers microhardness testing. The experimental results demonstrated that the model accurately predicts carbon diffusion behavior, with the maximum absolute deviation in carbon concentration within 0.013 wt. % and a relative error of less than 5%. The post-quenching hardness predictions showed strong agreement with experimental data, exhibiting a characteristic gradient hardening effect. The relative error between simulated and measured hardness values remained within 3.77%, confirming the robustness of the hardness estimation model. In addition to numerical validation, metallographic characterization of the carburized layer was performed using optical microscopy and scanning electron microscopy. The microstructural observations revealed typical hardened martensitic structures at the surface, a decreasing carbon gradient toward the core, and fine carbide precipitation at grain boundaries, consistent with model assumptions. These findings further corroborate the accuracy of the developed simulation framework. The presented model provides a computational tool for analyzing the coupled thermal-diffusion-mechanical behavior of G13Cr4Ni4Mo4V steel during heat treatment. It enables informed process design and optimization, thereby reducing dependence on empirical methods and minimizing time and material costs associated with traditional trial-and-error approaches. This framework has significant implications for improving process efficiency and ensuring performance reliability in the heat treatment of critical aerospace components.

Keywords: G13Cr4Mo4Ni4V bearing steel; vacuum carburizing; hardness; numerical simulation; carbon concentration

Grain Growth and Deformation Mechanisms in RENE41 Alloy Ring Forgings During Solution Treatment

Xiao Feng, Deng Xiaohu*, Wang Guangyu, Ju Dongying*

[1] School of Mechanical Engineering, TianjinUniversity of Technology and Education, Tianjin 300222, China

[2] Guizhou Aviation Technical Development CO., LTD, Guiyang 550081, China

[3] Zhejiang XCC Group Co., Ltd., 312500, Xinchang, Zhejiang, China

E-mail address: dengxh@ tute. edu. cn (Deng Xiaohu)

E-mail address: dyju. sitec@ gmail. com (Ju Dongying)

Abstract: This study employed a combined experimental and numerical simulation approach to investigate grain coarsening, stress, and deformation evolution during solution heat treatment of RENE41 superalloy ring forgings. To understand the influence of solution parameters on grain growth behavior, solution treatment experiments were conducted on RENE41 alloy across different heating temperatures (1050 ~ 1120 ℃) and holding times (60 ~ 240 min). The corresponding average grain sizes were measured. Results showed that grain growth in RENE41 alloy remained relatively slow at solution temperatures below 1066 ℃. Significant grain coarsening occurred when the solution temperature exceeded 1100 ℃. Furthermore, grain size increased rapidly with longer holding times, but the growth rate tended to stabilize after holding times exceeded 120 min. A grain growth kinetics model for RENE41 alloy during solution treatment was established using nonlinear regression. The model calculated grain sizes under various solution process conditions, and the calculated results agreed well with experimental data. Cooling curves on the ring forging surface were obtained using Computational Fluid Dynamics (CFD). The inverse heat transfer method was then used to calculate the comprehensive heat transfer coefficients at different characteristic locations. These heat transfer coefficients were applied as cooling boundary conditions to establish a thermo-mechanical coupled model for a single test ring (Φ610 mm × Φ560 mm × 90 mm) during solution treatment. The model calculated the post-cooling temperature field, stress field, and deformation distribution. Results indicated that residual stresses exhibited a typical distribution featuring compressive stresses on the surface and tensile stresses at the core. The maximum stress occurred near the circumference and center, likely due to significant plastic deformation in these areas. Simulated deformation results matched well with measurements, validating the accuracy of the RENE41 material's thermophysical properties and constitutive model. To account for loading configuration effects, CFD was used to obtain cooling curves for production rings (Φ700 mm × Φ575 mm × 134 mm) at different positions under single-layer flat loading and three-layer stacked loading. Heat transfer coefficients for different workpiece positions and their characteristic locations during cooling were calculated. Based on these results, three rings from single-layer loading and nine rings from three-layer loading were selected for analysis. Using the calculated heat transfer coefficients as boundary conditions, a three-dimensional thermo-mechanical coupled model for these twelve production rings was established. This model studied the temperature field, stress, and deformation magnitude/distribution in the rings and analyzed factors influencing deformation. Results revealed that rings in central positions cooled slower but more uniformly than peripheral rings. Peripheral rings also exhibited higher core tensile stresses than central rings. This work clarifies the deformation mechanisms in RENE41 alloy ring forgings during solution heat treatment, providing a reference for selecting solution process parameters.

Keywords: RENE41 alloy; solid solution treatment; numerical simulation; deformation; grain grow

Thermo-Fluid-Structural Coupling Modeling and Simulation of Oil Quenching for RENE41 Superalloy Rings

Xu Wangli[1], Deng Xiaohu[1*], Xiao Feng[1], Ju Dongying[2*]

[1] School of Mechanical Engineering, Tianjin University of Technology and Education, Tianjin 300222, China

[2] Zhejiang XCC Group Co., Ltd., 312500, Xinchang, Zhejiang, China

E-mail address: dengxh@tute.edu.cn (Deng Xiaohu)

E-mail address: dyju.sitec@gmail.com (Ju Dongying)

Abstract: To address deformation control challenges during oil quenching of RENE41 superalloy ring forgings, this study employed a combined approach of numerical simulation, theoretical analysis, and experimental validation. It systematically investigated how loading configurations affect the evolution of temperature and flow fields during the oil cooling process. First, a thermal-fluid-solid coupling model for a single ring forging was established. The full model incorporated the oil tank (5000 mm × 3000 mm × 3000 mm), the test ring positioned at the tray center, the tray (2000 mm × 2000 mm × 55 mm), and an agitation fan. A mixture multiphase flow model simulated quench oil vaporization under cooling conditions, while the k-ω SST (Shear Stress Transport) eddy viscosity model resolved the flow field characteristics within the tank. Simulated core cooling curves matched well with measurement results. Simulation results revealed non-uniform cooling where the ring's top face cooled fastest, followed by the side walls, with the bottom face cooling slowest. Correspondingly, flow velocity decayed first near the top face, then the outer wall, and last near the inner wall. Further analysis identified a typical three-stage cooling process comprising slow cooling (film boiling), rapid cooling (nucleate boiling), and slow equilibrium (natural convection), consistent with phase-change heat transfer theory. Furthermore, simulations examined the cooling process for production rings (Φ700 mm × Φ575 mm × 90 mm) under two loading configurations using trays sized 2300 mm × 2300 mm × 50 mm. The configurations were single-layer flat loading (9 rings per tray) and three-layer stacked loading (27 rings on 3 trays). Analysis showed that during single-layer loading, rings located at the tray corners cooled faster than central rings. Higher flow velocity persisted longer in the regions between rings. Stacked loading significantly reduced overall cooling efficiency and increased temperature non-uniformity. This occurred because, during the initial cooling stage, flow velocity near the rings was higher than near the tank bottom, exhibiting a decreasing trend from the top layer to the middle layer and then to the bottom layer (top > middle > bottom). Within each layer, flow velocity was higher in the center than at the sides. Later in the cooling process, flow velocity near the rings decreased, while the agitated zone maintained a higher velocity. This study clarifies the cooling characteristics of RENE41 ring forgings under the combined influence of temperature and flow fields during oil quenching. It provides a theoretical basis for optimizing loading strategies in the solution heat treatment process for superalloy ring forgings.

Keywords: thermo-fluid-solid coupling; superalloy; computational fluid dynamics; oil cooling

Thermal Barrier Coating R&D Driven by the Synergy of Computing, Data, and AI

Chong Xiaoyu

Kunming University of Science and Technology

Abstract: Thermal barrier coatings (TBCs) are critical for ensuring the safe, stable, and long-term service of hot-section components in aerospace vehicles. However, one of the primary causes of TBC failure is the thermal stress generated during preparation and service, which exceeds the interfacial bonding strength and damage tolerance. Given the current lack of efficient and accurate experimental methods to quantify thermal stress in TBCs, we adopted the concept and methodology of materials genome engineering. By integrating multiscale approaches—including first-principles calculations, genetic algorithm-support vector regression (GA-SVR), multiphysics-coupled finite element simulations, and thermal shock testing—we developed core software to enable high-throughput, intelligent computation of key data. We established parameter databases, experimental databases, and model databases, achieving automated data flow transfer and correlation modeling. This work has preliminarily developed a rapid quantification and evaluation method for thermal stress in TBCs, providing theoretical, methodological, and technical support for: Screening long-life TBC materials, Optimizing coating structures, Enhancing coating processes. This integrated approach bridges computational prediction and experimental validation, accelerating the development of next-generation thermal barrier coatings.

Digital Twin Modeling of Friction Stir Welding: Predicting Temperature, Material Properties and Bonding Quality

Chen Gaoqiang, Fan Yumo, Liu Xin, Yang Chengle, Shi Qingyu

State Key Laboratory of Clean and Efficient Turbomachinery Power Equipment, Department of Mechanical Engineering, Tsinghua University, Beijing, 100084, China

E-mail address: chengl@ tsinghua. edu. cn (Chen Gaoqiang)

Abstract: The digital twin of the welding process is the mirror image of the physical welding process in digital space. Its technical value depends on its accuracy and ability in real-time prediction of the welding process and weld quality. This study proposes a digital twin-based method for real-time quality monitoring in friction stir welding (FSW), addressing the challenge of rapid calibration for multi-parameter models under practical welding conditions. To overcome the limitations of existing models, such as high parameter sensitivity and slow computation, we developed a finite difference model with about 20 key parameters (e. g. , friction coefficient and precipitates evolution parameters) achieve faster-than-real-time computation. Based on experimental data from 7 welding trials (including temperature measurements, hardness tests, and joint tensile tests), the model parameters were calibrated using genetic algorithms. The results demonstrate the model's strong predictive capability for untrained welding tool parameters: peak temperature prediction errors were less than 5%, and microhardness prediction errors were less than 10%. The results show that, changing the time step has almost no effect on the accuracy of temperature field solving, unless the time step exceeds 4. 5 ms. as the time step increases, the time consumed for solving decreases. More remarkably, when the time step is more than 2 ms, the time consumed for simulation calculation is less than that of the actual welding process, and the real-time simulation and analysis of digital twins can be accomplished. For 6-mm-thick 6061-T6 aluminum alloy welding, the model completed simulations in 38sec on a personal computer, 68% faster than the physical welding time (118sec). This work provides an efficient tool for FSW process monitoring and optimization, particularly valuable for manufacturing applications such as aerospace components and electric vehicle battery trays, where rapid parameter adjustments are critical.

Keywords: friction stir welding; digital twin; weld properties prediction; defects prediction

Research on COSMAP Quenching Heat Treatment Simulation of GCr15 Thin-walled Bearing Rings

An Xinyang[1], Wang Jiangang[1*], Li Xusheng[2,3], Deng Xiaohu[4], Yu Chengshuang[2], Ju Dongying[2*]

[1] Hebei University of Science and technology, 050018, School of Materials Science and Engineering, Shi Jiazhuang, Hebei, China

[2] Zhejiang XCC Group Co., Ltd., 312500, Xinchang, Zhejiang, China

[3] Zhejiang University, 310030, School of Mechanical Engineering, Hangzhou, Zhejiang, China

[4] Tianjin University of Technology and Education, 300222, School of Mechanical Engineering, Tianjin, China

E-mail address: jgwang123@hebust.edu.cn (Wang Jiangang)

E-mail address: dyju.sitec@gmail.com (Ju Dongying)

Abstract: Bearings are critical components in heavy industrial equipment, widely employed in aerospace, transportation engineering, and high-end miniature intelligent devices. Their demanding operating conditions—characterized by high speed, heavy load, and elevated temperatures—impose stringent requirements on bearing materials and heat treatment processes. High-carbon chromium bearing steel is globally prevalent in bearing manufacturing due to its excellent fatigue resistance, optimal elasticity and toughness, high and uniform achievable hardness post-quenching and tempering, controllable alloy composition, and well-established heat treatment methods. GCr15 steel, a representative high-carbon chromium bearing steel, is renowned for its high hardness, superior wear resistance, and strong contact fatigue strength. It exhibits good machinability and formability after spheroidizing annealing, making it well-suited for producing bearing rollers and rings, as well as manufacturing mechanical components that require high wear resistance, high elastic limit, and high contact fatigue strength. This study employs COSMAP, a heat treatment simulation software developed based on the " metal-thermo-mechanical" coupling theory, to numerically simulate the quenching process of a thin-walled spherical bearing ring made of GCr15 steel. First, key physical properties of GCr15 steel—including transformation-induced plasticity, latent heat of phase transformation, coefficient of linear expansion, CCT (continuous cooling transformation) curves, and stress-strain curves—were obtained using JMatPro software. Then the thin-walled bearing ring was meshed for finite element analysis using GID software. Material properties, quenching process parameters, and the finite element mesh were exported into input files for simulation. The actual quenching process involved heating the thin-walled bearing ring to the austenitizing temperature in a continuous mesh-belt furnace, holding at this temperature for 40 minutes after a 15-minute heating period, followed by oil quenching at 363.15 K. To validate the simulation results, the quenched bearing ring was characterized using X-ray diffraction (XRD), optical microscopy (OM), scanning electron microscopy (SEM), and micro-Vickers hardness testing to analyze its phase composition, microstructure, and hardness. The simulation results show that due to the thin wall of the workpiece, the martensite distribution is uniform after quenching, with martensite content reaching 89%, surface hardness at 814 HV, and maximum part deformation of 0.157 mm. The experimental results indicate that the microstructure of the actual treated parts is mainly composed of acicular martensite, retained austenite in small amounts, and fine dispersed carbides. The measured average surface hardness is 810 HV, with a 1% error relative to the simulated value. Deformation measurements of 30 actual production parts showed an average maximum diameter deformation of 0.18 mm, closely matching the simulated value of 0.157 mm. In summary, the simulation results are consistent with experimental results, verifying the reliability and accuracy of the numerical simulation method for predicting microstructure, hardness, and deformation of quenched GCr15 thin-walled bearing rings.

Keywords: GCr15 steel; thin-walled bearing components; quenching process; numerical simulation; experimental validation

Modeling and Simulation of Microstructure Evolution During Inter-Critical Annealing of Cold-Rolled Dual-Phase Steel

Shen Gang, Gu Jianfeng*

School ofMechanical Engineering, Suzhou University of Science and Technology, China

E-mail address: jackshen2018@ usts. edu. cn (Shen Gang)

E-mail address: gujf@ sjtu. edu. cn (Gu Jianfeng)

Abstract: Dual-phase (DP) steel, as the flagship of the first generation advanced high-strength steels, has been widely used in the automotive industry due to the combined properties of high strength and good formability. Continuous annealing is widely used to produce DP steels, during which a series of complex metallurgical phenomena happen, such as ferrite recrystallization, cementite spheroidization, austenite formation and austenite-to-ferrite transformation. They proceed concurrently or consecutively with complex local interactions between each other, making the resulting microstructure closely process-related. Through-process simulation of the microstructure evolution during inter-critical annealing of a cold-rolled DP steel was investigated by a coupled Crystal Plasticity Finite Element Method-Cellular Automaton (CPFEM-CA) approach. The CA model with dual-scale cells was developed to describe the ferrite and cementite within the pearlite so that carbon redistribution to the neighboring ferrite and austenite during cementite dissolution can be considered. Simulation results dynamically revealed the microstructure evolution kinetics and morphology under the interaction of ferrite recrystallization and austenite formation during annealing. Two-stage transformation kinetics was presented during austenite formation, which corresponds to fast pearlite-to-austenite and slow ferrite-to-austenite transformation, respectively. Austenite nuclei nucleated at ferrite-ferrite boundaries were found to grow either into necklace-type distributed grain boundary austenite or dispersed intragranular austenite, depending on the local interactions between ferrite recrystallization and austenite formation, which is strongly affected by the heating rates and annealing temperatures. Minor changes in these interactions can cause profound changes in the transformation kinetics and the resulting microstructure. The coupled CPFEM-CA modeling approach gave good predictions of the microstructure evolution for DP steel production during industrial inter-critical annealing cycle. Simulation results on transformation kinetics and microstructure morphology were found to be in good agreement with experimental observations.

Keywords: dual-phase steel; continuous annealing; microstructure evolution; digital material representation; crystal plasticity finite element method; cellular automata

Research Progress in Numerical Simulation and Software Development for Heat Treatment

Tong Daming, Xu Jun, Wang Jing, Gu Jianfeng*

School of Materials Science and Engineering, Shanghai Jiao Tong University, China

E-mail address: daming_ tong@ sjtu. edu. cn (Tong Daming)

Abstract: With the continuous advancement of machine learning, artificial intelligence and relative technologies, intelligent heat treatment has become one of the core innovative technologies of heat treatment, and it is also a strong demand for transformation and upgrading of heat treatment technology. Heat treatment, as a critical processing of key parts manufacturing technology, plays a decisive role in enhancing material performance and extending service life. The internal physical mechanisms during the heat treatment process are intricate, involving a multitude of influencing parameters. It is difficult to implement process monitoring and regulation in the heat treatment process by using empirical or trial-and-error methods. There is an urgent need to establish effective means to sense and reflect the " black-box environment" within the furnace. This paper reviews recent developments in heat treatment simulation, with a focus on multi-physics coupling approaches that integrate thermal, metallurgical, and mechanical phenomena. Furthermore, significant progress has been made in the development of specialized simulation software that integrates multi-physics models, material databases, and user-friendly interfaces. Special attention is given to the development and application of the VirtualImago software series—an industrial-grade system designed for heat treatment process development and equipment optimization. VirtualImago consists of three core modules: a digital simulation engine, a data-driven intelligent design engine, and a materials-process knowledge database. Together, these components support a wide range of key functions across the heat treatment workflow, including process design, equipment configuration, and data management. The integration of numerical models with empirical knowledge and intelligent algorithms enhances the accuracy, efficiency, and adaptability of heat treatment solutions. This paper also discusses emerging trends such as AI-assisted modeling and digital twins, and concludes by outlining current challenges and future directions in the field. This study provides new insights into the integrated and coordinated control of equipment and workpieces during the heat treatment process.

Keywords: heat treatment; digital twins; model reduction; industrial software; intelligentization

Current-Carrying Friction and Wear Behavior of Dual-Phase Reinforced Copper Matrix Composites

Yang Tian[1], Jiang Yihui[2], Liang Shuhua[2]

[1] School of Materials Science and Engineering, Xi'an University of Technology, China

[2] Shaanxi Province Key Laboratory of Electrical Materials andInffltration Technology, School of Materials Science and Engineering, Xi'an University of Technology, China

E-mail address: 17389565563@ 163. com (Tian Yang)

E-mail address: jiangyihui@ xaut. edu. cn (Jiang Yihui)

Abstract: Current-carrying friction components in service must simultaneously meet the dual requirements of being both load-bearing and electrically conductive, necessitating not only sufficient mechanical properties but also excellent electrical and thermal conductivity. Conventional copper matrix composites, typically fabricated with single-phase reinforcement or through ex-situ methods, suffer from an imbalance between friction-reduction and wear-resistance properties, as well as poor interfacial bonding. This commonly leads to an inverse relationship between wear resistance and current-carrying capacity. In this study, a novel $2TiB_2/Cu\text{-}3B$ composite was synthesized in-situ via vacuum induction melting to investigate its current-carrying tribological behavior and wear mechanisms under conditions of electrical sliding contact. The synergistic effect of the anti-friction phase B particles and wear-resistant TiB_2 particles significantly enhances the current-carrying wear resistance of the copper matrix composites. During the friction process, an anti-friction layer is continuously formed and renewed through a dynamic self-adaptive mechanism, effectively minimizing direct contact between the friction pairs. Concurrently, the uniform distribution of the wear-resistant phase inhibits the excessive accumulation of the anti-friction layer, thereby optimizing its thickness and mitigating its adverse effects on current-carrying efficiency. Based on this synergistic mechanism, the in-situ introduced anti-friction and wear-resistant phases enable the copper matrix composite to exhibit outstanding current-carrying and wear-resistant properties by minimizing interfacial friction while ensuring efficient current transmission. The findings of this research provide valuable guidance for the material design of core components intended for high-performance, current-carrying friction service conditions.

Keywords: current-carrying friction; copper matrix composites; wear-resistant phase; anti-friction phase

Unveiling the Structure-Property Relationships of Cu-based Nanocrystalline Alloys Through a Quantitative Atomic-Resolution Study

Zhang Z. L. [1*]

[1] Erich Schmid Institute of Materials Science, Austrian Academy of Sciences

E-mail address: Zaoli. zhang@ oeaw. ac. at (Zhang Z. L.)

Abstract: Cu-based nanocrystalline (NC) alloys have garnered significant attention due to their superior electrical and mechanical properties compared to pure Cu materials[1-4]. To produce the Cu NCs alloy, various methods have been employed so far, including severe plastic deformation (SPD) and electrochemical methods, among others. The properties of Cu-based NCs are determined mainly by their structural features, including grain sizes and growth defects. Recently, it has been discovered that nanotwin architecture structures have a significant impact on the mechanical and electrical properties of Cu alloys. Controlling the density of defects and the structural peculiarities of Cu NC alloys becomes increasingly essential, as this can tailor their properties and deformation behaviors. In this presentation, Cu-Fe, Cu-Cr, Cu-Ag, and Cu-O nanocrystalline alloys were produced by SPD. Atomic-resolution studies on the structures of Cu-based NC alloys and their correlation with properties, using aberration-corrected high-resolution transmission electron microscopy and associated techniques (such as quantitative analysis), will be presented. Firstly, an immiscible Cu-Fe system can be alloyed to form a single phase utilizing high-pressure torsion (HPT). Micrometer-sized Cu and Fe powders were directly used as starting materials to be mixed as $(1\text{-}x)$ at. % Cu-x at. % Fe ($x = 0, 5, 15, 25, 35$) series of compositions. The Cu-Fe alloys with different atomic percentages of constituents were systematically investigated to get insights into the structural evolution of nanocomposites and to determine the solid solubility limit of Fe atoms in the Cu matrix under HPT deformation. Finally, all investigated compositions of $(1\text{-}x)$ at. % Cu-x at. % Fe with a single face-centered cubic structure were successfully formed after two stages of deformation, as indicated by X-ray diffraction results. The exact limit of solid solubility for Fe atoms in a Cu matrix was extended to 35 at% Fe by the HPT method. Grain size statistics from a transmission electron microscope image show that nanostructured Cu-Fe crystalline was obtained with an average size of 68 nm for an 85 at. % Cu-15 at. % Fe composition. Typical 'saturated softening' of micro-hardness was observed in the Cu-Fe supersaturated area. The possible mechanism of Fe atom dissolution into the Cu matrix was also discussed. On the other hand, the effects of oxygen impurities in nanocrystalline materials were carefully explored. A comparative investigation of Cu-Fe nanocrystalline alloys generated from different initial materials with different oxygen contents reveals that oxygen atoms go either to GBs or dissolve into the matrix. The former effectively decreases the grain boundary mobility, facilitating grain refinement, while the latter leads to lattice expansion and interstitial strengthening. Secondly, the 57 wt% Cu-43 wt% Cr dual-phase alloys deformed by high-pressure torsion (HPT) under 5 and 25 rotations have been investigated. We found that: (i) an inverse grain-size effect on stacking fault (SF) and twinning was observed in Cu. (ii) During the early deformation stage, in a way, grain refinement in Cu was controlled by the interaction between Cr solutes (or clusters) that were forced to dissolve into the Cu matrix and extended dislocation motion. iii) Twins created during the highly strained state are associated with the grain boundary activities and the recovery procedure in the extremely fine-grain size region. Moreover, the interaction between dislocations and twins plays a crucial role in forming the GBs of nanocrystalline materials. Thirdly, for the immiscible Cu-Ag NC alloys, accumulative roll bonding (ARB) was applied to fabricate the NC alloys. The significant structural features include planar defects, such as nanotwins and stacking faults. Except for these, the coherent Cu/Ag interfaces are dominant in the CuAg NC structures. The interactions of these defects and their high density largely determine the mechanical properties of CuAg NC alloys. Last, for the Cu-O NC alloy, we found that different nanosized Cu oxides are formed and distributed mainly at GBs. Oxygen atoms are also found to dissolve in the lattice, causing the interstitial strengthening. Furthermore, we reveal that the oxygen behavior in pure Cu NC differs from that in CuFe NC alloys[4].

Regulation of Age Hardening on Evolution of MgZnAl Phase in Sc/Zr-Microalloyed Mg-10Zn-5Al Alloys: Coherent Interface Strengthening and Precipitation Mechanisms

Jiang Aoxue[1], You Zhiyong[1*], Han Peide[1], Wang Kaiying[2]

[1] College of Materials Science and Engineering, Taiyuan University of Technology, Taiyuan, 030024, China

[2] Institute of Energy Innovation, Taiyuan University of Technology, Taiyuan, 030024, China

E-mail address: 1837864352@ qq. com (Jiang Aoxue)

E-mail address: 13603511486@ 163. com (You Zhiyong)

Abstract: This study systematically elucidates the differential regulatory mechanisms of Sc/Zr microalloying on the morphology, interfacial characteristics, and mechanical properties of MgZnAl precipitates formed during aging in Mg-10Zn-5Al alloys. Based on first-principles calculations, multiple interface models between the α-Mg matrix (basal/prismatic planes) and MgZnAl phase (basal/prismatic planes) were constructed, identifying coherent interfacial configurations with minimal interfacial energies (0. 135 J/m^2 and 0. 080 J/m^2), specifically the $(11\text{-}20)_{\alpha\text{-Mg}}// (0001)_{MgZnAl}$ orientation. These results provide thermodynamic foundations for interface orientation design. Experimental characterization confirmes that the ZAZr alloy exhibits a fully coherent interface between α-Mg and MgZnAl precipitates ($(11\text{-}20)_{\alpha\text{-Mg}}// (10\text{-}10)_{MgZnAl}$), with an ultralow lattice mismatch of 0. 4% and 99. 6% atomic column alignment consistency, aligning precisely with the computationally predicted low interfacial energy characteristics. In contrast, the ZASc alloy displays a semi-coherent interface ($(1\text{-}210)_{MgZnAl}// (01\text{-}1\text{-}1)_{\alpha\text{-Mg}}$) with a 3. 11% lattice mismatch, deviating from ideal coherency due to Sc-enhanced diffusion. This bidirectional " computational prediction-experimental validation" framework clarifies the physical origin of coherent relationships between MgZnAl precipitates and the α-Mg matrix, establishing a theoretical basis for interfacial optimization. Through combined TEM and mechanical analysis, the study demonstrates that the low diffusivity of Zr in the α-Mg matrix (40% lower than that of Sc) inhibits precipitate coarsening, facilitating MgZnAl nanoscale twin formation (2. 46 ±0. 11 nm thick) . Dislocation strengthening model-based analysis reveals that in the ZAZr alloy, fine precipitates ($d = 60.59$ nm, $\lambda = 862.4$ nm) achieve a theoretical strengthening increment of 34. 02 MPa—dominated by the Orowan bypass mechanism (30. 09 MPa) with supplementary dislocation shearing (3. 93 MPa) —closely aligning with experimental values (33. 64 MPa) . The fully coherent interface synergistically enhances dislocation obstruction through high interfacial binding energy and nanoscale twin-induced strain fields, resulting in a 14. 57% increase in tensile strength (216. 86 MPa), 21. 96% elevation in yield strength (186. 8 MPa), and 127. 12% improvement in elongation. Conversely, the high diffusivity of Sc in the α-Mg matrix promotes precipitate coarsening ($d = 97.87$ nm, $\lambda = 1120.71$ nm), reducing contributions from the Orowan mechanism (25. 63 MPa) and shearing effects (3. 22 MPa) . The significant disparity between theoretical (28. 86 MPa) and experimental (19. 45 MPa) strengthening stems from dislocation channeling in non-coherent regions, performance degradation induced by coarsening, and measurement errors. First-principles computational-experimental integration reveals that the fully coherent interface in ZAZr creates dislocation pinning barriers through charge density overlap, while the semi-coherent interface in ZASc accelerates dislocation escape via periodic slip bands. By integrating computational and experimental validation, this work elucidates the intrinsic correlations among microalloying element diffusivity, interfacial coherency, and dislocation obstruction efficacy. It establishes precise design principles for advancing high-strength/high-ductility magnesium alloys through interfacial engineering.

Keywords: age hardening; first-principles calculations; coherent interface strengthening; precipitation mechanisms; magnesium alloys

Research on the Synergistic Deformation and DRX Behavior of Tip/Mg

Deng Kunkun*

Shanxi Key Laboratory of advanced magnesium-based materials, College of Materials Science and Engineering,

Taiyuan University of Technology, Taiyuan 030024, China

Abstract: The spherical Ti particle (Ti_p) reinforced Mg-5Zn-0.5Ca (Ti_p/ZX50) composite was prepared via the semi-solid stirring casting process and the effects of Ti_p on the hot deformation and hot processing behavior of matrix alloy were investigated through uniaxial hot compression testing. The results indicate that a particle deformation zone (PDZ) forms around the Ti_p with the deformation of the Ti_p/ZX50 composite, which is propitious to the dynamic recrystallization (DRX) of the matrix alloy. The range of the PDZ and the promoting effect of the Ti_p on DRXed nucleation are inversely related to the deformation degree of the Ti_p. Moreover, the deformation of Ti_p alleviates the high stress in the matrix alloy during deformation, expanding the processing range and reducing the average deformation activation energy of the matrix alloy. Notably, the minimum processing temperature (493 K) of the Ti_p/ZX50 composite is significantly lower than that of hardened particle reinforced magnesium matrix composites. The hot deformation mechanism of the Ti_p/ZX50 composite is dislocation climb controlled by both lattice diffusion and pipe diffusion.

Keywords: deformable Ti_p; particle deformation zone; dynamic recrystallization; hot deformation mechanism

New Synthetic Semiconductor for Advanced Thermal Management—Cubic Boron Arsenide

Tian Fei*

School of Materials Science and Engineering, Sun Yat-sen University, China

E-mail address: tianf9@ mail. sysu. edu. cn (Tian Fei)

Abstract: High thermal conductivity materials have emerged as critical components for advanced thermal management for next-generation semiconductor devices. Among these, cubic boron arsenide (BAs) —long overlooked in semiconductor research—has garnered renewed interest owing to its ultrahigh thermal conductivity ($>1000\ W \cdot m^{-1} \cdot K^{-1}$ at room temperature, only second to diamond), which defies the widely-accepted Slack rule governing conventional materials. Recent advances further reveal its simultaneously high hole and electron mobilities with a rare room temperature ambipolar mobility value of $>1550\ cm^2 \cdot V^{-1} \cdot s^{-1}$, positioning it as a viable candidate for high-power and high-frequency electronics. Beyond these anomalies originating from the unique phonon and electron scattering dynamics, other interesting characteristics, including some unexpected optical and mechanical responses in BAs, have also been recently discovered. Preliminary demonstrations in integrated heat spreaders and BAs-based heterojunction devices underscore its transformative potential in realizing high-performance semiconductors that reconcile thermal-structural-electrical synergy. This presentation aims to elucidate recent breakthroughs in BAs semiconductor research, and catalyze interdisciplinary collaboration by highlighting its transformative potential in overcoming fundamental limits of conventional semiconductors.

Keywords: single crystal; thermal management; BAs

Strengthening-Toughening Mechanisms and Applications of Laminated Metal Composites

Zheng Shijian *

Tianjin Key Laboratory of Materials Laminating Fabrication and Interface Control Technology, School of Materials Science and Engineering, Hebei University of Technology, Tianjin, 300401, China

Abstract: Metallic materials exhibit pronounced size-dependent mechanical properties. Extensive studies since the 1950s have confirmed the Hall-Petch relationship, where metal strength increases with decreasing grain size. However, plastic deformability shows more complex size dependence: While the transition from coarse to fine grains typically enhances both tensile strength and elongation, the plasticity response further correlates with intrinsic crystal structure. Specifically, FCC and BCC metals display distinct critical grain sizes for the uniform-to-necking transition. Here, we utilize FCC/BCC size complementarity to achieve concurrent high strength and ductility in heterogeneous multilayers, while systematically investigating layer-thickness effects on deformation mechanisms and fracture behavior. In micrometer-scale multilayers, SUS304 shows layer-thickness-strengthening but suffers from shear band-induced ductility loss. Conversely, Q235 develops single-crystal-like characteristics with high-ductilityγ-fiber textures (< 111 >//ND) at reduced thicknesses. Strong interfacial bonding enables complementary FCC (SUS304) -BCC (Q235) effects: SUS304 provides strength via rapid strain hardening, whereas Q235's γ-texture enhances uniform strain accommodation, mitigating SUS304's stress concentration and promoting periodic (vs localized) necking. This synergy optimizes strength-toughness at intermediate thicknesses (~20 μm), beyond which SUS304's excessive strain localization degrades ductility.

Keywords: strengthening and toughening; bedded composite; component

Preparation and Microstructure of New Copper Matrix Composites with Current-Carrying Friction

Guo Xiuhua, Song Kexing, Li Shaolin, Zhou Yanjun, huang Tao

[a] School of Materials Science and Engineering, Henan University of Science and Technology, Luoyang, 471023, China

[b] Henan Academy of Sciences, Zhengzhou, Henan 450052, China

[c] Key Laboratory of Materials Science & Processing Technology for Non-ferrous Metals of Henan, Luoyang, 471023, China

Abstract: Achieving high strength and ductility for copper matrix composites reinforced by carbon nanofillers remains challenging due to secondary agglomeration, few intragranular distributions, and poor interfacial bonding in carbon/copper system. Here, we report an internal-oxidation-based fabrication strategy to simultaneously facilitate high-density in-situ nanoprecipitation into matrix grain interior as well as interfacial optimization between carbon nanotubes (CNTs) and copper matrix. Comprehensive experimental results show that in-situ generation of high-density coherent γ-Al_2O_3 nanoparticles with intragranular distribution and formation of amorphous copper oxides (Cu_xO_y) between CNTs and Cu are attributed to oxygen diffusion and in-situ solid reaction ascribing to thermodynamic driving force during internal oxidation. The co-strengthening effect of high density γ-Al_2O_3 nanoparticles and interface optimized CNTs is corroborated to contribute to the high mechanical properties of the composite, which exhibits a high tensile strength of up to 510 MPa and excellent ductility of 20.2%. Mechanism analysis indicates that both high-density intragranular γ-Al_2O_3 nanoparticles and interface optimized CNTs co-render high strength of the hybrid composite with regard to Orowan strengthening and highly effective load transfer strengthening respectively, which greatly agrees with theoretical quantitative and experimental values of corresponding strength contributions. Meanwhile, cooperation of the dislocation accumulation caused by intragranular γ-Al_2O_3 nanoparticles and cracking resistance enhanced by interfacial bonding optimized CNTs collectively maintains tensile ductility. The present design concept may provide a promising strategy for achieving the remarkable co-strengthening effect of nanoscale dual-reinforcements, contributing to high strength and ductility in other carbon nanofillers reinforced metal matrix composites.

Keywords: copper matrix composites; in-situ intragranular alumina nanoparticles; interface optimized carbon nanotubes; strength and ductility; oxygen diffusion and in-situ solid-reaction

Effect of Heat Treatment on the Structure and Properties of Amorphous Wrapped Nanocrystalline Duplex Alloy Coatings

Xiao Lilli

Institute of Materials, Henan Academy of Sciences/

E-mail address: xiaolili375@ 163. com (Xiao Lilli)

Abstract: High-entropy alloy is a new type of multi-principal alloy material composed of many main elements with equal or nearly equal atomic ratio. High-entropy alloy has simple solid solution structure, coupled with large lattice distortion, sluggish diffusion and other characteristics, showing excellent fracture toughness, good ductility, high fatigue strength, radiation resistance, high temperature softening resistance and other excellent properties. Since its discovery in 2004, it has received extensive attention and research in the fields of materials science, mechanics and applied physics. Taking CoCrFeMnNi (Cantor) alloy with face-centered cubic lattice structure as an example, although tremendous research results have been achieved in the microstructure adjusting and mechanical properties of this alloy and its derivative alloy system. However, the mechanisms behind its excellent fracture toughness and ductility and the difference between it and the traditional face-centered cubic metal are still unclear and debated. For example, to what extent can lattice distortion and local stalking fault energy affect the grain-size dependent strengthening effect of high-entropy alloys? Is the classical quantitative analysis of thermally activation parameters based on face-centered cubic materials applicable to high-entropy alloys? Does the creep behavior of high-entropy alloys with sluggish diffusion have better creep resistance? How do high-entropy alloys behave in terms of corrosion properties affecting their service life as potential metallic materials for engineering applications? The research and solution of the above problems will help to reveal the mechanical properties and service ability of high-entropy alloys deeply and comprehensively, and provide experimental basis and theoretical guidance for the design and engineering application of high-entropy alloys. Based on this, the intrinsic correlation and physical mechanism of the strength, plasticity, creep and corrosion properties of CoCrFeMnNi high-entropy alloy with the characteristics of structure and scale were systematically studied by altering grain size, phase composition, phase volume fraction and related characteristic microstructure size. The main research and results include the following aspects

Keywords: high-entropy alloys; strengthening mechanism; creep; strain rate sensitivity; corrosion

Microarc Oxidation Coating on Aluminum Matrix with Low Energy Consumption and High Wear-resistance Property

Zhang Xinwen, Ma Dongying, Pi Minghui, Sun Piaopiao, Xiao Lili, Li Rui, Zhang Zhaoqi, Zhang Guoshang*

Institute of Materials, Henan Academy of Sciences, Henan, China

E-mail address: xwzhang@ hnas. ac. cn (Zhang Xinwen)

E-mail address: getotten@ gmail. com (Zhang Guoshang)

Abstract: Microarc oxidation (MAO) technology is frequently employed to improved wearability and corrosion-resistance of aluminum and its alloys, however, the high energy consumption of MAO treatment remains a pressing problem to be urgently addressed. In this paper, the MAO coating on aluminum matrix with high wear-resistance is fabricated by controlled the duty ratio and frequency during MAO process, and the energy consumption can be reduced based on the synergistic effect for size of discharge center and duration of discharge process. The relationship between electrical parameters and the possible mechanism of MAO coating is sunmmerized to further evulate the energy and anti-wear property, and it is anticipated to provide potential idea for resolving the energy consumption issue of MAO technology.

Keywords: discharge; energy consumption; microarc oxidationn; wear-resistance

Achieving Excellent Creep Resistance of Mg-Zn-Ca-Gd Alloy via Dense G. P. Zones and Hierarchical Distribution of Phases

Liu Manping[1*], Cui Zhuang[1], Ying Zeng[2], Liu Xiaochun[3], Xu Shiwei[4]

[1] School of Materials Science and Engineering, Jiangsu University, Zhenjiang, China

[2] School of Material Science and Engineering, Southwest Jiaotong University, Chengdu, China

[3] College of Material Science and Engineering, Changsha University of Science & Technology, Changsha, China

[4] College of Mechanical and Vehicle Engineering, Hunan University, Changsha, China

E-mail address: manpingliu@ ujs. edu. cn (Liu Manping)

E-mail address: cuizhuang116@ 163. com (Cui Zhuang)

Abstract: Enhancing the creep resistance of magnesium (Mg) alloys at elevated temperatures remains a key challenge for structural applications. This study investigates the effect of Gd on the high-temperature creep behaviour of Mg-Zn-Ca alloys through alloy design and heat treatment. The addition of Gd reduces the minimum creep rate by an order of magnitude compared to Gd-free counterparts. Microstructural analysis reveals that Gd promotes the formation of W-phase ($Mg_3Zn_3Gd_2$) and nano-precipitates, leading to a more refined and hierarchical precipitate distribution. Notably, Gd induces unique basal/prismatic combined G. P. zones during creep, which act as effective barriers to dislocation motion. Solution treatment further enhances creep resistance by refining and densifying G. P. zones and nano-precipitates. For the first time, the atomic structure of the G. P. zones is directly visualized via aberration-corrected transmission electron microscopy. Dislocation-precipitate interaction analysis confirms that nano-precipitates impose strong Orowan-type barriers, significantly reducing dislocation mobility. This work demonstrates a precipitation-strengthening strategy via dynamic G. P. zones, offering valuable insights for the design of heat-resistant Mg alloys.

Keywords: mg alloy; creep resistance; G. P. zones; nano phase; dislocation; hierarchical distribution

Joining of Dissimilar Materials and Complex Structures: Technologies and Applications

Song Xiaoguo, Bian Hong

State Key Laboratory of Precision Welding & Joining of Materials and Structures, Harbin Institute of Technology, Harbin, China

Shandong Institute of Shipbuilding Technology, Weihai, China

E-mail address: xgsong@ hitwh. edu. cn (Song Xiaoguo)

E-mail address: bianhong@ hit. edu. cn (Bian Hong)

Abstract: Heterogeneous material components, by enabling the combination of superior properties from different materials, significantly enhance the flexibility of design and production for engineering structures. They meet the functional and performance requirements of modern engineering structures, offering higher technological and economic value, and possess broad application prospects across various fields. Consequently, reliable joining of dissimilar materials is particularly crucial. However, when joining dissimilar materials, significant differences in their physical, chemical, and mechanical properties lead to the following major scientific challenges in welding: (1) Large differences in wettability make it difficult to simultaneously wet both materials; (2) The joining interface is prone to immiscibility without reaction, or complex interfacial reactions, excessive formation of brittle compounds, and difficulties in controlling the reaction process; (3) Differences in thermal expansion coefficients result in substantial residual stresses at the interface difficult to alleviate. This report summarizes our team's research progress on brazing, laser welding, electron beam welding, ultrasonic joining, and friction stir welding of dissimilar materials, aiming to provide references for future research directions and technological breakthroughs in dissimilar material joining.

Keywords: welding and joining; heterogeneous materials; research progress; microstructure; mechanical properties

Effect of Heat Treatment on the Microstructure and Properties of CoCrNiNbx Medium Entropy Alloys

Xu Yitong[1], Li Bo[1*], Li Cong[1,2*], Wu Da[1], Shi Yuzhi[1]

[1] State Key Laboratory for MechanicalBehaviour of Materials, School of Materials Science and Engineering, Xi'an Jiaotong University, Xi'an, People's Republic of China

[2] School of Materials Science and Engineering, Nanyang Technological University, 639798, Singapore

E-mail address: libo616@ mail. xjtu. edu. cn (Li Bo)

E-mail address: licong0369@ 163. com (Li Cong)

Abstract: In this study, a series of $CoCrNiNb_x$ (x = 0. 2, 0. 3, 0. 4, and 0. 5) MEAs alloys were prepared by vacuum arc melting. The $CoCrNiNb_{0.4}$ alloy, which exhibited optimal comprehensive mechanical properties, underwent heat treatment at 800-1000 ℃ for 10 h. In the as-cast state, the addition of Nb caused the alloy to precipitate Laves phases, forming a two-phase structure with the FCC matrix phase. The MEA exhibited hypoeutectic behavior when x < 0. 4, hypereutectic behavior when x > 0. 4, and eutectic behavior when x = 0. 4. Among these, the fully eutectic structure of $CoCrNiNb_{0.4}$ exhibited excellent comprehensive mechanical properties, with a yield strength of 1151. 46 MPa, a compressive strength of 2219. 00 MPa, and plasticity maintained at 20. 2%. Following heat treatment of $CoCrNiNb_{0.4}$, it was observed that the FCC phase in the alloy structure transformed into the HCP phase, the Laves phase transformed into the σ phase, and small amounts of $NbNi_3$ and $NbCo_3$ phases precipitated. The yield strength and compressive strength of the alloy were significantly improved after heat treatment. At 900 ℃, the highest values were achieved, with a yield strength of 2480. 90 MPa and a compressive strength of 2859. 94 MPa, representing increases of 59. 91% and 28. 88%, respectively.

Keywords: medium entropy alloys; eutectic structure; heat treatment; mechanical properties

Proceedings of IFHTSE 2025

The 30th World Congress of International Federation for Heat Treatment and Surface Engineering
August, 19-22, 2025, Suzhou, China

Mapping Precipitate-Slip/twin Interactions in Mg-Sn Alloys: Insights From Strain Partitioning via HRDIC-EBSD Multimodal Analysis

Zeng Ying*, He Siyu, Yin Dongdi
School of Material Science and Engineering, SouthwestJiaotong University, Chengdu, China
E-mail address: clzy@swjtu.edu.cn (Zeng Ying)
E-mail address: 15508113181@163.com (He Siyu)

Abstract: Although precipitation strengthening has made significant progress for Mg alloys recently, quantitative understanding of the strain partitioning induced by precipitate-slip/twin interactions remains unclear. In this study, a multimodal analysis combining high-resolution digital image correlation (HRDIC) and electron backscatter diffraction (EBSD) was performed based on approximately 12 million measured strain data to map these interactions during compression at the nano-scale for a precipitate-strengthened Mg-5Sn alloy. Results show that strain localization for slip bands was reduced by 56-87% during interactions with coarse precipitates (>500 nm), markedly exceeding the ~20% reduction induced by fine precipitates (<500 nm). Similarly, strain localization for twins was reduced by 80-92% during interactions with coarse precipitates, inhibiting further twin propagation. Fine precipitates caused only a 30-38% reduction for strain localization. This work provides quantitative mapping of strain partitioning due to precipitate-slip/twin interactions at the nano-scale for the first time, offering new insights into understanding of precipitation strengthening in Mg alloys.

Keywords: Mg-Sn Alloy; HRDIC; Precipitate-Slip/Twinning interactions; Strain partitioning

Effect of Microalloying on the Microstructure and Properties of Cu-Cr-X Alloys for High-End Connectors

Du Yibo
Luoyang Institute of Science and Technology, Luoyang, China
E-mail address: dyb1408@126.com (Du Yibo)

Abstract: As a key component of electronic systems to transmit current and signals, connectors are widely used in the fields such as aerospace, electronic communications, and new energy vehicles. With the rapid development of the aforementioned fields, connector structures are becoming increasingly miniaturized and precision-oriented, and their service conditions are becoming more stringent. Therefore, copper alloy materials used for connectors are required to have high strength, high electrical conductivity, and better stress relaxation resistance. Precipitation sequence of Cu-Cr alloy was: supersaturated solid solution → fcc Cr-rich phase → bcc Cr-rich phase. The interaction energies between Si, Ti and Cr were −0.27 eV and −0.47 eV, respectively, formation of Si-Cr and Ti-Cr clusters served as heterogeneous nucleation particles of Cr-rich phase, and the bcc Cr-rich phase were fined. Because the interaction energies between Si, Ti and Cr were −0.27 eV and −0.47 eV, respectively, formation of Si-Cr and Ti-Cr clusters served as heterogeneous nucleation particles of Cr-rich phase, and the bcc Cr-rich phase were fined. Zr element promoted the fcc→bcc structure transition of the Cr-rich phase, and segregated at the interface between the Cr-rich phase and the matrix to hinder the coarsening of the bcc Cr-rich phase along the long axis direction, so the precipitates were refined by Zr element. The stress relaxation curve of the Cu-Cr alloy shows that in the stage of a sharp stress decrease, the strain rate decreases as the stress decreases. The obtained stress exponent ranges from 9.24 to 16.74, which reveal that dislocation creep dominates the stress relaxation behavior of the Cu-Cr alloy. Further analysis show that the bcc Cr-rich phase is the dominant factor affecting the stress relaxation resistance of the alloy. The significant improvement in stress relaxation resistance by Si + Ti is mainly attributed to the fact that Si exerts a dragging effect on dislocation movement, while Ti refines the bcc Cr-rich phase and enhances the blocking effect on dislocation movement. Zr promotes the structural transformation of the Cr-rich phase and inhibits its coarsening, thereby increasing the blocking effect of the bcc Cr-rich phase on dislocation movement and improving the stress relaxation resistance of the alloy.

Keywords: connector; microalloying; precipitated phase; stress relaxation resistance; high-strength and high conductivity copper alloy

Microstructures and Properties of NiTiTa Complex Materials for Biomedical Applications

Zhou Shirui, Li Wei *

School of Materials and Chemistry, University of Shanghai for Science and Technology, Shanghai, China

E-mail address: zhoushirui2022@163.com (Zhou Shirui)

E-mail address: liwei176@usst.edu.cn (Li Wei)

Abstract: This work demonstrates the comprehensive investigation and optimization of NiTiTa shape memory alloys for biomedical applications, focusing on enhancing X-ray visibility, mechanical properties, and functional performance through precise control of Ta content (ranging from 5 to 10 at. %), Ni/Ti ratio (varied from 1.05 to 1.2), and heat treatment processes such as solution treatment and annealing. Key findings reveal that increasing Ta content significantly refines the microstructure, reducing average grain size from 61.6 μm to 7.5 μm, while also improving X-ray visibility, with brightness increasing by 46.8% at 5 at. % Ta and 79.1% at 10 at. % Ta compared to pure NiTi. Additionally, Ta content elevates martensitic start temperature (Ms) from 98 ℃ to 109.7 ℃ at a rate of 2.34 ℃ per at. % Ta, and optimizes mechanical properties, such as achieving peak ultimate tensile strength (UTS) of 777.4 MPa at 6 at. % Ta. Furthermore, heat treatment optimization, particularly at 550 ℃ annealing for Ni/Ti ratio of 1.1, enhances strength to 951 MPa and elongation to 16%, with minimal irreversible strain in superelasticity tests (<0.17% after 10 loading-unloading cycles). The study highlights the synergistic effects of β-Ta precipitation pinning and Orowan strengthening for improved ductility and functionality, addressing challenges like brittleness at high Ta levels (>7 at. %) and nickel ion release risks, thereby advancing the development of high-performance medical devices such as vascular stents and orthopedic implants while outlining future directions for material refinement and industrial scalability.

Keywords: Shape-Memory Alloys (SMAs); Nickel-Titanium-Tantalum (NiTiTa); Heat treatment; X-ray visibility; Microstructures; Mechanical properties

Proceedings of IFHTSE 2025

The 30th World Congress of International Federation for Heat
Treatment and Surface Engineering
August, 19-22, 2025, Suzhou, China

Experimental and Simulation Study on Multiple Necking Phenomena in SP700 Titanium Alloy During Superplastic Deformation

Wei Ke, Wang Fangbin, Wang Gaochao, Sun Qianjiang

School of Materials Science and Engineering, Nanchang Hangkong University, 330063, P. R. China

E-mail address: weike@ nchu. edu. cn (Wei Ke)

Abstract: As a β-phase enriched α + β dual-phase titanium alloy, SP700 titanium alloy exhibits a superplastic forming temperature approximately 120 ℃ lower than that of TC4 titanium alloy, opening a new pathway for the low-temperature forming of aerospace titanium alloy components. This study investigated the superplastic deformation behavior of SP700 titanium alloy using a combined approach of high-temperature tensile experiments and finite element simulation. Based on the constant strain rate method, key superplastic tensile data, including true stress-true strain curves and elongation rate, were obtained under four deformation temperatures (755 ℃, 775 ℃, 795 ℃, and 815 ℃) and three strain rates ($1 \times 10^{-2}\ s^{-1}$, $1 \times 10^{-3}\ s^{-1}$, and $1 \times 10^{-4}\ s^{-1}$). The results indicate that a maximum elongation of up to 792% was achieved under the conditions of 755 ℃ and $1 \times 10^{-4}\ s^{-1}$. This high superplastic elongation is attributed to a deformation mechanism involving distributed instability induced by the low strain rate, leading to multiple necking and re-extension phenomena. The extended dwell time at high temperature during low-speed deformation promotes grain coarsening, which weakens grain boundary sliding. Thereby, the continuous migration of the necking position delays the formation of crack. A superplastic constitutive model for the SP700 alloy was established based on the Backofen equation. Concurrently, the critical damage parameters under different conditions were calculated using the C&L damage model. These models were then implemented in the DEFORM-3D finite element software to simulate the high-temperature tensile process. The simulation results successfully reproduced the multiple necking phenomena and strain localization distribution processes that are difficult to capture experimentally. Furthermore, the simulated elongation rate results showed good agreement with the experimental data, validating the high predictive accuracy of both the constitutive model and the C&L damage model. This research provides reliable theoretical support for the precise low-temperature forming and engineering application of SP700 titanium alloy components.

Keywords: SP700 titanium alloy; high-temperature tensile; superplastic deformation; multiple necking; finite element simulation

Digitally Enabled Cold Spray Additive Manufacturing and Remanufacturing

LI Wenbo[1], Deng Sihao[2]

[1] National Engineering Laboratory for Modern Materials Surface Engineering Technology, Institute of New Materials, Guangdong Academy of Sciences, Guangzhou, 510650, China

[2] Université Marie et Louis Pasteur, UTBM, CNRS, Laboratoire Interdisciplinaire Carnot de Bourgogne ICB UMR 6303, 90010 Belfort cedex, France

Abstract: Gas dynamic cold spraying (CS), a member of the thermal spray family, distinguishes itself as a solid-state material deposition process that bypasses the melting phase of feedstock particles. During operation, pressurized inert gas (e. g., helium or nitrogen) is accelerated through a converging-diverging nozzle to supersonic velocities (300-1200 m/s), entraining metallic or composite powders (typically 5-50 μm in diameter). These particles impact the substrate at kinetic energies exceeding their plastic deformation threshold (critical velocity), resulting in severe localized shear instability at the particle-substrate interface. This mechanism induces mechanical interlocking and metallurgical bonding through adiabatic shear instabilities, preserving the original microstructure and phase composition of the feedstock. Compared to conventional thermal spray techniques, CS eliminates issues such as oxidation, thermal stress-induced cracking, and intermetallic phase segregation, making it particularly advantageous for deposition of oxygen-sensitive materials. The non-melting nature of CS enables exceptional deposition efficiency (>90% for ductile metals) and precise localized material addition, positioning it as a promising candidate for additive manufacturing (AM) and in-situ repair of high-value components. However, inherent limitations constrain its broader adoption. The minimum feature size achievable with commercial CS systems is limited to ~1-2 mm due to particle jet divergence and nozzle diameter constraints, restricting applications to low-precision geometries (e. g., axisymmetric rotors or bulk coatings). The continuous powder feeding and the static parameter control requirements complicate spraying path planning, making the manufacturing/repair of complex structural components highly challenging. Additionally, poor path planning generates excessive coatings, necessitating extensive post-processing, which undermines the inherent near-net-shape advantage of CS process. To solve these problems, weproposed a digitally enabled CSAM and remanufacturing concept, which can manufacture parts with many geometric features and repair complex shape defects in batches. This research can use CS to build complex parts layer by layer, greatly shortening preparation time and lowering costs. The new method's combined deposition-checking-finishing process uses real spray equipment to make near-net-shape samples.

Keywords: cold spray; additive manufacturing; repair; path planning

The Effect of Post Heat Treatment on the Microstructure and Mechanical Properties of IN718 Fabricated by Laser Direct Energy Deposition

Zhao Zhao[1], Jin Yinman[1], Liu Shanxiu[1], Chen Zheng[1], Massimo Pellizzari[2], Xuewei Yan[1*]

[1] Zhengzhou University of Aeronautics, 450046, School of Aero Engine, No. 15, Wenyuan West Road, Zhengzhou, HN, China

University of Trento, 38123, Department of Industrial Engineering 9, via Sommarive, Trento, TN, Italy

E-mail address: zhaozhao@zua.edu.cn (Zhao Zhao)

E-mail address: jym13223916610@163.com (Jin Yinman)

Abstract: Laser Direct Energy Deposition (L-DED), an important metal additive manufacturing technology, enables the design and fabrication of complex, large-scale components, drawing significant attention from industries such as aerospace, automotive, and marine engineering. Inconel 718, a typical γ'' precipitation-strengthened nickel-based superalloy, has been extensively studied for L-DED processing. However, the mechanical properties of as-built Inconel 718 are inferior to those of its wrought counterparts due to the formation of harmful embrittling phases caused by strong microsegregation during the rapid solidification inherent to L-DED. Therefore, post-heat treatment is essential to tailor the microstructure and enhance mechanical performance. In this work, tailored post-heat treatment strategies were applied, and their effects on the microstructure, phase constitution, microhardness, and mechanical properties were systematically investigated. The results offer insights into the microstructural evolution induced by post-heat treatment and help identify an appropriate heat treatment window for L-DED Inconel 718, thereby supporting its broader application in advanced engineering fields.

Keywords: laser direct energy deposition; Ni superalloy; heat treatment; microstructure; mechanical properties

Innovations in Laser-Directed Energy Deposition of High-Strength Aluminum Alloys

Tan Chaolin

Soochow University

E-mail address: cltan@ suda. edu. cn (Tan Chaolin)

Abstract: High-strength aluminum (Al) alloys (e. g. , AA7000 series alloys) are widely used in aerospace due to their high strength-to-density ratio and superior ductility; however, Additive manufacturing (AM) high-strength Al alloys are still challenging due to the intrinsic properties. Although some success in processing high-strength Al alloys has been achieved with laser powder bed fusion (LPBF), these solutions have not yet been translated effectively to laser-directed energy deposition (LDED) process due to evident differences in deposition environment and thermal histories. State-of-art LDED Al alloys suggest a bottleneck in achieving yield strength higher than 350 MPa. This work will introduce materials and process innovations in LDED Al alloys to obtain high strength and high ductility. Several novel high-strength Al alloys customized for LDED will be introduced in terms of printability examination, microstructure evolution and mechanical properties. Innovative LDED processes, such as green laser deposition and coaxial wire deposition, are also investigated to understand the effect of process on the achievable mechanical properties. The synergic invocations in materials and processes achieved an excellent combination of strength-ductility in LDED Al alloys (with yield strength higher than 500 MPa), superior to a wide range of Al alloys processed by LDED

Keywords: additive manufacturing, 7075 aluminum alloy; ceramics reinforced metal composite; high strength aluminum alloy

Research on Residual Stress Control of Laser Forging Repair for Welding Cracks in Marine Energy Equipment

Zhang Yongkang[1], Lu Yi[2]

[1] School of Mechanical and Electrical Engineering, Guangdong University of Technology, China

[2] School of Mechanical and Electronic Engineering, Nanjing Forestry University, China

Abstract: The development of deep-sea crude oil is a national strategy, and deep-sea oil drilling platforms, crude oil transfer equipment, and offshore floating production and storage vessels are key equipment for deep-sea crude oil drilling, generation, and transfer, and are national treasures. These equipment have been in service at sea for a long time and are difficult to maintain. The conventional damage welding repair method only considers whether the cracks and pores exceed the standard, without considering the residual tensile stress after repair. In harsh service sea conditions, residual tensile stress can easily lead to fatigue crack damage, resulting in catastrophic accidents. In order to address the safety hazards of residual tensile stress in welding repair, this paper conducted research on laser forging fatigue repair, established a model of laser forging stress evolution, and obtained the evolution laws of residual stress, porosity, cracks, grain size, and fatigue life of laser forging. By utilizing AI technology, non-destructive X-ray residual stress measurement points were established based on limited surfaces, and the three-dimensional residual stress distribution along the depth direction of the weld repair area was reconstructed, providing a theoretical basis for the evaluation of fatigue life.

Keywords: laser forging; repair; fatigue life; drilling platform; floating production storage and offloading vessels (FPSO); crude oil transfer equipment (CTV)

High-Speed Laser-Directed Energy Deposition of Ti/Al Multi-Material Component

Wu Dongsheng[1*], Hisaya Komen[2], Manabu Tanaka[2], Sun Junhao[3], Li Zhuguo[3], Chen Ji[1], Wu Chuansong[1]

[1] Key Laboratory for Liquid-Solid Structure Evolution and Materials Processing, Institute of Materials Joining, Shandong University, China

[2] Joining and Welding Research Institute (JWRI), Osaka University, Japan

[3] Shanghai Key Laboratory of Material Laser Processing and Modification, Shanghai Jiao Tong University, China

E-mail address: dongsheng.wu@sdu.edu.cn (Wu Dongsheng)

E-mail address: h.komen.jwri@osaka-u.ac.jp (Hisaya Komen)

Abstract: A crack-free titanium/aluminum (Ti/Al) multi-material component (MMC) was fabricated by a high-speed laser-directed energy deposition process, in which the Ti and intermetallic compound (IMC) coatings improved the microhardness, electrochemical corrosion and wear performances, and the bottom Al alloy substrate reduced the material cost. The element content and microstructure were investigated by electron-probe microanalyzer (EPMA), electron backscatter diffraction (EBSD), and transmission electron microscopy (TEM). The formation enthalpies of Ti-Al IMCs were calculated by a first-principles model. The formation mechanism of the dominant α2-Ti3Al IMC was elucidated. The low Al content and high cooling rate caused the dominant α2-Ti3Al formation in the IMC coating. The Mg element promoted the α2-Ti3Al IMC formation, while the Si element had a minor influence on it. The study would provide the basis for manufacturing crack-free, anti-corrosion and wear-resistant synergistic reinforced coatings.

Keywords: high-speed laser-directed energy deposition; synergistic reinforcing coatings; Ti3Al phase; intermetallic compound

Research on Microstructure Evolution of Laser Forging Repair for Welding Cracks in MarineEnergy Equipment

LIAO Xinlong[1], ZHANG Yongkang[1*]

[1] Key Laboratory of Advanced Manufacturing Technology for Marine Energy Facilities in Guangdong Province, School of Mechanical and Electrical Engineering, Guangdong University of Technology, Guangzhou 510006, China

Abstract: Crude oil transfer equipment and Floating Production Storage and Offloading (FPSO) vessels are critical assets for deep-sea oil exploration, production, and transfer. Welded joints, serving as key connection units in marine energy equipment, have their service life at sea fundamentally determined by the performance of their microstructure. As a novel low-stress, fatigue-resistant welding repair technique, Laser Shock Forging (LSF) holds significant promise for substantially improving the microstructural characteristics and mechanical properties of welded joints in marine energy equipment. The application of LSF technology to repair critical components of marine structures effectively addresses the problem of crack initiation and subsequent fatigue failure under harsh marine service conditions. This study investigates the microstructural evolution during LSF repair. The microstructure of LSF-repaired specimens was characterized using Optical Microscopy (OM) and Field Emission Scanning Electron Microscopy (FE-SEM), while crystallographic features were analyzed via Electron Backscatter Diffraction (EBSD). This approach elucidated the evolution mechanisms governing defects such as pores, cracks, and coarse grains induced by the laser forging process. The results demonstrate that the high-energy shock waves introduced significant plastic deformation within the weld zone, leading to grain refinement and the formation of a graded microstructure. These findings provide a crucial theoretical basis for subsequent fatigue life assessment.

Keywords: laser forging; repair; microstructure; drilling platform; floating production storage and offloading vessels (FPSO); crude oil transfer equipment (CTV)

Research on Additive Manufacturing and Hot Isostatic Pressing for Biomedical Pure Tantalum

Liu Xin*, Qin Feng, Tan Chong

Institute of New Materials, Guangdong Academy of Sciences, Guangzhou, China

E-mail address: liuxin@gdinm.com (Liu Xin)

E-mail address: qinfeng0926@163.com (Qin Feng)

Abstract: In recent years, with the growing demand for bone repair materials in the biomedical field, the development of metallic implant materials with bone-like properties has become a research hotspot. Traditional biomedical metallic materials face challenges in meeting long-term implantation requirements due to their high modulus and potential cytotoxicity. Tantalum, with its excellent biocompatibility and osteogenic properties, is regarded as an ideal next-generation bone implant material. However, as a refractory metal, tantalum's high melting point makes it difficult to achieve precise fabrication of complex tantalum implant structures using conventional processing methods, limiting its ability to meet personalized customization needs for bone implants. This study employs laser powder bed fusion (LPBF) technology to systematically establish a full-process control methodology encompassing "powder characteristics-forming process-heat treatment," successfully optimizing the performance of both solid and porous tantalum materials. Through the optimization of radiofrequency plasma atomization, high sphericity and controlled oxygen content (as low as 150 ppm) were achieved in the 3D-printed tantalum powder. It was found that high-oxygen powder leads to non-equilibrium grain boundary oxygen segregation and intergranular brittle fracture in LPBF-processed tantalum samples. Based on the regulation of an appropriate oxygen content window, an in-situ heat treatment process utilizing a double-layer scanning strategy was employed to refine the grain structure of LPBF-processed tantalum samples while simultaneously enhancing their strength and ductility. Furthermore, an innovative low-temperature hot isostatic pressing (HIP) process was developed. Both 850 ℃ and 1350 ℃ HIP treatments effectively closed internal micropores in the tantalum samples without causing microstructural coarsening, leading to a simultaneous improvement in strength-ductility balance and high-cycle fatigue performance. Notably, the 1350 ℃ HIP process increased the fatigue life of tantalum samples from 10^5 to 10^7 cycles under a stress of 300 MPa. For porous tantalum materials, the 850 ℃ HIP process significantly enhanced static mechanical properties and high-cycle fatigue performance while maintaining structural integrity. In contrast, severe oxidation and performance degradation occurred during the 1350 ℃ HIP treatment of porous tantalum samples.

Keywords: pure tantalum; laser powder bed fusion; hot isostatic pressing; mechanical properties; biomedical applications

Prediction of the Short-Range Order in Fe-Ni-Cr Alloys

Kirill Osintsev*, Chen Xizhang

Wenzhou University, Wenzhou, China

E-mail address: kirilloss@yandex.ru (Kirill Osintsev)

E-mail address: chenxizhang@wzu.edu.cn (Chen Xizhang)

Abstract: In this study, we investigated the local chemical ordering of ternary Fe-Ni-Cr alloys. Short-range order (SRO) parameters, reflecting the probability for unlike pairs of atoms to have correlated site occupations, were calculated in the temperature range of 600 to 1200 K using the non-interacting molecule method. Each composition of FexNiyCrz alloys (where $20 \leqslant x, y, z \leqslant 60$ at. %, with the step of 1 at. %) was represented as a face-centered cubic crystal consisting of non-interacting pair clusters. The pair probabilities, dependent on the effective pair interaction energies, were calculated by molecular dynamics simulations at the 1st nearest neighbour distance. The results were validated against experimental SRO parameters obtained for a Fe56Ni23Cr21 single-crystal alloy annealed at 1273 K for 1 hour, and at 773 K for 10 hours, followed by quenching. The same ordering tendency as observed experimentally was found for Ni-Cr pairs (attractive interactions), while Fe-Ni and Fe-Cr pairs showed a low clustering tendency (repulsive interactions). The relative errors for Fe-Ni and Fe-Cr pairs were less than 5%, demonstrating a strong correlation with the experimental values. However, although the calculated SRO parameters for Ni-Cr pairs reflected the same tendency for attraction, they exhibited a relative error of 60%.

Keywords: short-range order; Fe-Ni-Cr; molecular dynamics; non-interacting molecule method

The $D0_{22}$ Precipitation Strengthens Behavior of $CoCrFe_{0.25}Ni_{1.75}Nb_{0.2}$ High Entropy Alloy by Laser Additive Manufactured

Liao Tianhai[1], Liu Qibin[1*]

Guizhou university

Abstract: In this study, the precipitation strengthening behavior of a Nb-bearing $CoCrFe_{0.25}Ni_{1.75}$-based high-entropy alloy (HEA) was investigated, focusing on coherent tetragonal $D0_{22}$ nanoprecipitates with the I4/mmm space group. The samples were fabricated via laser additive manufacturing (LAM) without metallurgical defects and subsequently subjected to post-heat treatment to induce nanoscale precipitation. The material was homogenized at 1200 ℃ for 2 h and aged at 650 ℃ for 24 h and 48 h. Heat treatment transformed the as-deposited dendritic microstructure into large equiaxed grains, accompanied by the precipitation of nanoscale, olive-like $D0_{22}$ phases. Aging for 48 h significantly enhanced the mechanical properties of HEA, increasing its yield strength to ~976.7 MPa and tensile strength to ~1133.1 MPa. Density functional theory (DFT) calculations revealed that while Nb addition increased the stacking fault energy, it also reduced the energy barrier for dislocation formation. Furthermore, the incorporation of 6 at% Fe improved the mechanical performance of the nano-precipitated phase by replacing Ni atoms in Ni_3Nb. These findings provide valuable insights into the role of superlattice $D0_{22}$ structures in HEAs, contributing to the development of high-strength, precipitation-hardened alloys.

Deposited Ductile-GPa CoCrNi-based FCC Medium Entropy Alloy with Continuously Precipitation by Directed Energy Deposition-Arc

Chen Xizhang

Wenzhou University

E-mail address: chenxizhang@ wzu. edu. cn (Chen Xizhang)

Abstract: An $L1_2$-strengthened Co30Cr18Ni42Al5Ti5 medium-entropy alloy was fabricated via directed energy deposition (DED) -Arc technique, focusing on investigating the modulation mechanism of the process on $L1_2$ phase precipitation behavior. The results showed that the high heat input and moderate cooling rate features of DED-Arc process effectively suppressed the discontinuous precipitation (DP) behavior: the coarse columnar crystal structure significantly reduces the number of grain boundaries; moderate cooling rate promotes homogeneous distribution of Al/Ti elements and eliminates grain boundary segregation; This "coarse grain-element homogenization" synergy results in a high-density distribution of the $L1_2$ phase within the grain through a continuous precipitation (CP) behavior. Directly deposited alloys exhibit gigapascal strength (~ 1090 MPa) and high uniform elongation (~ 28.4 %) . Furthermore, subsequent heat treatment of the deposited alloys confirmed the thermal stability of the continuous $L1_2$ precipitation, with increased $L1_2$ phase fraction while maintaining "FCC + $L1_2$" structure. This work provides guidance for the fabrication of L_{12}-strengthened high-entropy alloys and medium-entropy alloys with excellent mechanical properties by additive manufacturing.

Keywords: medium-entropy alloys; additive manufacturing; L12 phase; continuous precipitation

Laser Additive Manufacturing, Powder Production, Structure Design, Electrochemical Polishing, Blue Laser

Wang Hongze

Shanghai Jiao Tong University, China

E-mail address: hongzewang1@163.com (Wang Hongze)

Abstract: Laser additive manufacturing technology provides the possibility for rapid manufacturing of lightweight and high-strength aluminum alloy structures. Facing the urgent demand for batch manufacturing of high-performance aluminum alloy components in aviation, robotics and other fields, a series of work has been carried out around aluminum alloy powder, additive manufacturing process and structure. Through composition design, a series of in-situ mixed TiB2 particle-reinforced aluminum alloy powders were invented, which successfully transformed highly crack-sensitive materials into printable materials, and optimization of equipment and process for gas atomization powder production has been achieved based on in-situ monitoring and continuous-discrete coupling simulation ; through in-situ monitoring of the additive manufacturing process through synchrotron radiation imaging and multi-physics field coupling modeling, the laser additive manufacturing process was optimized, and the defects of the laser additive manufacturing process were significantly reduced; through the development of additive manufacturing-oriented Structural design and electrochemical polishing process research on the inner surface of complex flow channels for additive manufacturing, and high-performance complex structures such as lattice-filled large aircraft fan blades and bionic robot feet have been obtained. At the same time, a 2000W blue laser powder deposition additive manufacturing, repair and welding equipment has been developed, which has significantly improved the efficiency and precision of aluminum alloy laser manufacturing. This report proposes a technical route for mass-scale additive manufacturing of high-performance aluminum alloy components, and provides ideas for the large-scale application of aluminum alloy additive manufacturing technology.

Keywords: laser additive manufacturing; blue laser; laser cladding; laser repairment

Computational Analysis of Residual Stress Development in Laser Additive Manufacturing of γ-TiAl Alloys

Zhou Zheng[a], Tong Daming[a*], Gu Jianfeng[a,b]

[a] Institute of Materials Modification and Modelling, School of Materials Science and Engineering, Shanghai Jiao Tong University, Shanghai 200240, China

[b] Shanghai Key Laboratory of Materials Laser Processing and Modification, Shanghai Jiao Tong University, Shanghai 200240, China

E-mail addresses: daming_ tong@ sjtu. edu. cn (Daming Tong)

Abstract: Predicting residual stresses in directed laser deposition (DLD) additive manufacturing is essential for mitigating component cracking and improving mechanical integrity. Conventional experimental methods for assessing residual stresses across different processing conditions are resource-intensive. This study employed a coupled thermo-mechanical finite element model, integrating temperature-dependent material properties and element activation/deactivation strategies, to simulate the multi-layer DLD fabrication of γ-TiAl alloys (nominal composition: Ti-48Al-2Cr-2Nb). Validation against experimental thermocouple measurements and X-ray diffraction residual stress analysis confirmed the model's accuracy, with acceptable prediction errors for both temperature and stress fields. Simulations identified that the most significant residual stresses were localized at the wall-substrate interface along the primary laser scanning direction, primarily induced by the substantial thermal gradient between deposited material and the substrate during initial layer formation. Parametric investigations demonstrated that increasing laser power significantly reduced interfacial stresses by promoting enhanced substrate melting and stress distribution uniformity. Conversely, decreasing scanning speed markedly lowered internal stress variations. Furthermore, substrate preheating substantially diminished interfacial stresses by alleviating thermal mismatch.

Keywords: additive manufacturing; γ-TiAl alloy; residual stress; temperature-stress coupling; numerical simulation

Enhanced Strength-Thermophysical Synergy of Directed Energy Deposition-Arc Manufactured Inconel-Copper Bimetallic via Cooperative Heat Treatment

Liu Kun

Wenzhou university

E-mail address: liukunwz@163.com (Liu Kun)

Abstract: Achieving a synergistic balance between strength and service-related thermophysical properties remains a significant challenge in structural material. This study addresses this issue by employing cooperative heat treatment strategies to optimise Inconel-copper bimetallic structures fabricated via Directed Energy Deposition-Arc (DED-Arc) . Phase diagram calculations guided the design of these heat treatments. Comparative analyses revealed that the solution plus aging treatment (STA2#) achieved a balance between high strength and ductility with a tensile strength of 342.4 MPa and an elongation of 22.2%, while maintaining excellent thermal conductivity. Digital image correlation revealed localised deformation in the C18150 region, and the improved mechanical properties were attributed to grain boundary and precipitation strengthening. The thermophysical properties were significantly improved by the designed cooperative heat treatment, with the thermal conductivity of the STA2# bimetallic ranging from 41.16 to 56.08 W/ (mk) over a temperature range of 30 to 310 ℃, approximately 13% improvement than the as-fabricated sample. This improvement in thermal conductivity resulted from the decomposition of the supersaturated solid solution. This study provides a promising heat treatment strategy for advancing Inconel-copper bimetallic in next-generation rocket engine potential application.

Keywords: additive manufacturing; directed energy deposition; inconel-copper

Key Technologies for Laser Surface Modification and Life Extension of Metal Components in Complex Service Environments

Huang Shu

Jiangsu University, China

E-mail address: huangshu@ ujs. edu. cn (Huang Shu)

Abstract: Key components of aviation and marine engineering equipment face severe challenges such as fatigue damage, strength mismatch, and hydrogen induced fracture in extreme service environments such as high temperature, heavy load, and marine corrosion. Aiming at different material systems, usage environments, and failure modes, innovative methods of warm laser peening and new processes of ultrasonic laser composite shock processing strengthening have been proposed. These methods have overcome technical challenges such as controllable loading of shock wave pressure, stable stress/tissue strengthening effects, matching of strength and toughness properties, and uniform processing of complex structures. Original basic theory, engineering technology, and complete set of equipment for multi energy field composite laser peening for life extension have been formed, achieving high-temperature fatigue life extension and strength toughness matching of metal components, greatly improving the service life of aviation equipment. At the same time, the application of laser peening technology in hydrogen rich marine environments, has revealed the mechanism of hydrogen permeation and diffusion in the surface of ultra-high strain rate plastic deformation suppression materials, elucidated the stress/tissue synergistic anti hydrogen induced cracking and life extension mechanism, and effectively improved the service performance of key components of marine engineering equipment.

Keywords: laser fybrid hardening; shape regulation; property regulation; industrial applications

Composition Optimization and Performance Regulation of Hastelloy-X Alloys Fabricated by Laser Powder Bed Fusion

Chen Xiaohui

ShanghaiJiao Tong university, China

E-mail address: chenxiaohui0310@ sjtu. edu. cn (Chen Xiaohui)

Abstract: Laser Powder Bed Fusion (LPBF) has significant potential for fabricating Hastelloy-X nickel-based alloy for hot-end components in aerospace industry. However, the high temperature properties of LPBF Hastelloy-X alloy is insufficient due to detrimental phases precipitation and grain boundary weakening, greatly limiting its further applications. This project aims to enhance the solid solution strengthening effect and regulate the precipitation phase strengthening at grain boundaries by Composition Optimization. Based on such compositional optimization design, a new type of Hastelloy-X alloy with excellent comprehensive high temperature mechanical properties will be developed. The results will provide theoretical and technical support for the high-reliability and high performance manufacturing of hot-end components.

Keywords: hastelloy-X alloys; laser powder bed fusion; elevated temperature property

Microstructure Toughening Design and Mechanical Properties of Refractory High-Entropy Alloys for Laser Additive Manufacturing

Feng Kai

School of Materials Science and Engineering, Shanghai Jiao Tong University, Shanghai 200240, China

E-mail address: fengkai@ sjtu. edu. cn (Feng Kai)

Abstract: NbMoTaW refractory high entropy alloy (RHEA) exhibits excellent mechanical properties at ultra-high temperatures, but its room temperature brittleness limits its application in aerospace and other fields. In this study, two refractory high entropy alloys, $NbMoTaW_{100-x}C_x$ and $NbMoTaWTi_x$, were prepared by selective laser melting and the brittleness resistance of NbMoTaW alloy at room temperature was improved by alloying method. The results showed that the addition of 0. 5 at. % C significantly improved the formability and mechanical properties of NbMoTaW RHEA at room temperature. The yield strength, ultimate compressive strength and strain rate of $(NbMoTaW)_{99.5}C_{0.5}$ RHEA were increased to 1695 MPa, 1751 MPa and 6. 9%, respectively. With the increase of Ti content, the strength and plasticity of $NbMoTaWTi_x$ RHEA were also increased simultaneously. This study provides a new research idea for the fabrication of NbMoTaW refractory high entropy alloys with high strength and toughness.

Keywords: refractory high entropy alloy; selective laser melting; alloying method; microstructure; mechanical properties

Hot Compressive Deformation of W-modified $FeCr_2V$-Based Medium Entropy Alloys: Microstructure and Strengthening Mechanism Studies

Wang Yangfan[a], Wang Zhiyang[b*], Zhu Hanliang[a,b*], Qiu Zhijun[a,b], Pan Zengxi[a], Zhu Hongtao[a], David Wexler[a], Chen Xizhang[c], Zhang Mingxing[d], Li Huijun[a]

[a] School of Mechanical, Materials, Mechatronic and Biomedical Engineering, University of Wollongong, Northfields Avenue, Wollongong, NSW 2522, Australia

[b] Australian Nuclear Science and Technology Organisation (ANSTO), Sydney, NSW 2234, Australia

[c] School of Mechanical and Electrical Engineering, Wenzhou University, Wenzhou, 325035, China

[d] Division of Materials, School of Engineering, University of Queensland, St Lucia, Brisbane, Qld 4072, Australia

Abstract: The microstructure and mechanical properties of as-cast $FeCr_2VWx$ (x = 0, 0.1, 0.3, 0.5) medium-entropy alloys (MEAs) were investigated. The samples were prepared using the arc-melting technique under an argon atmosphere. $FeCr_2VWx$ (x = 0, 0.1, 0.3) MEAs exhibit a dual body-centred-cubic (BCC) structure. With the further addition of W, a new W-rich phase was observed in the microstructure of the $FeCr_2VW_{0.5}$ sample. Ambient and elevated temperature compressive test results revealed that the compressive performance of the MEAs improved with the introduction of W. Among them, $FeCr_2VW_{0.3}$ demonstrated an exceptional combination of enhanced strength and ductility, comparable to existing low activation HEAs at ambient temperature and 1223 K, evidenced by a yield strength of 1380 MPa and 586 MPa and an elongation exceeding 30% for both measurements. The excellent high-temperature performance was attributed to strengthening effects of precipitates and solid solution resulting from W addition. Moreover, it was found during the hot deformation process that $FeCr_2VWx$ MEAs with higher W content (above 0.3) resulted in more work hardening than that of the lower W contents, and required more dynamic recrystallization to achieve the dynamic equilibrium. These results provide valuable insights for the future development and microstructural engineering of new MEAs.

Keywords: Medium entropy alloy (MEAs); microstructure; high temperature mechanical properties; strengthening mechanism, low activation alloys

Laser Cleaning of Paint Coating From Aluminum Alloys and Composites As a Base Material

Szymon Tofil

Kielce University of Technology, Poland

E-mail address: L. Wang@ mail. sic. ac. cn (Szymon Tofil)

Abstract: This paper presents a comparative study on laser cleaning of paint coatings from aluminum alloys and composite substrates using different pulsed laser systems. The research aimed to determine optimal treatment parameters by varying the laser beam power density. Two laser sources were employed: the TruMicro 5325c (343 nm, UV) and the SPI G3. 1 SP40P (near-infrared). Samples with various technological coatings underwent multiple laser treatment variants, differing in pulse energy and number of pulses. Surface characterization was conducted using a Hirox KH-8700 digital microscope and a JEOL JSM-7100F scanning electron microscope (SEM) with energy-dispersive X-ray spectroscopy (EDS) for micro-area chemical composition analysis. The cleaned surfaces showed no evidence of melting; only coating removal effects were observed. A high degree of surface cleanliness was achieved after treatment. For thick coatings, repeated passes or the use of a higher average power laser is recommended to enhance cleaning efficiency.

Keywords: laser; SEM; UV; IR; cleaning; composites; aluminum alloys

Effect of Al Element on the Microstructure and Corrosion Behavior of Laser-Cladded Inconel 625 Alloy Coatings in SCO_2 Environment

Zhu Junjie, Wu Guolong*, Wang Jie, Wang Ye, Zhang Qunli, Yao Jianhua*

Zhejiang University of Technology, Hangzhou, China

E-mail address: glwu@ zjut. edu. cn (Wu Guolong)

E-mail address: lam@ zjut. edu. cn (Yao Jianhua)

Abstract: With the rapid development of economy and the growth of global energy demand, efficient and clean energy technology has attracted much attention. The Brayton cycle system, which uses supercritical carbon dioxide (SCO2) as the circulating medium, has great potential for application in the energy field due to its high circulating efficiency, simple structure and small size. However, SCO2 can cause severe corrosion on key structural components of the circulation system, resulting in a significant reduction in their service life. Nickel-based alloys usually show better overall performance in SCO2 environment, among which Inconel 625 is widely used in various extreme environments due to its excellent corrosion resistance and good high-temperature performance. Al, as an alloying element, can form a protective oxide layer, which plays an important role in improving the corrosion resistance to SCO2. Laser cladding technology is a widely used means of surface modification, characterized by high flexibility, low heat input and good economy, and is a potentially effective means to improve the service life of structural components in SCO2 environments. Based on this, this paper uses laser cladding technology to prepare Al-containing nickel-based alloy cladding layers with excellent SCO2 corrosion resistance by adding Al elements to Inconel 625 alloy. The effects of Al on the formation, microstructure, properties and corrosion behavior of Inconel 625 cladding in supercritical carbon dioxide environment are also investigated.

Keywords: laser cladding; Al element; microstructure; supercritical carbon dioxide; corrosion behavior

Mechanical Behavior and Deformation Mechanism of Ni50Cr20Fe20Al5Ti5 Medium-Entropy Alloys

Ma Shengguo

Taiyuan University of Technology, China

E-mail address: mashengguo@tyut.edu.cn (Ma Shengguo)

Abstract: A Co-free non-equiatomic $Ni_{2.5}CrFeAl_{0.25}Ti_{0.25}$ medium-entropy alloy (MEA) with an excellent strength-ductility synergy was fabricated, which shows a three-level phase structure composed of face-centered cubic (FCC), $L1_2$ (ordered FCC), and Cr-rich body-centered cubic (BCC) phase by thermomechanical processing. Specifically, the aged sample displays the outstanding yield tensile strength (YTS, ~1.2 GPa), ultimate tensile strength (UTS, ~1.6 GPa) and work-hardening rate (WHR, ~4.5 GPa) values as well as an acceptable tensile plasticity of ~16.6%. Theoretical calculations suggest that precipitation strengthening significantly contributes to achieving the fascinating tensile strength among various strengthening contributors. Further analyses reveal that multiple nanoscale stacking-fault (SF) networks are activated during plastic deformation in the aged alloy. Accordingly, the dual effects consisting of the three-level phase structure and SF networks lead to the combination of excellent tensile strength and strain-hardening capacity. Furthermore, the $Ni_{2.5}CrFeAl_{0.25}Ti_{0.25}$ MEA also shows an excellent strength-ductility combination at cryogenic temperature, in which the low-temperature tensile yield strength and ultimate tensile strength were respectively about 1.4 GPa and 2 GPa, followed by about 21% tensile plasticity.

Keywords: high/medium entropy alloys; heterogeneous structure; mechanical property; microstructure

Research on Shape-Property Regulation in Laser Hybrid Solid-State Phase Transformation Strengthening

Zhang Qunli
Zhejiang University of Technology
E-mail address: 30983343@ qq. com (Zhang Qunli)

Abstract: Driven by China's " Carbon peak, Carbon neutrality" strategy demanding high-performance, low-energy-consumption, and low-emission manufacturing, advanced heat treatment technologies are critical. Laser solid-state phase transformation, as a key surface engineering technology, offers inherent advantages like rapid heating/cooling, minimal distortion, elimination of quenching media, high efficiency, precise localization, low energy consumption, and zero pollution. However, achieving precise geometry-property regulation remains a challenge with conventional single-source laser quenching. Limitations include constrained and inconsistent hardening depth/width, uncontrollable microstructure, and difficulty in synergistically enhancing key properties like wear resistance, fatigue strength, and strength-toughness balance. Laser hybrid solid-state phase transformation strengthening technology addresses these limitations head-on. By leveraging synergistic multi-heat source, multi-energy field, and multi-process interactions, this approach enables precise regulation of both the geometry and properties of the strengthened zone. Specifically, it achieves: controlled and extended hardening depth and width, tailored to component design requirements. Enhanced and balanced properties including wear resistance, fatigue strength, core strength, toughness, and minimized distortion. It has become the core technology of high-performance surface manufacturing of key components of high-end equipment under complex working conditions, and provides technical support for the large-scale industrial applications of laser heat treatment. This report summarizes research progress in the fundamental mechanisms and hybrid process development for laser hybrid solid-state phase transformation strengthening. Through extensive analysis of industrial application cases, it demonstrates the successful implementation of this technology in achieving targeted shape-property regulation for critical components across energy power, rail transportation, and tool/die industries.

Keywords: laser Hybrid hardening; shape regulation; property regulation; industrial applications

Effect of Annealing, Quenching and Tempering Temperatures on the Microstructure and Properties of LPBF 3D-Printed Ti6Al4V

VanTrung Trinh[1*], HoangAnh Pham[2]

[1] School of Materials Science and Engineering, Hanoi University of Science and Technology, Vietnam

[2] Faculty of Materials for Energy, Shimane University, Japan

E-mail address: trung. trinhvan@ hust. edu. vn (VanTrung Trinh)

E-mail address: anhpham@ riko. shimane-u. ac. jp (HoangAnh Pham)

Abstract: Post-fabrication heat treatments are considered essential for AM Ti-6Al-4V parts to modify the as-fabricated microstructure, relieve residual stresses, improve mechanical properties, and ensure uniformity. This paper aims to investigate the effects of various heat treatment temperatures, including annealing temperature, quenching temperature, and tempering temperature on the microstructure and mechanical properties of Ti-6Al-4V alloy fabricated by LPBF 3D printing. Specimens were fabricated using LPBF and subjected to different heat treatment routes. These included annealing (furnace cooling) at 950 ℃ and 1010 ℃, and solution treatment followed by tempering. Solution treatment involved heating to 950 ℃ or 1010 ℃, holding for 30 minutes, and water quenching. Subsequent tempering was performed at 350, 450, and 550 ℃ for 4 hours on samples quenched from 950 ℃. Microstructural evolution was analyzed using Optical Microscopy, Scanning Electron Microscopy, X-ray Diffraction, Electron Backscatter Diffraction, and Electron Probe Microanalysis. Mechanical properties were evaluated by Vickers micro-hardness measurements. Results showed that the as-printed sample exhibited high hardness (365. 2 $HV_{0.1}$) primarily attributable to the presence of very fine α′ martensite. Annealing at 950 ℃ led to the decomposition of metastable α′ into the equilibrium α and β phases and a hardness of 329 $HV_{0.1}$. Annealing at 1010 ℃ resulted in the formation of coarser lamellar α + β with high β concentration of 1. 5 %, typically leading to the lowest hardness of 319 $HV_{0.1}$. Quenching from 950 ℃ resulted in a microstructure comprising the primary α' phase and transformed β regions, with a hardness of 332. 8 $HV_{0.1}$. Quenching from 1010 ℃ produced a microstructure dominated by transformed β regions, primarily α′ martensite, leading to a higher hardness of 371. 6 $HV_{0.1}$. Tempering the quenched samples (from 950 ℃) induced the progressive decomposition of metastable phases and the formation of the α + β microstructure. Hardness significantly increased from 361. 4 $HV_{0.1}$ (tempered at 350 ℃) to 382. 6 $HV_{0.1}$ (tempered at 450 ℃), and then reduced to 364. 4 $HV_{0.1}$ (tempered at 550 ℃), demonstrating that tempering temperature critically influences the final microstructure and hardness by affecting phase transformation and coarsening kinetics.

Keywords: Ti6Al4V alloy; LPBF 3D-Printing; annealing; quenching; tempering

Microstructure Evolution and Properties of Fe-Based Coatings Prepared by Extreme High-Speed Laser Cladding

Liu Yan*, Liu Jin, Wei Ziqi, Lv Hang, Chen Hui*

School of Materials Science and Engineering, SouthwestJiaotong University, China

E-mail address: liuyanzt@163.com (Liu Yan)

E-mail address: xnrpt@swjtu.edu.cn (Chen Hui)

Abstract: Under high-speed and heavy-haul operating conditions, railway axles are subjected to significant dynamic impact, bending, torsional, and vibrational loading over prolonged service periods. After long-term operation, various defects such as scratches and foreign object damage (FOD) tend to develop on the axle surface, compromising structural integrity and reducing service life, thereby directly threatening the operational safety and reliability. Once discarding damaged axles not only increases operational costs but also results in substantial resource wastage, wherein repair and remanufacturing technologies emerge as an effective solution to address these challenges. Conventional techniques such as overlay welding and thermal spraying often exhibit limitations including excessive residual stresses or inadequate interfacial bonding strength. In contrast, extreme high-speed laser cladding (EHLA), a variant of traditional laser cladding technology, demonstrates superior deposition efficiency, higher powder utilization, exceptional coating surface precision, and a reduced heat-affected zone in the substrate, making it particularly suitable for high-precision repair of wheel-axle components. In this study, EHLA was employed to fabricate an iron-based coating on LZ50 axle steel, investigating its potential for repair and remanufacturing applications. The research aims to reduce axle rejection rates and conserve resources through advanced surface engineering. A single-factor experimental design was implemented to investigate the regulatory mechanisms of key processing parameters (e.g., laser power and scanning speed) on the coatings' microstructure and mechanical properties. The results demonstrate that increasing laser power leads to greater coating thickness and heat-affected zone width, whereas higher scanning speeds effectively reduce both coating thickness and heat-affected zone width. Systematic characterization revealed gradient microstructural evolution in the remanufactured specimens: planar crystals nucleated preferentially near the fusion line, followed by epitaxial growth of columnar dendrites along the maximum heat dissipation direction, while equiaxed grains dominated the upper cladding region. Notably, interdiffusion of Fe and Cr elements at the coating-substrate interface confirmed the formation of sound metallurgicalbonding. Under optimized parameters condition, the cladding layer exhibited comparable microhardness and bonding strength to the substrate material, while the remanufacturing specimens' tensile properties meeting axle service requirements. Above all, these findings provide critical theoretical guidance and technical support for axle repair applications, offering significant potential for enhancing the service performance and extending the operational lifespan of railway transportation equipment in China.

Keywords: railway axles; EHLA; remanufacturing; microstructure and properties

Austenite Decomposition Kinetics During Multi-phase Microstructure Formation by Laser-induction Hybrid Solid-stare Phase Transformation

Abstract: Traditional homogeneous microstructures such as martensite, bainite, and pearlite struggle to balance strength and ductility in carbon steel. To solve this problem, a laser-induction hybrid phase transformation (LHPT) process was used to achieve the heterogeneous microstructure of pearlite and martensite during the continuous cooling transformation of supercooled austenite. A kinetic model of far-equilibrium austenite decomposition has been developed for rapid heating and cooling processes due to the combined action of multiple heat sources. During LHPT, martensite transformation follows the Koistinen-Marburger kinetics, while pearlite transformation is described by an adapted Avrami model where the conventional parameters (isothermal temperature, austenite grain size, and transformation time) are replaced with LHPT-specific variables including peak temperature, cooling rate, transformation duration, in order to . To determine the model parameters under the LHPT process, thermal expansion experiments were conducted under simulated processing conditions. Finally, a multiphysics-coupled numerical model for LHPT was established based on the phase transformation kinetics model. Numerical simulations analyzed the transient temperature field and microstructural evolution during heterogeneous phase transformation. Finally, comparison with experimental results confirmed the accuracy of the proposed austenite decomposition kinetics model, providing guidance for process parameter optimization.
Keywords: Heterogeneous microstructure, Laser, pearlite and martensite, kinetics model, numerical model

Excellent Thermal Shock Resistance of NiCrAlY Coatings on Copper Substrate via Laser Cladding

Liu Jiabin*

State Key Laboratory of Silicon Materials, School of Materials Science and Engineering, Zhejiang University, Hangzhou 310027, China

E-mail address: liujiabin@zju.edu.cn (Liu Jiabin)

Abstract: Coatings are widely used to protect substrate in extreme thermal environments (e.g., arc heaters), andthe thermal shock resistance is a crucial parameter for the coatings, which requires tight interlayer bonding betweencoatings and substrates. In this work, NiCrAlY coatings were highly required for the pure copper substrate torestrict the electric arc in arc heaters. To overcome the bonding difficulty of coating on copper surface, the NiCrAlYcoatings were prepared by two laser cladding methods: conventional laser cladding (CLC) and high-speed lasercladding (HSLC). The microstructure, composition, and thermal shock resistance of NiCrAlY cladding layersprepared by both methods were investigated. Benefitting from the high cooling rate and high energy density, theHSLC-layer has better composition uniformity and tighter interlayer bonding than the CLC-layer, achieving a30%-45% improvement in thermal cycling lifetime. Besides, the NiCrAlY layers prepared on copper substrate byboth laser cladding methods exhibit 3-10 times better thermal shock resistance than those NiCrAlY layers preparedby conventional spraying methods. It further confirms great effects of metallurgical bonding and compositionuniformity on thermal shock resistance of coatings. The NiCrAlY layers fail in the form of internal cracking andinterface peeling, and the corresponding failure mechanism is discussed.

Keywords: laser cladding; NiCrAlY; thermal shock resistance; microstructure

Optimisation of Laser Hardening Modes Using Nonlinear Programming and Computational Experiments

Mykola Anyakin[1*], Zhang Qunli[2], Yao Jianhua[2]

[1] National Technical University of Ukraine " Igor Sikorsky Kyiv Polytechnic Institute", Kyiv, Ukraine

[2] Institute of Laser Advanced Manufacturing, Zhejiang University of Technology, Hangzhou, China

E-mail address: anyakin@ukr.net (Mykola Anyakin)

E-mail address: zql@zjut.edu.cn (Zhang Qunli)

Abstract: The paper deals with thermal processes occurring in the body of a workpiece under the influence of focused laser radiation. In the design space, an adequate description of the heat affected zone formed in a 45 steel billet during laser treatment is established by the dependencies obtained as a result of numerical modelling of the laser surface treatment process. It is shown that when designing laser hardening operations, attention should be paid not only to achieving a given temperature interval in the heat affected zone, but also to its cooling rate, which should ensure the specified structural transformations with a minimum transition zone (between the hardening zone and the base metal). For example, for steel 45, the cooling rate should exceed 500 ℃/s. Due to the application of nonlinear programming methods, using the multidimensional nonlinear nonstationary thermal conductivity equation as an equation of connection between the treatment modes and the size of the hardened zone in the thermal impact zone, the treatment modes and the power density distribution in the focus spot were established, which allow almost 4 times to increase the area of the hardened zone (compared to the treatment with the usual Gaussian distribution of laser power density in the focus spot).

Keywords: laser surface hardening; optimisation; microhardness; numerical analysis; thermal cycle; cooling rate

Wear Behavior and High-temperature Oxidation Resistance of Laser Cladding $Al_{0.3}$CoCrFeNi-X High-entropy Alloy Coatings

Liu Hongxi

Kunming University of Science and Technology, Kunming, China

E-mail address: liuhx@ kust. edu. cn; piiiliuhx@ sina. com (Liu Hongxi)

Abstract: This work presents the effect of doping refractory elements on the microstructure, microhardness, wear resistance, and high-temperature oxidation resistance of laser cladding $Al_{0.3}$CoCrFeNi high-entropy alloy coatings (HEAC). $Al_{0.3}$CoCrFeNi has attracted extensive research interest due to its excellent processability and superior high-temperature oxidation resistance, making it a promising candidate for high-temperature structural materials. However, its mechanical properties and wear resistance still present room for improvement. Previous studies have shown that the addition of refractory elements such as Ti, Mo, Nb, and Ta can significantly enhance the mechanical properties and wear resistance of HEA without notably compromising high-temperature oxidation resistance. However, limited comparative studies are available regarding the specific effects of these elements on the microstructure, mechanical properties, wear resistance, and high-temperature oxidation resistance of the $Al_{0.3}$CoCrFeNi HEAC. This study aims to show the microstructure, wear resistance, and high-temperature oxidation resistance of laser cladding HEAC before and after being doped with refractory elements of Ti, Mo, Nb, and Ta. The incorporation of Ti, Mo, and Nb promotes the formation of AB_2-type Laves phases within the matrix, while the addition of Ta induces the precipitation of nanoscale amorphous phases. The results among HEAC doped with different refractory elements were compared through microhardness testing, wear testing, and high-temperature oxidation tests. The wear mechanism of the $Al_{0.3}$CoCrFeNi, $Al_{0.3}$CoCrFeNiTi$_{0.1}$, and $Al_{0.3}$CoCrFeNiMo$_{0.1}$ HEAC is primarily adhesive wear and oxidative wear, whereas the incorporation of Nb and Ta transforms the wear mechanism into abrasive and oxidative wear. The oxidation surface production of the $Al_{0.3}$CoCrFeNi HEAC is mainly composed of Cr_2O_3 and spinel oxides. Cr_2O_3 tends to volatilize at 1000 ℃, thereby facilitating the outward diffusion of Fe, Co, and Ni from the HEAC. These elements react at the oxidation surface to form spinel oxides. The introduction of Ti and Mo deteriorates the high-temperature oxidation resistance of the $Al_{0.3}$CoCrFeNi HEAC. In contrast, the addition of Nb and Ta promotes the formation of a composite oxide scale consisting of Al_2O_3 and Cr_2O_3 on the oxidation surface, which effectively suppresses the volatilization of Cr_2O_3. In particular, the $Al_{0.3}$CoCrFeNiTa$_{0.1}$ HEAC forms a $Cr_2O_3/Al_2O_3 + CrTaO_4$ double oxide scale that effectively inhibits internal oxidation of the HEAC compared to the $Al_{0.3}$CoCrFeNi and $Al_{0.3}$CoCrFeNiNb$_{0.1}$ HEAC. Thus, the $Al_{0.3}$CoCrFeNiTa$_{0.1}$ HEAC exhibits the best combination of microhardness, wear resistance, and high-temperature oxidation resistance among HEAC, and its oxidation behavior at different temperatures is further investigated. The $Al_{0.3}$CoCrFeNiTa$_{0.1}$ HEAC exhibits similar oxidation behavior at 900 ℃ and 1000 ℃. However, a double oxide scale composed of corundum derivative oxides and Cr_2O_3 at 1100 ℃ is formed, leading to a significant degradation in high-temperature oxidation resistance.

Keywords: laser cladding; high entropy alloy (HEA); wear behavior; high-temperature oxidation resistance

Experimental Investigation on Adhesive Bonding Performance of 5083 Aluminum Alloy via Laser Surface Texturing

Xu Yanlong, Li Wenge*, Zhao Yantao
School of Merchant Marine, Shanghai Maritime University, Shanghai 201306, China
E-mail address: lyxu@ shmtu. edu. cn (Xu Yanlong)
E-mail address: wgli@ shmtu. edu. cn (Li Wenge)

Abstract: Laser surface texturing (LST) is a highly effective technique for generating high-quality micro-textured patterns of various shapes and sizes on aluminum alloys. Textured surfaces can enhance the interaction between aluminum alloys and other materials by increasing the surface contact area. The goal of this study was to explore Nd: YAG laser potential for improving the adhesive bonding performance of 5083 aluminum alloys with epoxy resin coating. First, using an orthogonal experimental method, the study explores the effects of key process parameters—such as average power, scanning speed, pulse width, and pulse frequency—on the quality of LST. The relative importance of each process parameter was evaluated, leading to the identification of optimal processing conditions. Additionally, the impact of laser power, line spacing, and the number of scans on the surface characteristics was systematically examined. Experimental results show that the laser-textured surfaces exhibit enhanced hydrophilicity, which facilitates adhesive penetration and micromechanical interlocking. Shear strength tests on single-lap joints demonstrate a significant improvement in bonding performance compared to sandblasted specimens. Further analysis of the pull-out tests reveals a nonlinear relationship between adhesive strength and surface roughness. While adhesive strength gradually increased with surface roughness, beyond a certain threshold, excessive roughness resulted in a certain degree of decrease in strength. This study provides valuable insights for optimizing LST processes to achieve superior adhesive bonding performance in aluminum alloy applications.

Keywords: laser surface texturing; adhesive bonding; hydrophilicity; surface roughness; orthogonal experimental method

Coupling Analysis Model of High-speed Roller/Guide Rail System Dynamics and Thermodynamics Considering Surface Roughness

Xu Jian, Yang Zhen, Chang Xuefang, Li Qiang

School of Mechanical and Electrical Engineering, North University of China, Taiyuan, 030051, PR China

Abstract: Considering rotation and not considering rotation, a calculation was conducted on the friction and wear between the sliding pair in a high-speed rotating machine using a plane of Ra 6. 3 and Ra3. 2, indicating that Ra3. 2 has advantages. In higher firing rate Gatling guns, the guide rail should be processed more finely and have a smaller roughness. The results demonstrate that the stress increases a lot when the bolt is surface rough, which is 11. 5% higher than the flat condition. The temperature of Ra 6. 3 is about 100 degrees higher than the flat condition. It plays an important role in improving the service life of friction surfaces.

Keywords: reverse engineering; thermodynamic; sliding pair; surface roughness; rotation; sliding wear; rpm (round per minute)

Tribological Performance of Detonation-Sprayed Duplex $CrFeNiAl_{0.3}Ti_{0.3}$ High-Entropy Alloy-Matrix Self-Lubricating Coatings from 25 to 1000 ℃

Yang Dongsheng, Geng Yushan, Cheng Jun, Yang Jun*, La Peiqing*

[1] High-Temperature Tribology Research Group, State Key Laboratory of Solid Lubrication, Lanzhou Institute of Chemical Physics, Chinese Academy of Sciences, 18 Tianshui Middle Road, Chengguan District, Lanzhou, Gansu Province, 730000, China

[2] State Key Laboratory of Advanced Processing and Recycling of Nonferrous Metals, Provincial-Ministerial Co-construction, Lanzhou University of Technology, 730050, China

E-mail address: jyang@licp.cas.cn (Yang Jun)

E-mail address: pqla@lut.edu.cn (La Peiqing)

Abstract: To overcome the limitations of inadequate high-temperature wear resistance and frictional degradation in piston rings, cylinder liners, and air foil bearings of aerospace Stirling engines, NASA has successively developed four generations of PS-series high-temperature self-lubricating coatings tailored to the rigorous demands of aerospace propulsion systems operating under extreme thermal conditions. Despite the advanced PS400 coating demonstrating superior dimensional stability and reduced friction coefficient and wear at elevated temperatures, its tribological performance at room temperature remains suboptimal, with both friction coefficient and wear rate persisting at relatively high values. This underscores the necessity for continued investigation into long-life, high-temperature solid lubricating coatings that exhibit self-lubricating behavior, enhanced wear and oxidation resistance, and robust interfacial adhesion. In this study, a dual-phase $CrFeNiAl_{0.3}Ti_{0.3}$ high-entropy alloy (HEA) with outstanding thermal stability and a robust combination of strength and ductility across a wide temperature range was selected as the matrix. Lubricating phases such as Ag, (Ba, Ca) F_2, and $SrSO_4$ were incorporated to successfully fabricate $CrFeNiAl_{0.3}Ti_{0.3}$-Ag-(Ba, Ca) F_2 and $CrFeNiAl_{0.3}Ti_{0.3}$-Ag-$SrSO_4$ self-lubricating coatings operable over a broad temperature range (25 ~ 1000 ℃). From room temperature to 400 ℃, the dual-phase $CrFeNiAl_{0.3}Ti_{0.3}$ HEA-matrix self-lubricating coatings exploit the intrinsically low shear strength of Ag and the tribologically induced formation of a lubricating Ag-rich tribofilm to alleviate strain localization and stress concentration at the friction interface. This tribofilm serves as an energy dissipation medium for plastic deformation, thereby enhancing resistance to three-body abrasive wear. Between 400 and 1000 ℃, these coatings form a dual-glaze structure on the worn surfaces of tribo-pairs, the formation of a dense glaze layer effectively protects the exposed worn surface of the tribo-pairs and significantly suppresses high-temperature adhesive wear. Specifically, the high-temperature self-lubricating coating with a composition of 80 wt.% $CrFeNiAl_{0.3}Ti_{0.3}$-10 wt.% Ag-10 wt.% (Ba, Ca) F_2 demonstrated a reduction in friction coefficient by 0.14 ~ 0.63 and a decrease in wear rate by 1.05 ~ 5.67 × 10^{-5} mm^3/Nm over the 25 ~ 800 ℃ range compared to the pure $CrFeNiAl_{0.3}Ti_{0.3}$ coating. Additionally, the coating with 80 wt.% $CrFeNiAl_{0.3}Ti_{0.3}$-10 wt.% Ag-10 wt.% $SrSO_4$ exhibited excellent tribological performance across 25 ~ 1000 ℃, with friction coefficients as low as 0.29 ~ 0.49 and wear rates stabilized at 2.68 ~ 11.82 × 10^{-5} mm^3/Nm.

Keywords: high-entropy alloys; high-temperature solid lubrication; tribological performance

Enhanced Mechanical and Wear Properties of a Modified H13 Hot Work Die Steel via Tailored M_6C Formation and M_2C Stability

Huang Xiedong, Shan Quan *

Faculty of Material Science and Engineering, Kunming University of Science and Technology, Kunming 650093, PR China

E-mail address: hxd218@ qq. com (Huang Xiedong)

E-mail address: quanshan. ms@ gmail. com (Shan Quan)

Abstract: A novel modified hot work die steel, H13-M, was developed to address the increasing demands of high-temperature applications, where conventional H13 steel fails to provide sufficient performance above 600 ℃. The H13-M steel was designed by reducing the contents of C and Cr while incorporating W and Ni to enhance tempering resistance and high-temperature wear performance. Thermodynamic computational simulations revealed significant changes in carbide types and stability due to the compositional modifications, including the replacement of MC carbides with finer M_6C carbides. Microstructural characterization using SEM, TEM, XRD, and EBSD showed that H13-M features a refined austenitic grain structure and a higher density of fine, thermally stable secondary M_2C carbides. These carbides are semi-coherent with the matrix and effectively hinder dislocation motion. Tensile testing conducted between 20 ℃ and 700 ℃ demonstrated that while H13-M exhibited slightly lower strength at room temperature, it significantly outperformed conventional H13 above 300 ℃. Notably, H13-M retained a tensile strength of 1240 MPa at 600 ℃, 25. 8% higher than that of H13. High-temperature wear tests performed between 20 ℃ and 700 ℃ revealed consistently lower wear rates and more stable friction coefficients for H13-M. At 700 ℃, the wear rate of H13-M was 2. 28 times lower than that of H13, attributed to the formation of a dense, protective Fe_3O_4-rich tribo-oxide layer. This layer served as a thermal barrier and minimized surface recrystallization, in contrast to H13, which formed a brittle Fe_2O_3-rich oxide that led to frequent delamination and surface degradation. EBSD analysis further confirmed that H13-M's subsurface microstructure exhibited larger plastic deformation zones and retained high dislocation densities, thus preventing recrystallization-induced softening. This study establishes a strong correlation between compositional tuning, carbide evolution, and high-temperature mechanical and tribological properties, providing both theoretical insights and practical pathways for the development of next-generation hot work die steels.

Keywords: H13 hot work die steel; high-temperature wear resistance; tempering softening resistance; tribo-oxide layer

Tailoring Martensite and Ferrite via Tempering to Improve Impact-Abrasive Wear Resistance in Lower Alloy Medium Carbon Steel

Cai Youxiao, Shan Quan*

Faculty of Material Science and Engineering, Kunming University of Science and Technology, Kunming, China

E-mail address: cyx000808@163.com (Cai Youxiao)

E-mail address: quanshan.ms@gmail.com (Shan Quan)

Abstract: Martensitic steels are widely used in applications where wear resistance is critical, such as in mining, construction, and heavy machinery industries. However, the performance of martensitic steels under different wear conditions, particularly under impact-abrasive environments, is limited by their inherent brittleness. While tempered martensite is a well-established matrix for wear resistance, it may not fully meet the requirements for applications subject to severe impact and abrasion. To address this, tailoring the microstructure by incorporating ferrite to improve impact toughness, while maintaining sufficient strength, is a potential solution. In this study, a low-alloy medium carbon steel was developed, incorporating a reduction in nickel (Ni) and molybdenum (Mo) content to optimize the material's cost-performance ratio, while achieving outstanding impact-abrasive wear resistance. The aim of this study was to investigate the effect of tempering temperature (350 ℃-500 ℃) on the microstructure and mechanical properties of the steel. By controlling the tempering process, the steel's microstructure transitioned from a pure martensite structure to a martensite-ferrite dual-phase microstructure. The impact of this transformation on hardness, tensile strength, and impact toughness was systematically evaluated, along with its influence on the wear resistance under impact-abrasive conditions. As the tempering temperature increased, the martensitic phase content gradually decreased, and the ferrite phase content increased, resulting in improved impact toughness. Compared to the widely used Hardox steel, the impact toughness of the optimized steel increased significantly from 48 J/cm^2 to 136 J/cm^2. However, the increase in ferrite content led to a reduction in tensile strength, which decreased from 1214 MPa to 875 MPa, and a decrease in hardness from 458 HB to 230 HB. This trade-off between toughness and strength was attributed to the softening effect of ferrite. Despite the reduction in strength and hardness, the steel exhibited better impact toughness and a more favorable balance of strength and ductility for impact-abrasive wear applications. Wear testing under impact-abrasive conditions revealed that the wear resistance of the steel was not improved by the ferrite content alone. Instead, the wear rate increased with higher tempering temperatures. The optimal wear resistance was achieved at a tempering temperature of 450 ℃, where the combination of martensite and ferrite provided the best balance of toughness and strength for resisting impact and abrasion. The findings suggest that achieving a controlled martensite-ferrite microstructure is essential for improving wear resistance under impact-abrasive conditions, and this strategy can be applied to the development of cost-effective, high-performance wear-resistant steels.

Keywords: martensite wear-resistant steel; quenching and tempering; martensite-ferrite microstructure; impact abrasive wear; wear resistance

Mechanism of Vibration-Induced Waviness Wear on the Shaft Washer of GCr15 Thrust Bearings

Wang Jianan[1], Wu Qiong[2], Zhang Guanzhen[3], Tang Hujiao[2], Ren Ruiming[1]*

[1] School of Materials Science and Engineering, Dalian Jiaotong University China

[2] Wafangdian Bearing Group Co., Ltd, Liaoning, China

[3] Metals and chemistry Research Institute, China Academy of Raiway Sciences Corporation Limited, Beijing, China

E-mail address: 2641356689@qq.com (Wang Jianan)

E-mail address: rmren@djtu.edu.cn (Ren Ruiming)

Abstract: A primary early failure mode of motor bearings in industrial applications is the development of waviness wear on the inner/outer rings, leading to increased vibration and temperature. One proposed mechanism attributes this periodic surface damage to electric discharge ablation caused by shaft currents. This study conducted a comparative analysis of the surface microstructure and morphology within the waviness-damaged regions of failed insulated bearing outer rings using optical microscopy and scanning electron microscopy (SEM). The formation mechanism of the waviness damage was experimentally investigated. Results indicate no significant difference in surface microstructure or composition between the wave peaks and troughs in the failed insulated bearings. Both regions exhibit relatively uniform traces of discharge ablation and a gradient microstructure. To simulate the influence of rolling element damage on raceway wear and rolling contact fatigue (RCF) surface characteristics under actual operating conditions, localized spherical cap-shaped defects were artificially introduced onto the balls. This induced vibration between the rolling elements and the shaft washer. Thrust bearing rolling contact fatigue tests, conducted without an applied electric field, successfully reproduced waviness morphology on the shaft washer specimen surfaces. This confirms that the waviness damage primarily originates from system vibration, rather than shaft current discharge ablation. Analysis of the surface morphology at wave peaks and troughs revealed distinct wear mechanisms. Wave peak surfaces showed negligible wear traces. In contrast, wave trough regions exhibited significant evidence of adhesive wear, abrasive wear, and fatigue wear, indicating pronounced sliding of the geometrically defective balls over the surface. Fatigue spalling on the raceway predominantly initiated within the wave troughs. Cracks nucleated at the interface between the surface plastic deformation layer and the substrate, preferentially propagating along this interface, ultimately leading to fatigue spalling and consequent early bearing failure. Electron backscatter diffraction (EBSD) and transmission electron microscopy (TEM) analyses demonstrated that the surface microstructure at wave peaks remained largely unchanged. However, the surface microstructure within the wave troughs displayed clear evidence of severe plastic deformation and grain refinement. The average grain size of the α-phase in the wave trough subsurface layer refined from the original 1.75 μm to 1.14 μm, with localized regions even reaching the nanometer scale. Retained austenite was also detected within this refined structure.

Keywords: insulated bearing; waviness wear; rolling contact fatigue; grain refinement

TiNiCuZrNbTa Refractory High Entropy Alloy Coatings by Laser Cladding: Microstructure and Wear Mechanism Exploration

Wang Jiawei, Wu Bintao

NingxiaUniversity, China

E-mail address: 3129556877@ qq. com (Wang Jiawei)

Abstract: A novel $Ti_{35}Ni_{34}Cu_{10}Zr_{10}Nb_8Ta_3$ refractory high-entropy alloy (RHEA) coating was fabricated on copper substrates via laser-directed energy deposition (L-DED) to enhance wear resistance. The coating's top region exhibited a uniform grid-like distribution of HCP phase surrounding the BCC phase matrix. Due to rapid laser-induced solidification, the coating consisted of fine equiaxed grains with preferred orientation of the HCP phase. The formation of TiCuNi phases and ZrCu layer in RHEA coating exhibited siginificantly enhanced hardness and wear resistance, attributed to high strength and self-adaptation during friction. Consequently, the average friction coefficient decreased from 0. 284 (Cu substrate) to 0. 062 (RHEA coating) . The wear mechanisms were primarily abrasive and oxidative wear. Theses findings confirm that laser-deposited RHEA coatings effectively address the inherent poor wear resistance of copper components surface.

Keywords: TiNiCuZrNbTa RHEA; copper substrate; microstructure; wear mechanism

Research of XFEL Diamond Focusing Lens Manufacturing Based on Femtosecond Laser Processing and Mechanical Polishing

Yang Ning*, Wang Jiateng

Institute of Machinery Manufacturing of CAEP, Chengdu, China

E-mail address: yangning0706@126.com (Yang Ning)

E-mail address: 1927613737@qq.com (Wang Jiateng)

Abstract: XFEL (X-ray free-electron laser) plays an important role in many disciplines such as life, materials, and physics. The XFEL diamond focusing lens is the key device for beam focusing. Diamond is super hard with strong chemical inertness, the lens size is in the order of hundreds of microns, and has concave complex surface shape, so machining of diamond focusing lens is extremely difficult. This paper describes our attempts in laser and mechanical machining of XFEL diamond focusing lenses. This method has been validated in the preparation of bicavitating two-dimensional diamond focusing lenses by laser ablation and mechanical polishing by the Rafael Celestro's team. So far, no breakthrough has been made in China. In this paper, the surface with roughness of RMS 2.4560 um and accuracy of PV 33.7669 um of the parabolic contour pit machined by femtosecond laser processing were studied, and the black amorphous layer was covered on the surface, which was unfavorable to the focusing effect of the lens. No significant changes were observed after 30 min of polishing with a special polishing head, with RMS and PV values changing to 3.4057 um and 30.2431 um after 4 hours of polishing. Two polishing heads (50 and 100 types) were designed and polished again for 30 min, with RMS and PV values of 2.5001 um and 16.6842 um for 50 type and 2.5274 um and 25.0292 um for 100 type respectively. In our current research, the amorphous layer on the lens surface was not completely removed and the polishing head was severely worn. The reason is that the polishing head material is soft, and the laser processing surface is rough and hard, so the polishing head was worn and torn. Changing the polishing head material and adding appropriate polishing liquid have become the key research direction for improving the process.

Keywords: XFEL (X-ray Free Electron Laser); diamond focusing lens; femtosecond laser processing; mechanical polishing; surface roughness

Effects of C2H2 Flow Rate on Structure and Properties of Amorphous Carbon Films Prepared by Reactive Sputtering Graphite

LIU Jingzhou, JU Pengfei, TANG Lina

Shanghai Aerospace Equipment Manufacturer Co., Ltd., Shanghai 200245, China

Abstract: Amorphous carbon film, which combines good chemical inertia, wear resistance and mechanical properties, has been applied in aerospace, automotive, magnetic storage, mechanical processing and other fields. sp3 ratio and H content significant influence microstructure and tribological properties of amorphous carbon. Especially, the tribological behavior of amorphous carbon with different sp3 ratio and H content varies greatly in run-in stage under vacuum and air, which means a lot for wear pair design in aerospace equipment components. This work aims at investigating relationship between film structure and properties, so as to finding the optimum processing window. Mirror-polished 9Cr18 pieces (diameter 35 mm × thickness 3 mm) were used as substrate to deposit film by means of pulsed DC reactive sputtering graphite in Ar atmosphere mixed with different amount of acetylene. All substrates are ultrasonic cleaned in alcohol for 20 minutes to remove oil on surface followed by hot air drying, and then fixed on substrate holder. Before deposition, degassing was carried out by vacuum heating at 200 ℃ until base pressure reached 3×10^{-3} Pa. After that, plasma source with current of 50 A was applied to do Ar+ cleaning for 20 min, to further remove contaminants and activate substrate surfaces. 500 nm Cr was deposited below amorphous carbon layer to improve adhesion strength between substrate and film. For amorphous carbon layer deposition, 4 kW were applied on graphite target while a-100 V negative bias was applied on substrate, argon flow rate are among 50-100 cm^3/min, and acetylene flow rate changed from 0 to 100 cm^3/min to deposit amorphous carbon with various structures. In order to investigate effects of acetylene flow on microstructure, mechanical properties and tribological properties of films fabricated above. X-ray photoelectron spectrometer and Raman spectrometer were used to analyze microstructure, such as sp2, sp3, ID/IG, position and FWHM of D peak and G peak. Mechanical properties were characterized by nanoindentor. The tribological properties were examined through ball-disc tester, with load of 5 N, rotational speed of 1 m/s in air condition. Wear track on film as well as wear scar on mated ball were tested through white light interferometer and optical microscope. The results show that dense and uniform amorphous carbon film were prepared by reactive sputtering. Element analyzing demonstrates all films contained a certain amount of O element (6.36% - 13.86%) on surface layer, O content of lower than 1% can be reached for most film after Ar+ etch. With the increase of acetylene flow rate, H, E and H3/E2 increased first and then decreased. When acetylene flow rate was 10cm3/min, the hardness and elastic modulus reached maximum, which were 27.93 GPa and 233.55 GPa, respectively. Tribological results show average friction coefficient among 0.09-0.11 for all films, and the friction coefficient for start-up stage decreased along with H content increase. Wear resistance of 5 cm3/min sample is the best, minimum wear rate is 0.72×10^{-16} m^3/Nm. The sp3/sp2 and H can be tailored by adjusting acetylene flow in reactive sputtering process, so as to regulating mechanical properties and tribological properties of amorphous carbon films.

Keywords: acetylene; reactive sputtering; amorphous carbon; microstructure; mechanical properties; tribological properties

Nano Modified Wear-Resistant Materials and Coatings

Wang You

Harbin Institute of Technology, China

E-mail address: wangyou@ hit. edu. cn (Wang You)

Abstract: This report introduces the research ofour research group in the field of nano modified wear-resistant materials and coatings in the past 20 years, mainly divided into the following five parts:

1. Basic principles of nano modification;
2. Nano-modified CrMoCu alloy cast iron;
3. Nano-modified carbidic austempered ductile iron (CADI);
4. Nano-modified metal-ceramic coatings;
5. Nanostructured ceramic coatings.

Keywords: Nano modified; wear-resistant materials; Coatings

Crack Failure Analysis of Hot-Stamping Die Insert

Wang Shuo, Dou Zhiyang, Zuo Pengpeng*

School of Materials Science and Engineering, Jiangsu University of Science and Technology, Zhenjiang 212100, Jiangsu, China

E-mail address: 222210605223@ stu. just. edu. cn (Wang Shuo)

E-mail address: zpp0513@ 163. com (Zuo Pengpeng)

Abstract: In the hot-stamping process, the die must withstand the stress caused by mechanical loading, metal-plastic flow and thermal fatigue of cycling during 200-400 °C. Therefore, most researchers believe that the mechanical stress and the thermal stress accumulate continuously during tens of thousands of stamping processes (the service life of the hot-stamping die is about 200, 000 cycles), causing damage and cracking of the die. In the actual engineering application of hot-stamping dies, in addition to normal wear failure, according to incomplete statistics, more than 90% of failed dies start to crack from the inner walls of the water channels and then extend to the working surface, causing water leakage and rendering them unusable. Therefore, as this situation frequently occurs in engineering, it is necessary to further analyze its root cause. Despite a lot of research on the cause of die failure, these studies do not provide much attention to the cracks on the non-working area of the die insert. In order to determine the failure reason for the non-working area of a cracked A-pillar hot-stamping die insert, various instruments were used to detect the properties and microstructures of the cracks and matrix. The results show that the cracks are located in the area where the oxidative corrosion is more serious, and the cracks do not appear in the pitting area, verifying that crack initiation is related to the stress concentration on the upper half of the inner wall of the cooling channel. Meanwhile, pores and cracks exist in the grain boundary and crystal, making the impact energy of the die steel poor. Therefore, crack initiation and propagation easily occur along the brittle oxide layer. In summary, the die insert is damaged by stress-induced corrosion. In engineering applications of hot-stamping dies, we should pay more attention to the cracking of the cooling channel caused by stress and corrosion. Regarding the cause of die failure, four precautions are proposed: (1) The cooling channel should be cleaned with a weak acid solvent or high-pressure water after the die is offline. (2) The cooling channel should be air-dried with compressed air, and the oil-based protective layer should be coated to reduce the corrosion risk of the cooling channel. (3) The die should be subjected to stress relief annealing to reduce the dislocation density of the die steel. (4) The phenomenon of excessive stress in a certain area should be avoided according to the die design.

Keywords: Failure analysis; hot-stamping die; cooling channel; stress-induced corrosion

Failure Mechanism of the Roller forRoll-Forming of Single-Piece Flywheels in Automobiles

Huang Yu[a], Musa Sheikh MD Abu[a], Chen Zhian[a], Derek O. Northwood[b], Liu Cheng[a*]

[a] College of Mechanical Engineering, Yangzhou University, China 225100

[b] Mechanical, Auto and Materials Engineering, University of Windsor, Windsor, Ontario, Canada

E-mail address: liuch@ yzu. edu. cn (Liu Cheng)

Abstract: Based on the observation of fracture surface and microstructure, hardness testing and finite element simulation, a failure analysis of a 18CrNiMo7-6 steel roller in cold roll-forming of single-piece flywheels in automobiles was conducted. Cracks, furrows and material deficiency were present at the bottom of the groove of the failed rollers. Macroscopic fracture observation and hardness testing showed that the macroscopic and microhardness values in some areas at the bottom of the working groove of the roller were relatively low, and the microhardness distribution was uneven. The thickness of the hard layer was relatively thin and did not meet the design specifications. Observation and analysis using OM, SEM and EBSD indicated that the crack initiated from the bottom of the groove, where there was an uneven distribution of the surrounding microstructure and a high proportion of small-angle grain boundaries (about 35%), which accelerated the internal roller propagation of the crack. These research results provide a theoretical basis for the subsequent optimization design of high-lifetime rollers.

Keywords: Failure analysis, 18CrNiMo7-6 steel rollers, Carburizing and quenching, Hardness, Microstructure, Crack

Proceedings of IFHTSE 2025
The 30th World Congress of International Federation for Heat Treatment and Surface Engineering
August, 19-22, 2025, Suzhou, China

Failure Analysis of Cr-Mo-V Steel Stamping Die for Manufacturing Automobile Clutch Gear Hub

Chen Zhian[a], Al-seriyhi Waleed[a], Hang Yun[a], Derek O. Northwood[b], Liu Cheng[a*]

[a] College of Mechanical Engineering, Yangzhou University, China 225100

[b] Mechanical, Auto and Materials Engineering, University of Windsor, Windsor, Ontario, Canada

E-mail address: liuch@ yzu. edu. cn (Liu Cheng)

Abstract: This paper describes the failure analysis of a Cr-Mo-V steel gear-shaped stamping die used for manufacturing automobile clutch gear hubs. The failed die showed surface cracks at the tooth root, and obvious material loss at the tooth top. OM and SEM observations reveal that its microstructure mainly consists of tempered martensite, carbides, and a small amount of retained austenite. Hardness testing indicates that although the macro-hardness meets design specifications, 12.5% of the microhardness values are lower than the design requirements, and are unevenly distributed. SEM and EBSD observations, and analyses of the failed die, show that the stamping die operates in a complex environment, including impact, wear, and fatigue, among other loading forms. The presence, and uneven distribution, of carbides with a size exceeding 5 μm in the microstructure are the main reasons for the early failure of the die.

Keywords: Cold working die steel, Microstructure, Carbide, Failure, Crack

Tribological Behavior of Metallic Materials at the Microscale

Xia Wenzhen

Institute of Microstructure and Micro/Nanomechanics, School of Metallurgical Engineering,
Anhui University of Technology, Maanshan 243032, PR China

E-mail address: w. xia@ ahut. edu. cn (Xia Wenzhen)

Abstract: Friction and wear account for approximately 2-7% of a nation's GDP in economic losses. Despite centuries of research on tribology, the fundamental physical mechanisms governing friction and wear remain incompletely understood. This is primarily due to two key challenges: (1) Even macroscopically smooth surfaces consist of randomly distributed, multiscale asperities at the microscale, making precise modeling difficult; and (2) Friction occurs at the interface between contacting surfaces, rendering direct observation of the underlying physical processes at the microscale highly challenging. Currently, in situ electron microscopy tribometry has emerged as a cutting-edge approach to investigate the relationship between the frictional behavior of metallic materials and their microstructural evolution under single-asperity contact. However, conventional scanning and transmission electron microscopy (SEM/TEM) in situ friction testing methods still exhibit inherent limitations. This report will focus on microscale tribological methodologies, providing a detailed discussion on: (1) The formation mechanism of subsurface dislocation walls during the initial stages of friction, (2) The development of grain refinement layers in the later stages of wear, and (3) The influence of twin boundaries on the plastic deformation behavior under microscale friction. These studies aim to elucidate the micromechanical and microstructural evolution mechanisms governing friction and wear in metallic systems.

Keywords: microscale tribology; dislocation wall; grain refinement; twin boundary

Unravelling the Strain Hardening and Wear Mechanisms of Worn Surface for Metallic Materials by Multi-Pass Dual Indenter Scratch Methodology

Xu Xiaojun[1], Wang Xinghong[1], Long Yu[1], Sybrand van der Zwaag[2], Xu Wei[3], Zhu Minhao[1]

[1] School of Materials Science and Engineering, Southwest Jiaotong University, Chengdu, 610031, China

[2] Novel Aerospace Materials group, Faculty of Aerospace Engineering, Delft University of Technology, 2629HS Delft, The Netherlands

[3] State Key Laboratory of Digital Steel, Northeastern University, 110819, Shenyang, China

E-mail address: xiaojunxu0617@126.com (Xu Xiaojun)

Abstract: The real-life abrasion process is that abrasive particles continuously move along the worn surface being strain hardened and abraded under local contact conditions. As a result, a subsurface plastic deformation layer with different degree of strain hardening was created, which is the actual component undergoing the continuous abrasion process and eventually determines the abrasion performance of the material. In order to mimic the real-life abrasion process, a multi-pass dual-indenter scratch test methodology was developed to approach the real abrasion condition, which not only probes damage formation during the actual scratching (abrasion) process but also probes its interaction with the damage and the strain hardening effect in the deformed surface layer caused by prior local scratch deformations. In the current studies, five steel grades with different work hardening capacities, i. e. Interstitial-Free ferritic steel (IF steel), Fully Martensitic steel (FM steel), Dual Phase steel (DP steel), Quench Partitioning steel (Q&P steel) and TWining Induced Plasticity steel (TWIP steel) were selected. Systematic scratch resistance experiments were performed to investigate the damage mechanisms, the strain hardening behavior and the development of subsurface deformation layers. Moreover, the scratch behaviour was correlated to the tensile behaviour and the corresponding strain hardening characteristics. A two-stage strain hardening model was applied to explain the scratch behaviour at different loads. It can give a good interpretation on the scratch behaviour in the different load conditions and may provide a new insight into correlate the strain hardening with abrasion resistance.

Keywords: Multi-pass dual indenter scratch test; abrasive wear; stain hardening; subsurface layer; abrasive wear mechanism

Nanostructured White Etching Matter Formation Mechanisms due to Rolling Contact Fatigue in Bearings

Ren Ruiming*, Wang Jianan, Wu Qiong, Zhang Guanzhen, Tang Hujiao

[1] School of Materials Science and Engineering, Dalian Jiaotong University China

[2] Wafangdian Bearing Group Co., Ltd, Liaoning, China

[3] Metals and chemistry Research Institute, China Academy of Raiway Sciences Corporation Limited, Beijing, China

E-mail address: rmren@ djtu. edu. cn (Ren Ruiming)

E-mail address: 2641356689@ qq. com (Wang Jianan)

Abstract: In this paper, scanning electron microscopy, transmission electron microscopy and nanoindentation were used to systematically analyze the nanostructured white etched structures in the insulated bearings of the high-speed rail in the lower line, and two types of nanostructured white etched structures with different microscopic characteristics and properties were found. Combined with the service conditions, microstructure and performance characteristics, the formation mechanism of the two types of white etched microstructure is discussed, and it is concluded that one of them is formed under a single large contact load and strain, and the other is formed under a lower load and strain and a high fatigue cycle, and the latter has finer grain size and higher hardness. Combined with the experimental observations, the uneven carbon content of ordinary high-Cr bearing steels, the characteristics of mixed martensitic microstructures and the martensitic phase transition theory, a " soft zone model" was proposed, which preferentially formed in the martensite region of the original austenite grain boundary slats. Through experimental measurement and finite element simulation, it is concluded that once the plastic deformation of the low-strength " soft zone" occurs under fatigue load, it will cause the microscopic residual stress change of the deformed area and the surrounding slightly high-strength matrix, and the residual stress and external load work together to make the " soft zone" continue to undergo plastic deformation, secondary carbide and tempered carbide dissolve, and martensitic slats continue to refine until they reach the nanometer level, and at the same time, its hardness can reach nearly 3 times the average hardness of the matrix. This theory can satisfactorily explain the experimental results such as the irregular distribution characteristics of white etched microstructure and cracks, the formation of white etched microstructure before cracks, and the hardness of white etched microstructure is significantly higher than that of the matrix, and provides a theoretical basis for slowing down or even inhibiting the formation of white etched cracks that cause early bearing failure.

Keywords: early peeling of bearings; white etching matter and white etching cracks; bearing microstructure; plastic deformation nanocrystallization; martensitic transformation in bearing steels

Nano Modified Wear-Resistant Materials and Coatings

Wang You, Zhang Xiaodong, Meng Junsheng

School of Materials Science and Engineering Harbin Institute of Technology, Harbin, China

E-mail address: wangyou@ hit. edu. cn (Wang You)

E-mail address: 222011@ sdjtu. edu. cn (Meng Junsheng)

Abstract: This report introduces the research of our research group in the field of nano modified wear-resistant materials and coatings in the past 20 years, mainly divided into the following five parts: 1-Basic principles of nano modification; 2-Nano-modified CrMoCu alloy cast iron; 3-Nano-modified carbidic austempered ductile iron (CADI); 4-Nano-modified metal-ceramic coatings; 5-Nanostructured ceramic coatings.

Keywords: nano modification; CrMoCu alloys; carbidic austempered ductile iron (CADI); metal-ceramic coatings; nanostructured ceramic coatings

CO_2 Reduction for Case Hardening

Volker Heuer*, Chen Xin

ALD Vacuum Technologies GmbH Otto-von-Guericke-Platz 163457 Hanau, Germany

E-mail address: dr. volker. heuer@ ald-vt. de

Abstract: Reducing CO_2-output is one of the top-priorities for today's manufacturing industry. Process heat is one of the major contributors to industry's CO_2-output. For the process of case hardening, electrically heated vacuum furnaces represent an ecological and economical alternative to gas-fired atmospheric furnaces. The main amount of energy during the vacuum case hardening process is consumed by idle losses, followed by heating up of the load and then followed by the energy needed to quench the load. Therefore, it is key to reduce idle losses by improving the insulation of the treatment chambers. It has been demonstrated that idle losses can be reduced by 55% with improved chamber insulation. In addition, high temperature carburizing offers the potential to reduce energy consumption during the case hardening process. When using a higher carburizing temperature, more energy is needed for heating up the load. But this is overcompensated by a significant reduction in idle losses due to the shorter treatment time. For example, when targeting a case hardening depth of 1, 5mm, the energy consumption is reduced by 21% when raising carburizing temperature from 930 ℃ to 1030 ℃. The paper shows practical examples on several ways and means to reduce the energy consumption of vacuum furnaces applied for the case hardening process.

Keywords: CO_2; energy efficiency; vacuum heat treatment; low pressure carburizing; gas quenching

Supersized Vertical Electric Furnace Assembling Unit with Protective Atmosphere

Li Xianjun[1], Wang Decheng[1], Yan Jiahong[2], Liu Junjie[1], Hou Junqing[1], Wang Zhanjun[1],
Zhang Wenliang[1], Zhang Minghao[1], Luo Ping[1]

[1] China Academy of Machinery Beijing Institute of Mechanical & Electrical Research Co., Ltd., Beijing, China

[2] Henan Jianghe Machinery Co., Ltd., Henan, China

E-mail address: imlixj@163.com (Li Xianjun)

Abstract: With the development of China's aerospace industry, the requirements for metal shells of large aerospace engines are also increasing. Precision heat treatment for large, ultra high strength steel thin-walled metal shells is a key process in the manufacturing of metal shells for large aerospace engines, which directly affects product performance. There are many problems with the common well furnace equipment for thin-walled cylindrical metal shells heat treatment. Firstly, the transfer time during quenching process cannot meet the requirements. Secondly, the degree of automation is low, and manual transfer operations are relied between each process, which cannot achieve the homogenization of product performance. Thirdly, the size of product is too large, and the effective size and uniform furnace temperature of conventional equipment's furnace are difficult to treat. Fourthly, uneven cooling is prone to occur during the quenching process, leading to excessive deformation of the product. The above issues directly constrain the manufacturing capability and level of metal shells for large aerospace engines. Based on the simulation analysis and research of quenching deformation problems of large-diameter thin-walled structural components, how to ensure the uniformity of furnace temperature for metal shells with different wall thicknesses, rapid heating of preheating furnaces and quenching furnaces, and the uniformity of protective atmosphere, China Academy of Machinery Beijing Institute of Mechanical & Electrical Research Co., Ltd. has innovatively developed a super large protective atmosphere combined well type electric furnace unit production line. The furnace unit has the technical characteristics of super large furnace gas sequential circulation stirring, large tonnage product hanging fast start and stop, quenching furnace automatic transfer with precise positioning, automatic hanging and detachment lifting equipment, uniform protection of atmosphere throughout the heating transfer and cooling process of mobile equipment, imitation fixed-point spraying at variable wall thickness, and overall cooling deformation control. The furnace unit broken through the key technologies such as atmosphere protection during the entire process of heating and cooling for large-diameter thin-walled cylindrical components, the hot air circulation stirring system of ultra large furnace, the minimal deformation control of aerospace large ultra high strength steel thin-walled metal shell during quenching process. The products treated by the furnace unit have uniform performance and small deformation, meeting the urgent need for quenching and tempering heat treatment of large aerospace components and solving the needs of major aerospace projects in China. At the same time, the furnace unit fills the gap in heat treatment equipment for large thin-walled shell components in China, obtains a series of independent intellectual property rights, enhances China's research and development capabilities for heat treatment processes and equipment for large thin-walled shell components. The performance and accuracy indicators of the constituent equipment achieve international advanced level, and elevating China's large thin-walled metal shell heat treatment equipment to the high-end equipment category.

Keywords: heat treatment; furnace; aerospace application; thin-walled structures

Salt Quenching: Innovations and Performance at SOLO Swiss

Christian LIDIN

Technical Director SOLO Swiss, Switzerland

E-mail address: lidin@ soloswiss. com (Christian LIDIN)

Abstract: Salt quenching, traditionally seen as outdated or environmentally problematic, has undergone significant innovation at SOLO Swiss. This technique now offers a reliable, high-performance, and sustainable alternative to oil and vacuum gas quenching. With its superior thermal conductivity and isothermal characteristics, salt quenching ensures uniform cooling, low residual stresses, and minimal deformation—particularly for parts requiring high dimensional accuracy and fatigue resistance. SOLO Swiss furnaces integrate advanced technologies such as dynamic salt level regulation, automated carousels, and closed-loop salt recycling. These features allow precise heat treatment cycles in the 180-550 ℃ range, suitable for both martensitic and bainitic transformations. Results show reduced distortion (up to 50% compared to oil quenching) and optimised microstructures with low retained austenite, eliminating the need for additional tempering in some cases. Case studies on ring-shaped and thin-walled parts demonstrate improved dimensional stability and mechanical performance. Furthermore, environmental and safety benefits are substantial: no fire risk, no toxic emissions, and complete salt recovery through water-based washing and evaporation. This paper presents the technical principles, performance data, and industrial applications of salt quenching as implemented by SOLO Swiss, highlighting its compatibility with modern manufacturing and sustainability requirements.

Keywords: salt quenching; martensitic transformation; bainitic structure; dimensional stability; isothermal cooling; sustainable heat treatment

Practice of Smart Manufacturing in Heat Treatment Enterprises

Lu Zejian

Guangdong Strong Metal Technology Co. Ltd., China

E-mail address: kinlu@strongmetal.com.cn (Lu Zejian)

Abstract: Heat treatment is a key link in the manufacturing industry, but traditional production methods face issues such as low levels of automation, high energy consumption, and environmental pollution. The development of smart manufacturing technology has provided opportunities for the transformation and upgrading of the heat treatment industry. The intelligent needs of heat treatment enterprises and introduces practical application achievements were introduced such as multi-task flexible heat treatment production lines, automated warehouse space allocation optimization, machine vision-based workpiece automatic grasping technology, and autonomous mobile AGVs.

Keywords: flexible job scheduling; heat treatment production line; intelligent operation and maintenance; intelligent robots; machine vision

Highly Reliable Hot Stamping Production Line with Protective Atmosphere Roller Hearth Furnace

Hou JunQing, Wang Decheng, Li XianJun, Zhang Wenliang, Jiang Chao, Shi YueZhang, Nie XiaoPeng
China Academy of Machinery Beijing Institute of Mechanical & Electrical Research Co. , Ltd. , Beijing, China
E-mail address: hjqofking@ 163. com (Hou JunQing)

Abstract: In recent years, China's new energy vehicle (NEV) industry has experienced rapid development, with a significant increase in the speed of model updates and iterations. In this situation, the hot stamping quenching integrated technology, as a core means to achieve automotive lightweighting, has been widely applied in different types of vehicle. The hot stamping production line for ultra-high strength steel, as the key equipment for the hot stamping quenching integrated technology, has a decisive impact on the product performance and promotion cost due to its furnace temperature uniformity, stability of the protective atmosphere, and operational reliability. At present, domestic hot stamping production line technology and equipment face three major core challenges. Firstly, the flat and elongated structural design of the roller hearth furnace, coupled with the susceptibility of the heating elements in the front end of the heating furnace to external damage, severely interferes with the furnace temperature uniformity, thereby affecting the austenite transformation and final quenching performance of the product. Secondly, the frequent opening of the furnace door during the production process (once every 15-30 seconds) causes repeated disturbances in the furnace atmosphere, making it difficult to maintain stable oxygen content and dew point. Thirdly, the production line involves the coordinated operation of dozens of pieces of equipment. Insufficient coordination among the equipment during mass production leads to extended production cycles and frequent interruptions. In response to the above-mentioned technical bottlenecks, China Academy of Machinery Beijing Institute of Mechanical & Electrical Research Co. , Ltd. has successfully overcome key technologies such as high-redundancy multi-gradient precise temperature control, adaptive precise control of the protective atmosphere, and coordinated control of multiple equipment. They have innovatively developed a highly reliable hot stamping production line for ultra-high strength steel. This production line has significant advantages, including excellent heating uniformity, strong anti-interference capability, stable protective atmosphere, efficient equipment coordination, short production cycle, and high output. The highly reliable hot stamping production line is shown in Figure 1.

Fig. 1 Highly Reliable Hot Stamping Production Line

After practical application verification of more than 20 production lines, the products manufactured using this production line have uniform performance and good consistency, fully meeting the automotive industry's demands for mass production and high consistency. It has effectively broken the long-term dependence on imports in the core equipment field of China's automotive parts manufacturing. Moreover, this technology and equipment have been successfully extended to the special vehicle field, filling the technical gap of hot stamping quenching integrated equipment for multi-curved and irregular components of 6-14 mm medium-thick plates. During the project development process, a series of independent intellectual property rights achievements have been obtained, significantly enhancing China's R & D capabilities in non-standard heat treatment processes and equipment. The performance and precision indicators of the related equipment have reached the international advanced level, propelling China's ultra-high strength steel hot stamping quenching integrated equipment into the high-end manufacturing category.

Keywords: hot stamping; roller hearth furnace; highly reliable; vehicle body components

Effect of Through-Thickness Cooling on the Microstructure Evolution for Medium Carbon Alloy Steels Using Gas Quenching

Ju Yulin[1*], Dixon Nathan[1], Matthrew Green[2], Oindrila Mukherjee[1], Theo Teasdale[1], Geoff West[1], Martin Strangwood[1], Claire Davis[1]

[1] Warwick Manufacturing Group, University of Warwick, Coventry, UK

[2] Research & Forensic Metallurgy Department, Liberty Steel Group, UK

E-mail address: Yulin. Ju@ warwick. ac. uk (Ju Yulin)

Abstract: Salt quenching, traditionally seen as outdated or environmentally problematic, has undergone significant innovation at SOLO Swiss. This technique now offers a reliable, high-performance, and sustainable alternative to oil and vacuum gas quenching. With its superior thermal conductivity and isothermal characteristics, salt quenching ensures uniform cooling, low residual stresses, and minimal deformation—particularly for parts requiring high dimensional accuracy and fatigue resistance. SOLO Swiss furnaces integrate advanced technologies such as dynamic salt level regulation, automated carousels, and closed-loop salt recycling. These features allow precise heat treatment cycles in the 180-550 ℃ range, suitable for both martensitic and bainitic transformations. Results show reduced distortion (up to 50% compared to oil quenching) and optimised microstructures with low retained austenite, eliminating the need for additional tempering in some cases. Case studies on ring-shaped and thin-walled parts demonstrate improved dimensional stability and mechanical performance. Furthermore, environmental and safety benefits are substantial: no fire risk, no toxic emissions, and complete salt recovery through water-based washing and evaporation. This paper presents the technical principles, performance data, and industrial applications of salt quenching as implemented by SOLO Swiss, highlighting its compatibility with modern manufacturing and sustainability requirements.

Keywords: salt quenching; martensitic transformation; bainitic structure; dimensional stability; isothermal cooling; sustainable heat treatment

Study on Surface Structure and Properties of GCr15 Bearing Treated by Gas Nitrocarburizing Quenching

Su Yang[1], Shi Yousen[1], Zhang Zhichong[1], Wang Qin[2]

[1] Jiangsu Fengdong Thermal Technology Co., Ltd.

[2] Jiangsu Fengdong Heat Treatment & Surface Engineering Research Co., Ltd.

E-mail address: su. yang@ fengdong. com (Su Yang)

E-mail address: shi. yousen@ fengdong. com (Shi Yousen)

Abstract: The GCr15 bearing parts were subjected to gas carbon-nitrogen co-permeation quenching treatment. Different processes were tested to investigate the effects on the surface carbon-nitrogen concentration distribution, microstructure, surface hardness, and hardening layer depth. The results show that the heat treatment process parameters have an impact on the structure and performance. Amog them, the ammonia gas flow rate has a significant influence. An appropriate ammonia gas flow rate is the key to obtaining a good surface structute and performance. After carbon-nitrogen co-permeation quenching treatrment, the surface handness of the product is higher than that of the core, especially the impact resistance after tempering has significantly improved.

Keywords: gas carbon and nitrogen co-permeation; carbide free layer

Study on High Concentration Deep Vacuum Carburizing-Nitriding Thermal Treatment Process

Zhang Zhichong, Shi Yousen, Su Yang, Li Yong, Mao Jun

JiangsuFengdong Thermal Technology Co. , Ltd.

E-mail address: zhang. zhichong@ fengdong. com (Zhang Zhichong)

E-mail address: shi. yousen@ fengdong. com (Shi Yousen)

Abstract: The surface microstructure, hardness, and effective hardening depth of 18CrNiMo7-6 camshafts after vacuum carburizing (970 ℃) with air cooling, nitriding, quenching, and low-temperature tempering were investigated using optical microscopy, electron probe microanalysis (EPMA), spectrometric composition analysis, residual austenite measurement, and hardness/hardening depth testing. The results showed that the surface microstructure after vacuum carburizing and nitriding followed by quenching primarily consists of martensite, residual austenite, and dispersed carbonitrides. The effective hardening depth reached approximately 1. 8 mm, with surface carbon content of 1. 2% and nitrogen content of 0. 76%. The surface hardness exceeded 60 HRC, and the residual austenite content was about 25%, successfully resolving the conflict between high surface hardness and high residual austenite content.

Keywords: 18CrNiMo7-6 steel; camshaft; vacuum carburizing and nitriding; residual austenite; fatigue life

Effect of Vacuum Annealing on Microstructure and Properties of 1J117 Soft Magnetic Alloy

Wu Xingping, Huang Zihuang, Jin Jing, Chen Xingxing, Tang Lina*
Shanghai Aerospace Equipment Manufacturer Co. , Ltd. , China
E-mail address: Wuxingping149@ 163. com (Wu Xingping)

Abstract: This work presents the effects of vacuum annealing process on the microstructure, mechanical properties and magnetic properties of 1J117 soft magnetic alloy were studied. The internal structure of the soft magnetic alloy can be reconstructed through annealing heat treatment, the grain orientation can be adjusted, the internal stress can be fully eliminated, and a certain degree of order can be controlled, so as to stimulate the magnetic properties and mechanical properties of the material. The larger the diameter, the lower the tensile strength of 1J117 after heat treatment, the coercivity and tensile strength of the alloy decrease, and the saturation magnetic induction strength increases with the increase of annealing temperature. When the annealing temperature is 1150 ℃, the holding time is 2h, and the fast cooling temperature is 450 ℃, the magnetic properties of 1J117 alloy are better, B4000 can reach 1. 397T, B8000 is 1. 526T, Hc is 29. 81A/m, the tensile strength is 375MPa, the yield limit is 280MPa, the elongation after fracture is 39% , and the shrinkage is 68%.

Keywords: 1J117 soft magnetic allot; vacuum annealing; microstructure; property

The Application of BBH-6000 Furnace in the Deep Carburizing Process for the Wind Power Industry

Yan Peiqing, Yang Liang, Xia Yuanhai, Liu hui

Yantai Fengdong Hot Technology Co., Ltd. Yantai Shandong 265500, China

Abstract: With the continuous development of the wind power new energy industry, the proportion of wind power in the market has been increasing, which has driven the vigorous development of the wind power component industry. The requirements for product quality in component manufacturing have been continuously raised, while at the same time, optimization of manufacturing costs needs to be carried out constantly. In response to the market situation, we conducted in-depth research on large pre-vacuum multi-purpose furnaces, and manufactured the large pre-vacuum controllable atmosphere multi-purpose furnace BBH-6000. It has improvements in internal oxidation control, carbide level control, and deformation control, and compared to small and medium-sized multi-purpose furnaces, it reduces the manufacturing cost of single products, achieving energy conservation and consumption reduction.

Keywords: Pre-vacuum controllable atmosphere multi-purpose furnace; BBH-6000; deep carburizing; effective hardening layer; internal oxidation; carbides; uniformity; deformation

Study on Cooling Capacity of High Pressure Nitrogen

LI Jingmin, Zhou Wenfeng, Lu Xiaoyao

Institute of Mechanical Manufacturing Technology, China Academy of Engineering Physics, Mianyang 621900, China

Abstract: In actual production the reference data of high pressure nitrogen is few. In this paper, the 12bar high pressure furnace was studied. The cooling rate of test samples which were made by different materials and had different size was measured at different nitrogen pressure. The results showed that the cooling capacity of high pressure nitrogen was less than standard quenching oil And as the effective thickness of samples increased, the cooling rate decreased obviously. So high pressure nitrogen quenching was only fit for medium alloy steel and high-alloy steel with high hardenability and the batch should be considered together.

Keywords: nitrogen; quenching; cooling character; nitrogen

Precision Controlled Atmosphere Heat Treatment and Its Emission Reduction Technology

Zhu Wenming, Zhu Xiaojun, Han Boqun

Jiangsu Fengdong Thermal Technology Co., Ltd., Yancheng Jiangsu, China

Abstract: This article introduces a computer-aided design system for temperature and flow fields in carburizing furnaces, improve the uniformity of carburizing furnace temperature and atmosphere, and achieve precise control of carburizing layer, structure, and performance; A large-scale pre-vacuum controlled atmosphere carburizing multi-purpose furnace adopts the method of pre-vacuum in the front chamber, can effectively reduce the air intake during the work piece entering the furnace, saving purging time, reducing internal oxidation, significantly improving product carburizing quality and production efficiency, and achieving a reduction in process gas consumption; The atmosphere recycling device achieves about 30% reduction in gas consumption for the carburizing furnace. Additionally, the waste heat recycling of heat treatment can get a comprehensive energy-saving effect of 5-10%; the LMS/FMS intelligent management system enhances heat treatment quality, reduce energy consumption and emissions, and promote intelligent development.

Keywords: temperature and flow field simulation; pre-vacuum multi-purpose furnace; atmosphere recycling device; the waste heat recycling of heat treatment; the LMS/FMS intelligent management system

Vacuum Carburizing for Industrial Robot Parts — Vacuum Carburizing Treatment of Cycloid Gears for RV Reducers

Yao Jun, Wang Songming

JiangsuIHI-Fengdong Vacuum Technology Limited Company, Jiangsu Province, China

E-mail address: yao. jun@ ihi-fengdong. cn (Yao Jun)

Abstract: This article introduces the vacuum carburizing process applied to cycloid gears used in RV reducers for industrial robots. The reducer serves as the core component of industrial robots, with the cycloid gear being the critical part of RV reducers. Due to severe mixed grain structures and stress concentration resulting from raw materials and machining processes, precision carburizing during heat treatment is essential. This ensures qualified metallographic structure and mechanical properties while achieving optimal dimensional deformation. Given the inherent advantages of low-pressure vacuum carburizing over atmosphere carburizing, it is widely adopted for cycloid gear carburizing. This paper reports process parameters suitable for a specific model of RV reducer cycloid gear. By implementing appropriate vacuum carburizing techniques, the optimal combination of performance metrics was achieved in mass production. The process has now been adopted by several RV reducer manufacturers for batch production.

Keywords: RV reducer; Cycloid gear; Vacuum carburizing and quenching; SCM420H; Precision carburizing

High-end Application of ion Nitriding Technology in key Components of Hydrogen Engine

Xu Zhibin[1], Liu Deyun[1], You Yitian[1], Zhang Xiangyu[1], Wu Maoyong[1], Lv zhidou[1]

[1] Qingdao Fengdong Thermal Technology Co., LTD, Shandong, China

Abstract: This study addresses the wear and hydrogen embrittlement issues faced by key components of hydrogen engines under in harsh operating conditions. It innovatively employs ion nitrogen-carbon co-infiltration technology to the surface strengthen martensitic stainless steel. The experimental results show that the wear resistance of the modified material is increased by up to 80%.

Keywords: martensitic stainless steel; ion nitrogen and carbon co-infiltration; hydrogen engine injection system

Research on the Application of Digital Twin System for Vacuum Heat Treatment Equipment

Zhang Jiarui

China Academy of Machinery Beijing Research Institute of Mechanical & Electrical Technology, Beijing, China

E-mail address: zhangjiaxiaorui@126.com (Zhang Jiarui)

Abstract: In order to promote the informatization and intelligence development of vacuum heat treatment equipment, this article proposes a method to build a digital twin system for vacuum heat treatment equipment and presents several typical application scenarios. To make the method have more reference value, this article selects the WZST type vacuum carburizing and oil quenching furnace as the research object, which has the common characteristics of multiple types of vacuum heat treatment equipment and can perform various heat treatment such as carburizing, carbonitriding, gas cooling, oil quench, annealing, etc. Based on the five-dimensional digital twin model which is the most mature and widely used in different industries, this article builds a digital twin system for vacuum heat treatment equipment from five dimensions: Physical Entity (PE), Virtual Entity (VE), Digital Twin Data (DD), Services (Ss), Connection (CN).

Keywords: vacuum heat treatment equipment; digital twin, quenching; carburizing

Design and Preparation of High Temperature Protective Coatings for Marine Environments

Li Xiaoya, Zhu Xun, Shi Qian, Lin Songsheng, Su Yifan, Deng Chunming

Institute of New Materials, Guangdong Academy of Sciences, National Engineering Laboratory of Modern Materials Surface Engineering Technology, Guangdong Provincial Key Laboratory of Modern Surface Engineering Technology, Guangzhou 510651, China

Abstract: The coastal or near-sea areas are influenced by the inland environment of our country, the salt components in the atmosphere are mainly composed of 75% Na_2SO_4 +25% NaCl. When aviation engines are in service in such environment, they are affected by high temperature, high humidity, high load and high salt fog, which will cause them to not only endure the effects of high-temperature oxidation, but also the impact of thermal corrosion, and leads to a 30-50% reduction in service life and seriously affects the safety, stability and lifespan of aviation engines. As the turbine inlet temperature of the aeroengine increases, the combined effects of high temperature, high humidity, high salt spray and other factors make the existing NiCrAlY high-temperature protective coating unable to meet the current needs of aeroengine development. Therefore, it is urgent to develop long-life high-temperature protective coatings suitable for marine environments. However, the research on marine thermal corrosion in China started late, and there are relatively few related studies and unclear corrosion mechanisms, resulting in the marine environmental thermal protection technology lagging far behind that of Western development countries. This study addresses the aforementioned problems of the NiCrAlY coating, proposes the modification by doping with Zr/Dy active elements, systematically investigates the influence of active element doping on the corrosion resistance of the NiCrAlY coating in seawater, and reveals the related corrosion mechanisms.

Keywords: thermal corrosion; marine environment

Preparation and the Corrosion Resistance of Surface MgAl Alloyed NdFeB Magnets

Xu Guangqing[1,2], Ran Renhao[1], Lv Jun[1,2]

[1] School of Materials Science and Engineering, Hefei University of Technology, Hefei 230009, China

[2] Key Laboratory of Advanced Functional Materials and Devices of Anhui Province, Hefei 230009, China

Abstract: Aiming at the problem of poor corrosion resistance ofNdFeB magnets, the surface layer of NdFeB magnets was alloyed with MgAl by magnetron sputtering combined with thermal diffusion to improve the corrosion resistance of the magnets, and the corrosion resistance mechanism was investigated. X-ray diffractometer, X-ray photoelectron spectrometer, scanning electron microscope and transmission electron microscope are used to characterize the composition, phase structure, morphology, and elemental distribution of the MgAl alloyed NdFeB magnet surface layer. The corrosion resistance of surface MgAl alloyed magnets is studied by static total immersion corrosion experiments, corrosion electrochemical experiments and accelerated aging experiments, and the preparation process parameters of surface alloying are optimized. The corrosion resistance mechanism of surface MgAl alloyed magnets was studied by atomic force microscopy to test the potential difference between phases of magnets and combined with the corrosion process. By combining magnetron sputtering and vacuum diffusion processes, anMgAl alloy layer is formed within a range of 4 to 22 μm on the surface of NdFeB magnets. Mg and Al elements mainly diffuse through the grain boundaries and exist in the form of ions and elemental substances. After surface MgAl alloying treatment, the corrosion resistance of NdFeB magnets is significantly improved. The optimized NdFeB (MgAl) -600-105 sample shows a red spot appearance time of 16 hours in the static full immersion corrosion test. Its self-corrosion potential shifts from −0.914 V to −0.725 V, the self-corrosion current density decreases from 6.896×10^{-5} A·cm^{-2} to 1.955×10^{-7} A·cm^{-2}, and the impedance value reachs as high as 27514 Ω·cm^2. MgAl alloying effectively stabilizes the structure of the Nd-rich phase at the grain boundaries, significantly reduces the potential difference between the main phase and the intergranular phase, directly weakens the local self-corrosion phenomenon caused by the potential difference, fundamentally lowers the self-corrosion tendency and corrosion rate of the magnet, and provides a solid guarantee for the long-term stability of the magnet in harsh environments.

Keywords: NdFeB; surface MgAl alloying; magnetron sputtering; grain boundary diffusion; corrosion resistance

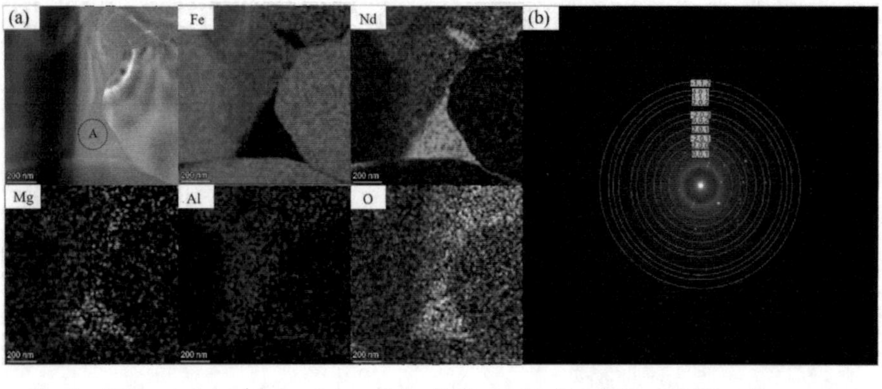

Fig. 1 TEM morphology (a), elemental distribution and selected electron diffraction patterns (b) of NdFeB (MgAl) -600-105 magnets

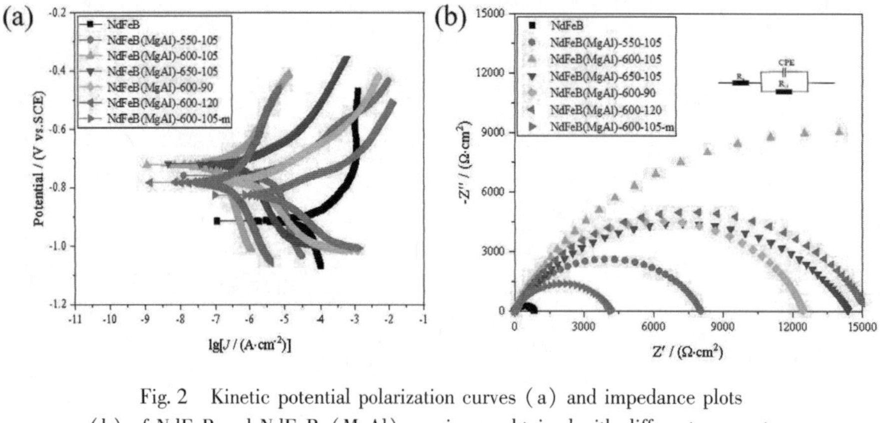

Fig. 2 Kinetic potential polarization curves (a) and impedance plots (b) of NdFeB and NdFeB (MgAl) specimens obtained with different parameters

PACVD and Nitriding Technologies as Sustainable Alternatives to Chrome VI Coatings for Hydraulic Applications

Zhang Liang, Thomas Bauer

RUBIG

E-mail address: davis. zhang@ rubig. com (Zhang Liang)

Abstract: Due to increasing restrictions under the REACH regulation, the use of Chromium (VI) -based coatings is becoming limited and calls for environmentally friendly and non-toxic alternatives. Especially in the hydraulics industry, where wear and corrosion resistance are critical, high-performance surface technologies are urgently needed. This presentation provides a comprehensive comparison between traditional Chromium (VI) coatings and advanced plasma-based technologies from RÜBIG: PLASOX ©, GASOX ©, and Duplex DLC Xtended ©. Using standardized test methods such as salt spray testing (DIN EN ISO 9227), tribological wear testing, and rock impact tests, these technologies were evaluated regarding corrosion resistance, friction behavior, and coating durability. The results demonstrate that PLASOX © and GASOX ©, when applied on properly prepared surfaces, offer comparable tribological properties to hard chrome. Duplex DLC Xtended © outperforms traditional hard chrome coatings, particularly in premium applications requiring superior wear and corrosion protection. The talk delves into the underlying physical and chemical principles and highlights industrial applications, capabilities, and limitations of each process. It aims to support decision-makers in transitioning toward sustainable surface treatment technologies.

Keywords: PVD; PACVD; nitride; plasma; coating

Digital & Intelligent Production Line of Low Pressure Carburizing

Shou Jiandong

Shanghai Advanced Metallurgical Technology Corporation, Shanghai, China

E-mail address: https: //www. samtc. com/ (Shou Jiandong)

Abstract: Heat treatment, as one of the core processes in material processing, plays a crucial role in the performance and lifecycle of products in industrial manufacturing. Vacuum low-pressure carburizing is a surface heat treatment strengthening technology that produces activated carbon atoms through high-temperature cracking reaction gas in a vacuum environment and diffuses them to the surface of materials. It improves the surface hardness and wear resistance of low carbon steel and other materials while maintaining the original toughness of the core. Compared with traditional carburizing processes, vacuum low-pressure carburizing has higher temperature, higher carburizing efficiency, better uniformity of carburizing layer, no intergranular oxidation and surface decarburization, and has advantages such as cleanliness, high efficiency, low-carbon energy saving, and precise controllability. It is an important direction for the development of carburizing technology. The vast majority of existing low-pressure carburizing furnaces on the market have low levels of digitization and intelligence, high requirements for on-site operators, low automation, inability to achieve fully automated operation with supporting equipment, unscientific production scheduling, and serious waste of equipment resources. SAMT integrates vacuum low-pressure carburizing furnaces with vacuum tempering furnaces, industrial cleaning machines, and other equipment through the use of manufacturing execution systems (MES), advanced scheduling systems (APS), enterprise resource planning systems (ERP), and visual digital twin systems, combined with 5G networks and cloud servers. Automated carts and loading and unloading platforms are used to form a fully automated digital and intelligent vacuum low-pressure production line, achieving more scientific and efficient scheduling and minimizing the impact of human operation errors and mistakes during equipment operation. The production process has higher controllability, reliability, and repeatability. Meanwhile, through the visualization of digital twin systems, 5G networks, and cloud servers, we can monitor the operation of factory equipment and the production process of products anytime and anywhere, effectively avoiding unexpected situations caused by equipment failures or other factors and making timely and effective interventions and adjustments.

Keywords: Digital; Intelligent; low pressure carburizing; vacuum heat treatment; environment protection; energy conservation; efficient

Heattreatment Process Combining Vacuum and Controlled Atmosphere

Zhang Minghao[1], Li Xianjun[1], Zhang Wenliang[1], Li Shiwang[1], Luo Ping[1]

[1] China Academy of Machinery Beijing Institute of Mechanical & Electrical Research Co., Ltd., Beijing, China

E-mail address: 1021431797@ qq. com (Zhang Minghao)

Abstract: Vacuum carburizing technology is a new technology developed in heat treatment industry in recent years. It has many incomparable advantages over conventional carburizing, including fast infiltration, environmental protection, energy conservation, easy operation and no gas explosion risk. However, even with many advantages, the vacuum carburizing technology has not been widely used, mainly because: ① Carburizing atmosphere is a kind of uncontrollable hydrocarbon cracking gas, which can not carry out decarburizing and carburizing treatment for carbon content of carburized layers on different surfaces of steel parts, resulting in too high carbon content in three-dimensional carburizing at sharp corners of workpieces, and the level of carbides exceeds the standard due to the impossibility of decarburizing through gas; In addition, if hydrocarbon gases and cracked carbon atoms are not ejected or splashed on the workpiece surface, there will be no carburized layer on the workpiece surface, i. e. the workpiece is scrapped. ②The uniformity of carburizing layer is restricted by the quantity, geometry, total surface area and stacking density of workpieces to be carburized in the furnace. If any condition changes, the uniformity of carburizing layer is difficult to be guaranteed, and even the carburizing leakage of individual workpieces may occur. Therefore, only a large number of tests can be carried out to ensure that the nozzle of the hydrocarbon gas set in the furnace can spread over the surface of each workpiece, so as to meet the requirements of the uniformity of the carburizing layer. ③A large amount of carbon black will be generated in the furnace, which will affect the normal use of the equipment in serious cases, increase the maintenance cost of the equipment and reduce the service life of the equipment. Compared with the vacuum carburizing technology, the controlled atmosphere carburizing technology has a longer history and wider application. After decades of development, this technology has become mature, with good carburizing effect and high equipment reliability. The carburizing atmosphere is a controllable water gas dynamic equilibrium reaction gas, which effectively solves the defects such as unqualified carbides at sharp corners, poor uniformity of carburizing layer and serious carbon black in the vacuum carburizing process. Compared with vacuum carburizing, the defects are also obvious: ① Internal oxidation of carburized layer surface. ② A large amount of hydrocarbons shall be filled into the furnace to maintain positive furnace pressure, so as to prevent air suction from causing carburizing gas explosion, which not only wastes a large amount of energy, but also causes environmental pollution. In view of the problems existing in the above existing vacuum carburizing and controlled atmosphere carburizing technologies, this paper develops a vacuum-controlled atmosphere heat treatment process. Combined with the advantages of vacuum heat treatment and controlled atmosphere carburizing process, this process not only solves the problems of oxidation in the carburized layer and explosion risk in the controlled atmosphere carburizing process, but also solves the problems of the vacuum carburizing process, such as substandard carbide level at the sharp corner, poor uniformity of the carburized layer, serious carbon black, etc.

Keywords: heat treatment; furnace; aerospace application; thin-walled structures

The Influence of Carbon Potential and Ammonia Ratio on the Morphology of Surface Carbides in 16Cr3NiWMoVNbE Gear Steel

Song Jian[1]

Weifang Fengdong heat treatment Co. , LTD, China

E-mail address: song. jian@ fengdong. com (Song Jian)

Abstract: This paper studies the influence of different ammonia ratios and different carbon potentials on the morphology of carbides on the surface of 16Cr3NiWMoVNbE gear steel during gas carbon-nitrogen co-permeation. The experiments show that when the carbon potential is low or the ammonia ratio is high, the carbide morphology is mostly fine granular; when the carbon potential is high or the ammonia ratio is low, the carbides are mostly in the form of angular blocks and mesh-like. The optimized process with a carbon-permeation temperature of 870 N℃, a carbon potential of 0. 8% C, and an ammonia ratio of 6% can better meet the technical requirements.

Keywords: 16Cr3NiWMoVNbE; Gas nitrogen-carbon co-deposition process; Carbide

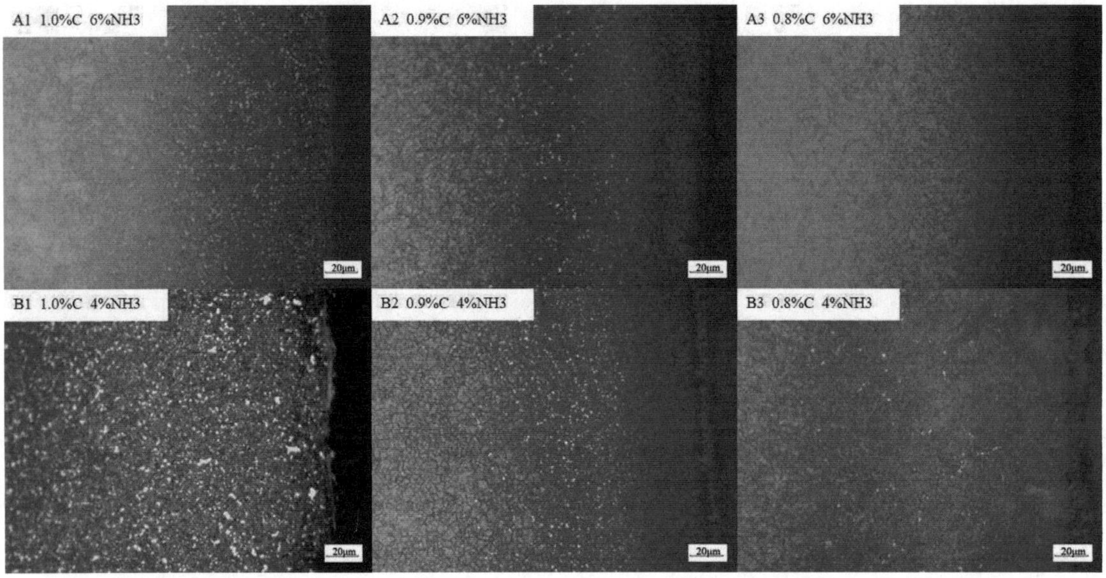

Fig. 1 The morphology of surface carbides in the products under different carbon potential and ammonia gas ratios

Study on Vacuum Low-Pressure Carburizing and High-Pressure Gas Quenching Process of TL4521 Automobile Gears

Du Chunhui, Chen Xuyang, Yang Guangwen

Beijing Research Institute of Mechanical and Electrical Technology Ltd., Beijing 100083, China

E-mail address: yangyangcxy@163.com (Chen Xuyang)

Abstract: The new low-pressure carburizing process based on the saturation adjustment method can obtain the diffuse carbide structure through the precise control of the carburizing process, which has low energy consumption and high efficiency. High-pressure gas quenching technology is clean and pollution-free, with small deformation and controllable cooling speed, which can replace oil quenching. The results show that when the depth of the permeable layer is required to be about 0.6mm, the carburizing process time at 960 ℃ is reduced by 88min compared with that of 930 ℃, and after high-pressure gas quenching, all the indicators meet the technical requirements, the surface hardness uniformity is not greater than ±1.5 HRC, the permeability depth uniformity is not greater than ±0.1 mm, and the deformation is reduced by 0.02 mm.

Keywords: low-pressure carburizing; high-pressure gas quenching; gear

Application & Technology of Electrospark Deposited High Performance Coating

Guo Ce'an

School of Equipment Engineering, Shengyang Ligong University, China

E-mail address: bigocean1979@163.com (Guo Ce'an)

Abstract: This work investigates the application and technology of high-performance coatings deposited by electrospark deposition (ESD). ESD is a process that utilizes high-temperature plasma generated by pulsed discharge to melt and vaporize electrode materials, which are then sprayed onto the workpiece surface to form a coating. This technique offers several advantages, including strong coating adhesion, low heat input to the substrate, a wide range of coating materials, local repair capability, simple equipment, easy operation, compatibility with other processes, low energy consumption, and environmental friendliness. Through ESD, high-performance coatings such as tantalum (Ta) coatings, high-entropy alloy coatings, self-lubricating coatings, and MCrAlY coatings can be deposited on various material surfaces. These coatings significantly enhance the hardness, wear resistance, and high-temperature oxidation resistance of the substrate materials. For instance, the ESD Ta coating exhibits 1.2% increase in hardness, 21% reduction in elasticity modulus, 1.6 times increase in the H/E ratio, and 14.7 times increase in the H^3/E^2 ratio compared to gun steel substrate. It also has a friction coefficient of less than 0.14 and reduces wear volume by 89% compared to gun steel substrate. The ESD technology has broad application prospects in aerospace, military, nuclear industry, mining machinery, petrochemicals, and power generation, effectively enhancing the service life and reliability of key components.

Keywords: electrospark deposition; high-performance coatings; tribological properties; tantalum coating; coating adhesion

High Temperature Coatings for SiC_f/SiC Composites

Wang Jingyang

Advanced Ceramics and Composites Division, Institute of Metal Research, CAS, China

Institute of Coating Technology for Hydrogen Gas Turbines, Liaoning Academy of Materials, China

E-mail address: jywang@ imr. ac. cn; jywang@ lam. ln. cn (Wang Jingyang)

Abstract: SiC_f/SiC composite is disruptive material for the hot-section components in new generation aviation engine. High temperature coatings, including environmental barrier coating, thermal/environmental barrier coating, as well as abradable coating, can protect various SiC_f/SiC components against harsh thermal and chemical attacks in combustion environment. Rare earth silicate high temperature coatings are reliable up to the surface temperature of 1350 ℃. The key technology depends on coating design, processing of thermal spray powders, and plasma spray technology. This talk presents the recent progresses of rare earth silicate high temperature coatings. The developments support the applications of SiC_f/SiC composite in combustion environment.

Keywords: High temperature Coating; ceramic-matrix-composite; data-driven design; processing and evaluation technology

Study of High-Temperature Oxidation and Diffusion Inhibition Mechanism of NiCrAlY Coatings with High-Entropy Oxide Diffusion Barrier

Liu He[1*], liang Zi[1], Zhou Jintao[1], Liu Qingqing[1]

[1] School of Materials Science and Engineering, Shenyang Ligong University, Shenyang, China

E-mail address: 1105333164@qq.com (Liu He)

Abstract: In high-temperature environments, inevitable element interdiffusion occurs between the coating and the substrate, which seriously threatens the long-term service performance of superalloy components. Traditional metallic and ceramic diffusion barriers have several drawbacks such as inadequate multi-element diffusion resistance and weak interfacial bonding. In this study, an (AlCoCrNiRe)O high-entropy oxide diffusion barrier was deposited between the René N5 substrate and the NiCrAlY coating via magnetron sputtering technology, creating a NiCrAlY coating with high-entropy oxide diffusion barrier. Compared with the traditional NiCrAlY coating, systematic isothermal oxidation experiments were conducted. The results demonstrate that the (AlCoCrNiRe)O high-entropy oxide diffusion barrier exhibits outstanding stability during oxidation. The multi-element synergy within the (AlCoCrNiRe)O diffusion barrier, coupled with the progressive formation of dense α-Al_2O_3 layers on both sides and the enrichment of Cr, collectively establishes a multidimensional diffusion barrier. This effectively inhibits element interdiffusion between the coating and the substrate. After 500 h of isothermal oxidation, the coating incorporating the barrier exhibited higher hardness and modulus. A dense, continuous, and exclusively α-Al_2O_3 oxide scale formed on the surface, resulting in lower oxidation weight gain and demonstrating superior oxidation resistance. The precipitation of topologically close-packed (TCP) phases within the substrate was significantly reduced and remained stable, preserving the mechanical properties of the superalloy. Limited element diffusion facilitated metallurgical bonding at the interface, enhancing interfacial adhesion. The (AlCoCrNiRe)O diffusion barrier offers advantages in both diffusion inhibition effectiveness and interfacial bonding strength, presenting promising application prospects. This research provides theoretical support for developing next generation high-temperature protective coatings for aero-engine applications.

Keywords: High entropy coating; diffusion barrier; NiCrAlY coating; High-temperature oxidation; diffusion inhibition

Oxidation and Hot Corrosion Behaviour of Pt Modified Aluminized NiCrAlYSi Coating on a Ni-based Single Crystal Superalloy

Sun Jian[1*], Su Mengjiang[2*]

[1] Advanced Research Institute for Multidisciplinary Science, Qilu University of Technology (Shandong Academy of Sciences), 3501Daxue Road, Jinan 250353, PR China

[2] Institute of Metal Research, Chinese Academy of Sciences, 72 Wenhua Road, Shenyang, 110016, China

E-mail address: sunjian@ qlu. edu. cn (Sun Jian)

E-mail address: smjiang@ imr. ac. cn (Su Mengjiang)

Abstract: In this work, the aluminized NiCrAlYSi coating and Pt modified aluminized NiCrAlYSi coating were prepared on Ni-based single crystal superalloy substrates. Isothermal and cyclic oxidation behaviour of both coatings were investigated. The Pt modified aluminized NiCrAlYSi coating showed an enhanced oxidation resistance due to the presence of Pt decreasing the oxidation rate and improving the adherence of alumina scale. The addition of Pt in aluminized NiCrAlYSi coating inhibited the phase transformation from β to γ'. Besides, the surface rumpling of Pt modified aluminized NiCrAlYSi coating was suppressed as compared to the aluminized NiCrAlYSi coating. The hot corrosion tests of Pt modified aluminized NiCrAlYSi coating coated with mixed salt films of Na_2SO_4 + K_2SO_4 and NaCl + Na_2SO_4 were carried out at 900 ℃. Compared to the aluminized NiCrAlYSi coating, the Pt modified aluminized NiCrAlYSi coating exhibited an enhanced hot corrosion resistance in these two types of mixed salts. The existence of Pt in the aluminized coating restrained the outward diffusion of substrate elements and effectively suppressed internal sulfidation and oxidation during hot corrosion test. The degradation mechanism of the coating in the corrosion condition was also discussed.

Keywords: metal coatings; superalloys; platinum; oxidation; hot corrosion

Heat Treatment and Surface Modification-Induced Microstructural Evolution Mechanisms and Their Correlation with Material Properties

Dong Shizhi

School of Materials Science and Engineering, Liaoning Technical University, China

E-mail address: dongshizhi@lntu.edu.cn (Dong Shizhi)

Abstract: In response to the service requirements for the preparation of coatings on engine blades, a gradient NiCoCrAlY coating with a Ni-Re diffusion barrier was fabricated on gas turbine engine blades via surface modification methods. This coating effectively reduces the precipitation thickness of SRZ by more than 50%. The oxidation life of the coupon-level bond coat under 1050 ℃ cyclic oxidation (1 hour per cycle) exceeds 700 cycles, and the coating exhibits a hot corrosion resistance life of more than 200 hours under 900 ℃ mixed salt conditions. Furthermore, the microstructure of materials was effectively regulated through heat treatment, constructing multi-scale mineral/nano-amorphous functional carriers. This enabled the controllable construction of single-atom mineral-amorphous functionalized carriers. Using first-principles calculations, the mechanism of the catalytic reaction process involving single atoms supported on mineral/amorphous oxides was analyzed. Three types of single-atom catalysts utilizing mineral/amorphous functionalized carriers were synthesized, laying a foundation for the industrial application of mineral-amorphous functionalized carrier single-atom catalysts, which demonstrate excellent water electrolysis performance.

Keywords: heat treatment; Ni-Re diffusion barriertensile; hot corrosion; mineral-amorphous; heterojunction structure; catalysis

Proceedings of IFHTSE 2025

The 30th World Congress of International Federation for Heat Treatment and Surface Engineering
August, 19-22, 2025, Suzhou, China

The Current Application Statusof Additive Manufacturing Technology in Superalloys

Liang Jingjing

Institute of Metal Research, Chinese Academy of Sciences, Shenyang, China

E-mail address: jjliang@ imr. ac. cn (Liang Jingjing)

Abstract: Additive manufacturing technology is an emerging advanced fabrication technique in recent years. This technology demonstrates significant advantages in the short-process preparation of complex-structured superalloy components. It has already found engineering applications in fields such as aviation, aerospace, and ground-based gas turbines. This paper introduces the current application status of additively manufactured superalloys, summarizes the microstructural characteristics of the produced components, as well as the types and formation mechanisms of metallurgical defects. Furthermore, it elaborates on the important achievements made by domestic and international researchers in two key aspects: superalloy composition design and additive manufacturing process optimization. Finally, the paper analyzes some challenges affecting the large-scale engineering application of additively manufactured superalloy components and provides an outlook on future development trends for these materials.

Keywords: additive manufacturing; superalloys; aerospace; ground-based gas turbines

Effect of the M/B Dual-Phase Microstructure Induced by Heat Treatment on the Microstructure and Properties of G13Cr4Mo4Ni4V Aerospace Bearing Steel

Yu Xingfu[1*], Wang Shijie[1], Zheng Dongyue[1], Su Yong[2]

[1] School of Mechanical Engineering, Shenyang University of Technology, Shenyang, China

[2] School of Mechanical and Power Engineering, Shenyang University of Chemical Technology, Shenyang, China

E-mail address: Yuxingfu@163.com (Yu Xingfu)

E-mail address: suyong_1979@126.com (Su Yong)

Abstract: G13Cr4Mo4Ni4V aerospace bearing steel was gas-carburised and the resulting carbon gradient in the case was quantified. Subsequent multi-step vacuum austempering at different temperatures was followed by microstructural characterisation and hardness profiling to elucidate the evolution of martensite/lower-bainite (M/B) duplex structures and their influence on mechanical properties. Carbon content decreases continuously from 1.42 wt% at the surface to 0.13 wt% in the core. After austempering, the surface layer exhibits an M/B duplex structure whose bainite fraction varies with austempering temperature, whereas the core comprises lath martensite. Surface hardness is governed by the volume fractions of carbides, martensite, bainite and retained austenite. Retained austenite persists in the carburised case; with increasing distance from the surface, progressively more retained austenite transforms into martensite and bainite during austempering, increasing hardness and changing the retained austenite morphology from blocky to film-like. Optimising the case microstructure by multi-step austempering markedly improves the impact toughness and rotating-bending fatigue performance of G13Cr4Mo4Ni4V steel.

Keywords: carburizing; aerospace bearing steel; vacuum multi-stage austempering; heat treatment; impact toughness; rotary-bending fatigue performance

Research on Molten Pool Characteristics during Laser Deposition

Yang Guang

College of Mechanical & Electrical Engineering, Shenyang Aerospace University, China

E-mail address: yangguang@ sau. edu. cn (Yang Guang)

Abstract: This work focuses on the interaction between laser and materials, which involves complex physicochemical interactions between high-energy-density laser beams and materials. Upon absorbing laser energy, materials rapidly melt or even vaporize, forming molten pools and keyhole structure. The Study results have revealed that the formation of laser molten pools initiates when the energy density in the laser-focused region exceeds the material's vaporization threshold. The recoil pressure of metal vapor pushes away the liquid metal, thereby forming the initial keyhole. Multiple reflections of the laser on the inner wall of the keyhole enhance the energy absorption rate, driving the molten pool to expand both in depth and breadth. The expansion stage is dominated by Marangoni convection, where the surface tension gradient, induced by temperature differences, drives the flow of liquid metal. During the stable stage of the molten pool, a specific convective structure is maintained by the balance of multiple forces, including recoil pressure and surface tension. Laser deposition repairing technology is widely applied in aviation materials, such as the repair of Ti-6Al-4V titanium alloy engine blades and the gradient composition design of Inconel 718 turbine disks. In the field of aviation material repair, laser deposition repair technology exhibits technical advantages of high precision and high reliability.

Keywords: laser molten pool; evolutionary process; porosity defect; laser deposition repairing technology

Pickling of 40Cr Hot-Rolled Steel Plates and the Structure of Oxide Scale

Jiang Haoran[1], Liu Xiaojiang[2*], Zhang Haoyu[1*]

[1] School of Materials Science and Engineering, Shenyang University of Technology, China

[2] School of Mechanical Engineering, Shenyang University of Technology, China

E-mail address: oxidescale@ sut. edu. cn (Liu Xiaojiang)

E-mail address: zhanghaoyu@ sut. edu. cn (Zhang Haoyu)

Abstract: This paper investigates the pickling behavior of 40Cr hot-rolled steel plates and the structural characteristics of their oxide scales, aiming to clarify the correlation between oxide scale structure and pickling efficiency. 40Cr steel is widely used in the field of mechanical manufacturing. During the hot-rolling process, a dense oxide scale is formed, which needs to be removed by pickling to ensure the quality of subsequent processing. The oxide scale of 40Cr hot-rolled steel plates presents a multi-layer structure, and the thickness and density of each layer have a significant impact on the pickling rate. In this study, various oxide scale structures of 40Cr steel generated under different conditions (heating temperature, cooling rate) were characterized to analyze their morphological features, elemental composition, and phase structure. Meanwhile, pickling experiments were conducted on oxide scales with different structures to explore the dissolution behavior of oxide scales and the influence of their own structure on pickling efficiency. Optimizing pickling parameters based on the oxide scale structure can effectively accelerate the dissolution of the inner layer and improve the surface quality of the steel plate.

Keywords: oxide scale; pickling; surface quality; hot-rolled steel plate; scale structure

Electropulsing Treatment Modification of GH4068 Nickel-Based Superalloy and M50 Steel: Microstructure Evolution and Performance Optimization

Li Jinguo[1*], Liu Jide[1*], Guo Jingdong[2]

[1] Shi-changxu Innovation Center for Advanced Materials, Institute of Metal Research, Chinese Academy of Sciences, China

[2] Shenyang National Laboratory for Materials Science, Institute of Metal Research, Chinese Academy of Sciences, China

E-mail address: jgli@ imr. ac. cn (Li Jinguo)

E-mail address: jdliu@ imr. ac. cn (Liu Jide)

Abstract: Electropulsing treatment (EPT), as a multi-field coupling technology integrating thermal, mechanical, and electrical effects, demonstrates significant advantages in regulating microstructure evolution and optimizing properties of metallic materials. This study systematically investigates the modification effects and underlying mechanisms of EPT on GH4068 nickel-based superalloy and M50 steel. For GH4068 alloy, compared with conventional heat treatment, EPT exhibits a targeted dissolution effect that not only significantly accelerates the dissolution of large γ' phases but also effectively reconciles the trade-off between grain boundary strengthening and precipitation strengthening, thereby enhancing both room-temperature and high-temperature strength-ductility synergy. In M50 steel, EPT remarkably reduces the dissolution temperature of M_2C carbides while suppressing the coarsening of MC carbides, leading to substantially improved corrosion resistance. Although elevated temperatures induce Kirkendall pore formation due to non-uniform interdiffusion between Mo and Fe atoms, EPT enables rapid pore healing compared to traditional heat treatment. Mechanism analysis reveals that the non-thermal effects of EPT, including reduced dissolution energy barriers and enhanced diffusion coefficients, facilitate low-temperature/short-duration phase dissolution, while accelerated atomic diffusion toward pores accounts for the rapid elimination of Kirkendall voids.

Keywords: electropulsing treatment; GH4068 nickel-based superalloy; M50 steel; microstructure evolution; property optimization

Research on Thermal Equipment Technology and Application of Ultra-High and Oversized Vertical High-Pressure Gas Quenching Furnaces

Ma WeiDong

ACME, Changsha, Hunan, 410100

Abstract: Vacuum heat treatment with high-pressure gas quenching equipment and process technology offers significant advantages for critical components, including oxidation-free processing, no pollution, excellent surface quality, and minimal distortion. Following the development trends in vacuum heat treatment thermal equipment and processes, there is an increasing focus on treating ultra-high and oversized components, long rod-shaped parts, and long thin-walled cylindrical parts, particularly in specialized and national defense sectors such as aerospace and nuclear industries. The application of vacuum heat treatment technology and processes for slender rod-shaped thin-walled parts in these fields is growing substantially, consequently driving demand for ultra-high and oversized vacuum heat treatment thermal equipment. In recent years, high-pressure gas quenching furnaces have gained widespread adoption in China, replacing a significant number of atmosphere furnaces and vacuum oil quenching furnaces. This paper elaborates on the research content and application prospects of ultra-high and oversized vertical high-pressure gas quenching furnace thermal equipment. It details the design structure and technical characteristics of such furnaces. Crucially, the paper emphasizes that ensuring uniform cooling rates (cooling rate synchronization) along the entire length (upper, middle, and lower sections) of ultra-long slender rod-shaped parts during high-pressure gas quenching is paramount, not just uniform heating. Only through cooling rate synchronization can simultaneous martensitic transformation be achieved throughout the part. The paper comprehensively introduces the proprietary core research technologies developed for these ultra-high and oversized vertical high-pressure gas quenching furnaces. The application of this equipment and its core technologies is vital for achieving uniform hardness throughout the entire length of ultra-slender rod-shaped parts after quenching and for minimizing quenching distortion along the full length.

Keywords: Ultra-High and Oversized Vertical High-Pressure Gas Quenching Furnace; Vacuum High-Pressure Gas Quenching of Slender Rod and Long Thin-Walled Cylindrical Parts; Cooling Rate Synchronization; Hardness Uniformity of Slender Rod Parts

No.	Data from ACME	
①	Operating Temperature range	400 ℃-1450 ℃
②	400 ℃ Furnace Temperature Uniformity	±5 ℃
③	950 ℃ Furnace Temperature Uniformity	±3 ℃
④	Thermal Field Design Principle	Six-Sided Heating + Multi-Zone Temperature Control
⑤	Effective Heating Zone Size	Diameter: Φ0.6 m——Φ2 m, Height: H3 m—H7 m
⑥	Flatness of the Loading Platform	±0.2mm
⑦	Workload Capacity	1000kg-5000kg
⑧	Ultimate Vacuum Level	5×10-4Pa
⑨	Selectable Gas Quenching Pressure Range	6~40bar
⑩	Gas Quenching Heat Transfer Mechanism	Multistage External Circulation Heat Transfer Mechanism
⑪	Initial Gas Jet Velocity during Quenching	40m/s
⑫	Pressure Vessel Shell Treatment	Stress relief annealing + X-ray
⑬	Furnace Aperture Dimensional Tolerance	CNC Machining

Fig. 1 Key Technical Parameters

Fig. 2 Model Vertical Gas Quenching Furnace at the Institute of Metal Research, Chinese Academy of Sciences (Effective Height 4.5 m)

Research Progress of Numerical Simulation of Induction Heat Treatment

Zhang Lun, Li Xianjun, Zhang Wenliang, Sun Lizhuang, Chen Baofeng, Luo Ping

China Academy of Machinery Beijing Research Institute of Mechanical & Electrical Technology Co. , Ltd. , Beijing 100083, China

Abstract: Induction heat treatment technology is to improve the performance of the workpiece through rapid heating by electromagnetic heat generation and rapid cooling with quenching, and is widely used in aerospace, automobile manufacturing, industrial machinery as well as other important fields. In this paper, the electromagnetic field, temperature field, phase change dynamics, stress/strain field and interaction coupling field involved in induction heat treatment process are systematically introduced, and the basic principle of numerical simulation of induction heat treatment process is expounded. By discussing the development history of numerical simulation of induction heat treatment process and analyzing typical numerical simulation cases of induction heat treatment process, the research progress and challenges are summarized, and the future development is prospected.

Keywords: induction hardening; heat treatment; numerical simulation

Carbide-Free Tempered Fe-C Martensitic Steels Fabricated by Fast Heating

Xiong Zhiping

[1] National Key Laboratory of Science and Technology on Materials under Shock and Impact, School of Materials Science and Engineering, Beijing Institute of Technology, Beijing 100081, China

[2] Tangshan Key Laboratory of High-Performance Metals and Ceramics, Tangshan Research Institute BIT, Tangshan 063000, China

E-mail address: zpxiong@ bit. edu. cn (Xiong Zhiping)

Abstract: Quenched martensite is very strong but brittle. Thus, tempering is necessary in order to improve the ductility. However, carbides are formed during tempering even with the addition of silicon and aluminum. These carbides can serve as the nucleation sites for voids and cracks, which are detrimental to mechanical performance. This work provides a strategy to completely prevent carbide precipitation in martensite during tempering through Mn heterogeneous distribution between Mn-enriched austenite and Mn-depleted martensite. This Mn-heterogeneity can be produced by fast heating Mn-enriched carbides and Mn-depleted ferrite. As a result, during tempering the carbide formation is completely inhibited in martensitic matrix where instead, the saturated carbon atoms diffuse into austenite. For these carbide-free Fe-C martensitic steels, the product of strength and ductility can be improved due to the absence of carbide precipitation and the existence of retained austenite. The strategy for completely avoiding carbide precipitation during tempering in the martensitic matrix will promote the development of a new series of carbide-free Fe-C martensitic steels.

Keywords: martensite; fast heating; mn enrichment; pearlite; quenching and tempering

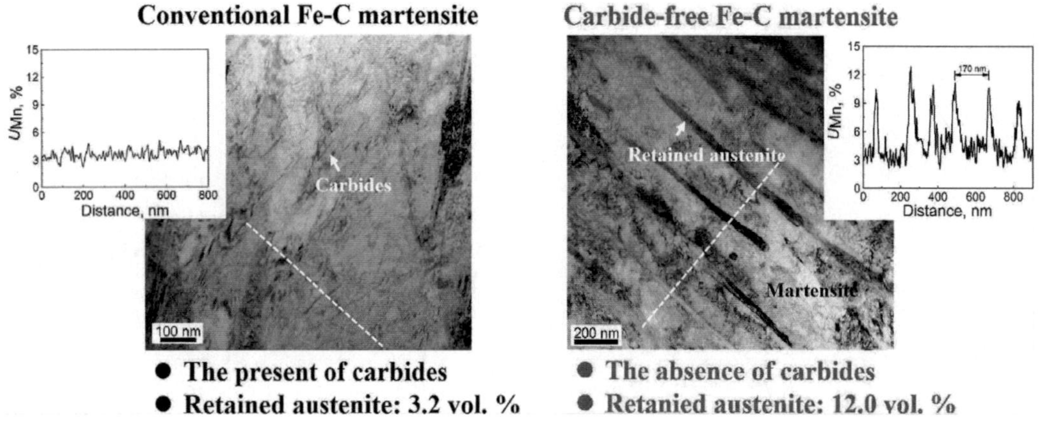

Proceedings of IFHTSE 2025

The 30th World Congress of International Federation for Heat Treatment and Surface Engineering
August, 19-22, 2025, Suzhou, China

Research on the Optimization of Induction Hardening Process for the Inner Wall of Gray Iron Material Piston Body

Xie Xuexing, Han Zhang, Wang Ting, Guan Minchao, Cao Liang, Yu Aiping

Haitian Plastic Machinery Group Co., Ltd., China

E-mail address: tianyi665650@163.com (Xie Xuexing)

Abstract: This paper focuses on the problem of easy cracking and scrapping of the inner wall of gray cast iron plunger bodies during induction hardening. The study mainly investigates the influence of the original matrix structure and graphite morphology of gray cast iron on the cracking of the inner wall of the plunger body during induction hardening. Through physical and chemical detection and analysis of the cracked inner wall of the gray cast iron plunger body, the existing inductor structure and quenching process parameters were optimized, and the plunger body was pre-treated by normalizing before induction hardening. This significantly reduced the content of phosphorus eutectics, and timely low-temperature tempering treatment after induction hardening successfully solved the problem of easy cracking and scrapping of the inner wall of the gray cast iron plunger body during induction hardening. The research results show that compared with gray cast iron with D-type graphite morphology, gray cast iron with A-type graphite morphology is more conducive to reducing the cracking risk during induction hardening, and the matrix structure after normalizing is also more conducive to reducing the cracking risk during induction hardening than the matrix structure without normalizing.

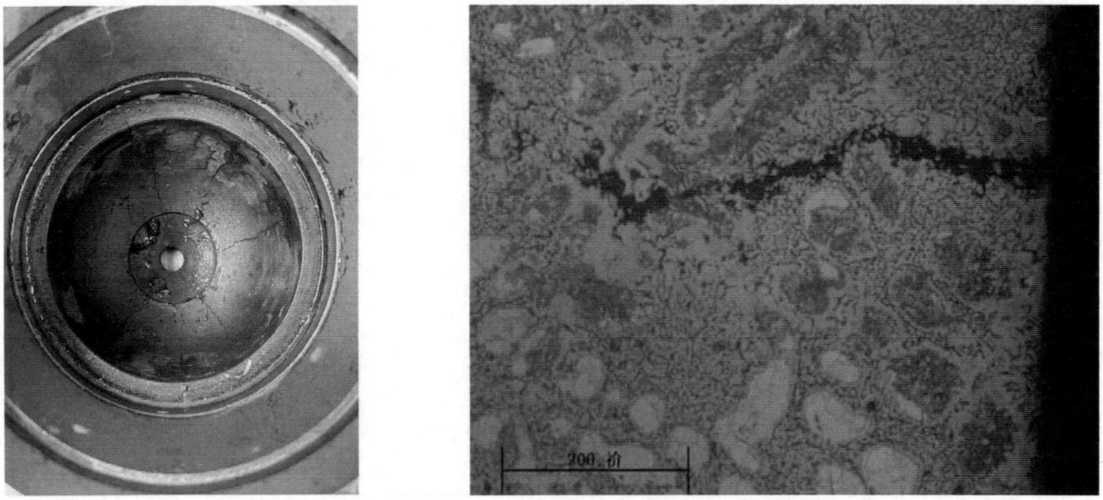

(a) The inner wall of the plunger cracked after induction quenching

(b) Cracks and fissures appeared on the inner wall of the plunger after induction quenching

Fig. 1 Microstructure of cracking cracks on the inner wall of the plunger after induction

Proceedings of IFHTSE 2025

The 30th World Congress of International Federation for Heat Treatment and Surface Engineering
August, 19-22, 2025, Suzhou, China

Effect of Microstructure and Dislocation Density on Hardness and Toughness in Induction Hardened Medium Carbon Steel

Kyosuke Matsuda[1], Tomomi Tsubakura[2], Masahiro Tsukahara[2], OsamuIdohara[2], Setsuo Takaki[3]

[1] Research and Development of Neturen Co., Ltd, 7-4-10, Tamura, Hiratsuka-city, Kanagawa, 254-0013, Japan

[2] Neturen Co., Ltd

[3] Emeritus professor of Kyushu University

E-mail address: kyosuke-matsuda@k-neturen-co.jp (Kyosuke Matsuda)

E-mail address: tomomi-tsubakura9026@k-neturen.co.jp (Tomomi Tsubakura)

Abstract: Induction hardening (IH) is a rapid and short-term heating process that makes possible to obtain fine prior austenite (γ) grains (PAGs) in martensitic steel; thus, IH is applied to structural steel materials to improve toughness and ductility. Many studies have reported about the effects of grain refinement on fatigue strength and mechanical properties in induction hardened steels. However, most reports focus on the relationships between the PAGs and the mechanical properties, and there are few discussions about the effect of the substructure change with the refinement of PAGs on the mechanical properties of carbon-containing martensitic steels. In this study, the changes in substructures and dislocation density accompanied with the refinement of PAGs were investigated, and their effects on hardness and toughness were clarified in induction-hardened Fe-0.40C-1.07Cr-0.15Mo alloy. A commercial SCM440 (Fe-0.40C-1.07Cr-0.15Mo alloy) with the shape of a round bar (φ18 mm) was subjected to homogenization, quenching and tempering, followed by IH and sub zero treatment. The quenching temperature of the IH was changed in the range between 1223 K and to 1523 K, and five types of samples were prepared. For these samples, measurements of retained austenite by saturation magnetization measurements, microstructural characterizations by optical microscopy and SEM-EBSD, Ms temperature measurements by dilatation, dislocation density analysis by the modified Williamson-Hall (mWH) method, and the Vickers hardness tests were conducted. In addition, low-temperature tempering at 473 K for 3.6 ks was performed for the samples and then the Charpy impact tests were conducted at room temperature (296 K). All the as-quenched samples contain no retained austenite. The PAG size become finer as the heating temperature decreased, and fully lath martensitic structure with a PAG size from 4 to 50 μm are obtained. The γ grain size has little effect on the Ms temperature, dislocation density, dislocation character but the substructure such as block and packet is refined with the refinement of PAG. The values of Vickers hardness were almost independent of the PAG size, and all samples showed similar hardness. On the other hand, the Charpy impact value at room temperature increased with the refinement of the PAGs. Fractograph reveals that cleavage and ductile fracture surfaces are mixed in the specimen with 50 μm PAGs, and cleavage fracture was suppressed as the PAG size became finer. The suppression of cleavage fracture is thought to be due to the complexity of the propagation path of brittle cracks within fine substructure of block and packet. In conclusion, the improvement of toughness produced by the refinement of PAGs is due to the suppression of the cleavage fracture, that attributes to the refinement of the substructure in lath martensitic steels.

Keywords: induction hardening; medium carbon steel; grain refinement; dislocation density; toughness

The 30th World Congress of International Federation for Heat Treatment and Surface Engineering
August, 19-22, 2025, Suzhou, China

Progress of Green Heat Treatment Technology

Yang Siyuan, Luo Ping, Zhang Wenliang

China Academy of Machinery Beijing Research Institute of Mechanical & Electrical TechnologyCo. , Ltd. , 100083, Beijing, China

E-mail address: m17822012378@ 163. com (Yang Siyuan)

Abstract: China's " dual carbon" policy mandates low-carbon transformation across industrial sectors, compelling the heat treatment industry to advance green technologies. Induction heat treatment provides a sustainable solution characterized by high efficiency, energy conservation, pollution-free operation, precise temperature control, minimal distortion, and automation compatibility. This process significantly enhances wear resistance and fatigue strength in surface-hardened components including gears, shafts, and crankshafts. The laser-induction hybrid quenching technique synergistically elevates surface hardening temperatures while extending case dimensions, overcoming power limitations inherent to standalone laser processes which focused energy enables rapid phase transformation and substrate-mediated self-quenching. This integrated approach concurrently mitigates distortion and cracking susceptibility, establishing a pioneering direction for eco-friendly heat treatment development. Collectively, these innovations provide foundational support for the industry's green transition toward carbon neutrality objectives.

Keywords: green development; induction heat treatment; laser heat treatment; numerical simulation

Induction Innovations in Hardening of Bearings and Raceways

Zeng Guoqing

Inductotherm Industries (Shanghai) Ltd. Shanghai, China

E-mail address: zengguoqing@ inductotherm. com. cn (Zeng Guoqing)

Abstract: Based on the geometry and heat-treating requirements of bearings and raceways, this paper introduces one new technology in induction hardening for bearings and raceways. According to the principle of electromagnetic field coupling, to bring the principle and technical features of active/passive coil. By analysing the conventional hardening pattern and its problems, to present the innovative active/passive coil for bearings and raceways induction hardening application: by heating the top and bottom raceways with separable split coupled top and bottom active/passive coil, it reduces the parts distortion, decrease the grinding, increase throughput, lower the energy consumption and extend the coil lifetime. With this innovative and laboratory proven technology, it will substantially improve the deficiency and the problems from conventional induction bearings and raceways hardening.

Keywords: hub bearing hardening; induction hardening pattern; electromagnetic field coupling; active/passive coupling coil; Induction hardening distortion

Research on Mechanical Model of Induction Heating Straightening Process for Large-Scale Complex Hull Welding Structures

Zhang Xuebiao*, Sun Baopeng
School of Naval Architecture & Ocean Engineering, Dalian University of Technology, China
Dalian Key Laboratory of Advanced Shipbuilding Technology
E-mail address: xbzhang@dlut.edu.cn (Zhang Xuebiao)
E-mail address: sunbaopeng371@mail.dlut.edu.cn (Sun Baopeng)

Abstract: Although the traditional elastic inherent strain method is widely used, it is incapable to deal with the complex effects of welding residual strains, as a result, the deformation prediction of the continuous process of welding-induction heating straightening is limited. So, this paper establishes a mechanical model for the thermal straightening process, which analyses the inherent strain value of the continuous process and its influence radius by quantifying the maximum cumulative plastic strain during the induction heating straightening process under the condition of welding residual strain as the initial input. Numerical simulations and experimental validation of the T-profiles and the stiffened plate structure demonstrate that the deformation prediction results of the continuous process achieved by the model through the elastic inherent strain method are basically consistent with the experimental measurements. The model significantly improves computational efficiency while maintaining prediction accuracy, providing an effective solution for the prediction of welding-straightening continuous deformation for large-scale complex structures of ship hulls.

Keywords: elastic inherent strain; welding residual strain; straightening; elastic-plastic analysis; deformation prediction

Research and Application of Low-Frequency Induction Hardening Technology for Large Backup Rolls

Chen Taihui, Zhao Xichun, Han Zhao, Shi Yuan

Tianjin Heavy Equipment Engineering Research Co., Ltd., China

E-mail address: snowffee@126.com (Chen Taihui)

Abstract: A large-scale backup roll induction quenching integrated equipment was developed with complete localization based on the one-shot method principle. The innovative application of high-power low-frequency (4 MW/25 Hz) digital power supply and multi-turn variable tap technology for induction coil ensures the uniformity of the temperature field on the surface of the backup roll body ≤ ±5 ℃. Combined with multi-furnace engineering temperature field tests, the induction heating simulation technology for large backup rolls was successfully developed, achieving full coverage of induction quenching technology applications for nearly 70 specifications of backup rolls (including φ1000 mm ~ φ1650 mm × 1200 mm ~ 2400 mm). A large number of industrial application practices have shown that the independently developed low-frequency static integral induction quenching integrated technology for large backup rolls can stably achieve the technical indicators of "roller body hardness uniformity △ ≤ ±1HSD" and "zero" hardness drop at 100mm in the radial direction of the working layer, with the hardness quality of the rolls reaching the international top level.

Keywords: large backup roll; high-power low-frequency; integral induction; simulation modeling; industrial application

Progress of Induction Hardening Technology for Large-modulus Wind Turbine Gear Rings

Yan Mangang, Wang Zhengbing, Zhao Xiuhua

Beijing e-Magnetic Induction Technology Co., Ltd., China

Abstract: This study comprehensively reviews the technological breakthroughs and industrial applications of induction hardening for large-modulus wind power gear rings. To meet the demands for high reliability and cost-effective manufacturing of wind turbine gearboxes, key challenges including full-tooth-width uniform hardening, deformation control, and batch stability were addressed. By developing dual-station tooth-by-tooth scanning quenching machines, profiled inductors, and full-process energy control, high-precision induction hardening was achieved for gear rings with modules of 7-32 mm and diameters of 1.3-3.8 m. The hardened layer exhibited uniform distribution along the tooth profile and width (100% coverage). A comprehensive quality control system was established, covering raw material, forgings, process monitoring, and non-destructive testing, enabling zero-failure mass production of over 100,000 gear rings. This technology signifies China's transition from technical follower to global leader in gear induction hardening, providing a paradigm for core component manufacturing in clean energy equipment.

Keywords: wind power gear ring; induction hardening; full-tooth-width hardening; dual-station quenching machine; quality control; large-modulus gear

Thermal Analysis of Variable Cross Section Cylinder Induction Heating by Half Open Anti-Profiling Coil: Simulation and Experimental Validation

Wang Zidong

University of Science and Technology Beijing, China

Abstract: Resistance heating for variable cross-section cylinders before quenching results in the coarsening of austenite grains in the thin-wall segment, and inevitably leads to nonhomogeneous hardness and inferior properties after quenching. In this study, a half open anti-profiling coil is designed for induction heating of the workpiece before hardening. A numerical model is established, couples the electromagnetic-thermal field and the rotary motion of the variable cross section cylinder to investigate the system heating efficiency and heating quality. The calculated data reveal the high efficiency of induction heating system, although the efficiency decreases significantly as the magnetic permeability rapidly drops at high temperature periods. Additionally, the temperature distribution along the axial, circumferential, and radial directions fluctuate in different parts at the end of heating time. Narrower air gap improves the heating efficiency and axial direction heating quality, albeit at the expense of reduced radial direction heating quality. An induction heating experiment validated the simulation results with an error range of 1.61% to 7.59%. These results suggest that to improve heating quality while ensuring high heating efficiency, it is crucial to design a suitable air gap that accommodates various wall thickness segments of the workpiece. Moreover, incorporating insulation measures is essential to reduce temperature distribution difference caused by surface heat dissipation. This study presents a promising technical solution for upgrading the current heat treatment process.

Keywords: variable cross-section cylinder; induction heating; temperature field; half open anti-profiling coil; heat treatment

The Construction and Practice of an Innovative System for the Entire Chain of Induction Heating Coils

Zhang Zongjie

Luoyang Shenghua Induction Heating Co., Ltd., China

Abstract: In the field of surface modification of mechanical parts, the application of induction heating is becoming increasingly widespread. The design and manufacturing level of induction heating coils directly affects the processing quality, production efficiency and energy-saving effect. Currently, induction heating coils face the problem that the design process relies on the scattered experience of experts, making it difficult to achieve efficient coupling of electromagnetic fields and thermal fields, as well as effective control of thermal stress and fatigue life. This paper aims to build a " design-manufacturing-application-maintenance" full-chain system. By re-decomposing the essence of the coupling between electromagnetic fields and thermal fields in induction heating, it reveals the mechanism of skin effect, proximity effect and ring effect in surface quenching processes such as gradient heating and moving heating, laying a theoretical foundation for breaking through the limitations of traditional design. The research proposes a paradigm shift in the design method from " empirical rules" to " digital twins", achieving the optimization of coil structure parameters through multi-physical field coupling modeling, and then intelligence. In terms of manufacturing processes, 3D printing technology is introduced to form complex-shaped coils, and a " additive manufacturing-precision processing-surface treatment" composite manufacturing process route is constructed, significantly improving the geometric accuracy, compactness and reliability of the coils. In response to the traditional passive maintenance mode, a predictive maintenance system based on the Internet of Things is constructed. By real-time collection of multi-dimensional data such as coil temperature rise and current harmonics, the service status of the heating coils can be accurately evaluated and fault warnings can be issued. The construction of the full-chain innovation system for induction heating coils can not only break through the technical barriers in the design and manufacturing of coils, but also promote their transformation from single devices to intelligent units.

Study on the Influence of Residual Stress on the Thermal Treatment Deformation Behavior of GCr15 Bearing Rings

Chen Zhenyu, Du Lianming, Lu Xiandong

School of Mechanical Engineering, University of Jinan, 250022, Jinan, China

E-mail address: 2129949778@ qq. com (Chen Zhenyu)

E-mail address: me_ dulm@ ujn. edu. cn (Du Lianming)

Abstract: With the development of manufacturing technology, the production efficiency of bearings has been raised to higher level. The rings are the key components of bearing which affect the quality and production efficiency of the bearing. During heat treatment, the bearing ring is prone to deformation, leading to high scrap rates and low grinding efficiency. The residual stress before heat treatment is one of the important facts which significantly affect the degree of deformation. Therefore, studying the relationship between residual stress and deformation is crucial for addressing the issue of deformation in the heat treatment of the bearing rings. This study focuses on thrust bearing rings made by GCr15 material. An X-ray diffraction technology was used to quantitatively analyze the residual stress on the surface of the ring after turning, the runout of the workpiece end face was accurately measure by a dial indicator, and the data collected before and after the treatment were plotted to line graphs when different heat treatment processes. The line graphs indicate a significant positive correlation between the deformation of the workpiece after quenching and the surface residual stress. Simulations with the Deform finite element analysis software were conducted to verify the consistency between the distribution of residual stress and the deformation pattern. The simulation results compared the stress conditions and deformation behavior of different areas of the ring before and after heat treatment, revealing that the deformation is particularly pronounced in the regions where residual stress is concentrated, especially on the inner and outer surfaces of the ring. It shows that the degree of quenching deformation can be effectively controlled, the grinding allowance can be reduced and the processing efficiency can be improved by controlling the residual stress in the machining process.

Keywords: thrust bearing ring; residual stress; heat treatment; deformation

Comparison of Residual Stress Elimination in Welded Steel Structures by Different Annealing Processes

Han Zhuang, Xie Xuexing, Hao Liu, Yu Aiping, Guan Minchao, Cao Liang

Haitian Plastic Machinery Group Co. , Ltd. , China

E-mail address: 823399985@ qq. com (Han Zhuang)

Abstract: The annealing process is the main method for eliminating welding residual stresses in large steel structures. The parameters of the annealing process, such as the holding temperature and holding time, are generally set based on empirical formulas related to heat transfer efficiency. This study utilized X-ray residual stress detection methods to compare the welding residual stress values under different holding temperatures and holding times, and analyzed to determine the most suitable annealing process parameters for this large steel structure. The test samples were large steel structures with a length greater than 5 meters and a width greater than 2. 5 meters, mainly composed of Q355B t30-100mm thick plates, square steel pipes, and I-beams. The main stress detection equipment was the Japanese Pulstec company-X-ray residual stress analyzer (μ-X360 s), and the auxiliary equipment was a circulating pump electrolytic polishing machine. Three groups of test samples were set up, with two welded steel structures in each group, and different process parameters were adopted for each group. The conditions of group A were a holding temperature of 550 ℃ and a holding time of 8 hours; group B had a holding temperature of 550 ℃ and a holding time of 4 hours; group C had a holding temperature of 500 ℃ and a holding time of 4 hours. The residual stress measurement points were all selected at the center of the weld seam, and the electrolytic polishing depth of the points was 0. 25 mm. The test results showed that the stress relief effect of group B was similar to that of group A, while the stress relief effect of group C was significantly different from the other two groups. Therefore, the holding temperature and holding time adopted by group B were the most suitable annealing process parameters for this steel structure among the three groups of experiments.

Fig. 1 Distribution of Main Stress Measurement Points

A Study of Bath Agitation and Distribution of Cooling Capability in a Commercial Quenching System

Ma Min, Zhang Xiaoyu, Yao Jihong, Sergio Gallegos

Quaker Houghton China

E-mail address: minma78@126.com (Ma Min)

Abstract: Different designs of the impeller or turbine and quenching tank can generate different flow around the workload. In this paper, the distribution of velocity, cooling performance and quenching severity with fast quenching oils in the effective working zone of quenching tank are studied in a commercial quenching system. The system was divided into three layers in the vertical direction and three circles in the horizontal direction. Each layer was labeled to be 13 points. The velocity, cooling performance and quenching severity of middle layer tested under 40 ℃, 70 ℃, 500 rpm and 1000 rpm. It is interesting to find the velocity, cooling performance and quenching severity changed with the points locations. Round rod Ø3.5 cm, Length 8 cm made of plant carbon steel AISI 1045 steel were put in these points for quenching. The surface hardness of the rods is consistent with the quench severity and velocities on each point.

Keywords: quench tank; distribution of cooling capability; agitation distribution; quenching severity

Research and Application Progress on Green Quenching Media

Wang Tao, Lv Bohao

Nanjing Kerun Industrial Media Co., Ltd., China

E-mail address: lvbh@njkerun.com (Lv Bohao)

Abstract: When the concept of green quenching media first emerged, it was defined as a medium that is safe, healthy, environmentally friendly and poses no harm to the environment or human health. With the advancement of society, the state's and enterprises' requirements for environmental protection have been continuously increasing. Now, a more comprehensive definition of green quenching media has been established: it must be safe, healthy, and environmentally friendly, with easily treatable waste liquid, offer superior cost and performance, and be capable of intelligent monitoring. In the traditional metal heat treatment industry, mineral oil-based quenching oil has been the main quenching media. It has advantages such as maturity and easy availability, but we cannot ignore its harm to the environment and human health. With the advancement of environmental protection policies and the continuous development of society, green quenching media are gradually gaining prominence, and their definitions are being continually refined. Currently, the common green quenching media mainly include water-based quenching media, plant oil-based quenching media, and gas quenching media. Water-based quenching media have been developed in China for over three decades and are relatively mature in application. Different cooling effects can be achieved by controlling the concentration of the refrigerant. Plant oil-based quenching media have the primary advantage of being environmentally friendly and biodegradable, addressing the issues of low flash points and poor environmental friendliness associated with mineral oil-based quenching media. They represent a new direction in the research and development of green quenching media. Gas quenching media, with their high cleanliness and controllability, have already replaced oil-based and salt-based quenching media in some industries. During the use of quenching media, there are inevitable reasons, such as aging, excessive impurities, and the inability to monitor in real-time, which lead to a decline in quenching medium performance and, consequently, insufficient product performance and additional losses. Therefore, online monitoring equipment for the quenching medium, centrifugal filter, and nitrate filter are also rapidly developing and upgrading rapidly. The green quenching medium is a product of social development, and intelligent monitoring is an urgent need for the future of the metal heat treatment industry. Both of them are facing different challenges and prospects. This paper will provide a detailed account of the application progress of these three green quenching media, the direction, path, and development trend of green quenching media research and development, as well as the future development direction of intelligent monitoring.

Keywords: quenching media; greening; gas quenching media; vegetable oil-based quenching media; water-based quenching media; intelligent monitoring

Effect of Primary Carbide on Heat Treatment Microscopic Residual Stress of 8Cr4Mo4V Bearing Outer Ring

Xia Tao[1], Tao Yuan[1], Song Tianpeng[1], Su Shuaiwei[1], Liu Yong[1], Zhu Jingchuan[1]

[1] Harbin Institute of Technology, Heilongjiang, China

E-mail address: 19112354078@163.com (Xia Tao)

Abstract: The micro residual tensile stress at the interface of primary carbides can accelerate the fatigue failure of bearings. Therefore, this study focuses on the influence of primary carbides on the interfacial micro residual stress. The Density functional theory (DFT) calculation was used to investigate the effects of alloying elements and carbon vacancy on the mechanical properties of primary carbides VC and Mo_2C. DFT calculation results show that alloying and introduction of carbon vacancy can reduce formation enthalpy of primary carbides. Furthermore, by combining the DFT calculation results with finite element simulations, the influence of the shape and mechanical properties of primary carbides on interfacial micro stress was explored. The results of the numerical simulation of heat treatment indicate that the maximum quenching residual tensile stress is 750 MPa, which decreases to 50 MPa after first tempering. Using macro tensile as the boundary condition for mesoscopic model, the influence of carbides on the interfacial micro residual stress was analyzed. Introducing carbon vacancy can effectively reduce the interfacial micro stress, while the effects of alloying on the interfacial stresses of VC and Mo_2C are opposite. This study provides a new idea for modifying primary carbides, reducing micro stress concentration in bearing rings at loading area, and increasing fatigue life.

Keywords: 8Cr4Mo4V steel bearing rings; heat treatment; micro residual stress; numerical simulation

The Influence of Carbides on the Strength for Q-P-T Steel

Yang Yuchen[1], Zheng Hongqing[1], Zuo Xunwei[2*], Wan Jianfeng[1*], Rong Yonghua[1], Chen Nailu[2]

[1] School of Materials Science and Engineering, Shanghai Jiao Tong University, Shanghai 200240, China

[2] Shanghai Key Laboratory of Materials Laser Processing and Modification, School of Materials Science and Engineering, Shanghai Jiao Tong University, Shanghai 200240, China

Abstract: A phase-field finite element model for Q-P-T (Quenching-Partitioning-Tempering) steel was employed to comprehensively investigate the evolution of microstructural and mechanical parameters during the heat treatment process. The study specifically focused on tracking the variations in carbon concentration, local stress and plastic strain, and the volume fraction of carbides precipitated during tempering. It also explored the mechanical and kinetic behavior of the carbide/martensite interface, which plays a crucial role in determining the steel's final properties. During the quenching stage, the rapid cooling of high-carbon steel leads to the formation of lenticular martensite, a hard and brittle phase characterized by a supersaturated carbon content. However, this transformation is often incomplete, leaving behind a certain amount of retained austenite. The subsequent partitioning process enables carbon atoms to redistribute from the supersaturated martensite into the retained austenite. This carbon enrichment of austenite improves its stability, reducing the likelihood of its transformation into martensite during further cooling or mechanical loading, which is beneficial for enhancing ductility and toughness. In the tempering stage, typically carried out at relatively low temperatures, carbides precipitate preferentially within the martensite matrix. The simulation results indicate that the volume fraction of carbides differs significantly between steels of different carbon contents. High-carbon steels exhibit a higher volume fraction of carbide precipitation compared to low-carbon steels, owing to the greater availability of excess carbon. This precipitation of carbides alters the local mechanical environment of the martensite. Specifically, as carbides nucleate and grow, the surrounding martensitic regions experience an increase in local stress levels. Moreover, the presence of carbides induces localized plastic strain in the adjacent martensite, which is notably greater than the plastic strain generated solely by the martensitic transformation during quenching. This plastic deformation arises due to the misfit between the carbides and the martensite lattice and the volume change associated with carbide precipitation. At the carbide/martensite interface, as carbides continue to grow, the carbon concentration at the interface rises progressively, leading to an increase in the maximum equilibrium stress and equivalent plastic strain in this region. These factors collectively influence the mechanical performance of Q-P-T steel, highlighting the significance of controlling carbide precipitation behavior to tailor the final properties of high-strength steels.

Keywords: Q-P-T steels; carbides; phase-field; strengthening mechanism

Proceedings of IFHTSE 2025

The 30th World Congress of International Federation for Heat Treatment and Surface Engineering
August, 19-22, 2025, Suzhou, China

The Interaction Between Cracks and Carbides in Q-P-T Steel

Yang Yuchen[1], Zheng Hongqing[1], Zuo Xunwei[2*], Wan Jianfeng[1*], Rong Yonghua[1], Chen Nailu[2]

[1] School of Materials Science and Engineering, Shanghai Jiao Tong University, Shanghai 200240, China

[2] Shanghai Key Laboratory of Materials Laser Processing and Modification, School of Materials Science and Engineering, Shanghai Jiao Tong University, Shanghai 200240, China

Abstract: Toughness is a fundamental mechanical property that describes a material's capacity to absorb energy through plastic deformation and fracture before ultimately failing. Materials with higher toughness can endure greater deformation and dissipate more energy without fracturing, which is essential for ensuring the reliability and safety of components used in demanding applications, such as structural, automotive, or aerospace engineering. One of the most effective approaches for evaluating toughness is through the study of crack initiation and propagation behaviors, as these phenomena directly relate to how materials respond under stress and how they ultimately fracture. In this context, heat treatment processes play a pivotal role in tailoring the microstructure and, consequently, the mechanical properties of steels. While the Quenching and Partitioning (Q&P) process is known for enhancing strength and ductility by stabilizing retained austenite, the subsequent tempering step in the Quenching-Partitioning-Tempering (Q-P-T) process offers an additional opportunity to optimize the balance between strength and toughness. Tempering allows for the controlled precipitation of carbides within the martensite matrix, which can influence crack propagation behavior and overall fracture resistance. In this study, a fracture phase-field finite element model was employed to simulate and analyze the mechanisms of crack propagation in Q-P-T steel. The model allowed detailed investigation into how cracks initiate and evolve within these phases, and how microstructural features such as carbides influence the crack path and resistance to fracture. The simulation results revealed that as the carbon content in steel increases, the volume fraction of carbides precipitated within martensite correspondingly rises. This higher carbide fraction alters the local microstructure, increasing the likelihood that propagating cracks, which typically originate from an initial edge notch or defect, will intersect carbide phases during their growth. Such interactions introduce significant obstacles to crack propagation because carbides disrupt the continuity of the martensitic matrix and locally increase the stress field, making crack growth more difficult. Furthermore, a comparative analysis between Q&P steel and Q-P-T steel indicated that the presence of carbides in martensite leads to more substantial hindrance of crack propagation within martensitic laths in Q-P-T steel. When cracks propagate from initial edge notches and traverse the martensitic microstructure, they encounter greater resistance in Q-P-T steel, attributable to the carbide-induced obstruction. This highlights the important role that carbide precipitation plays in improving the toughness of Q-P-T steel by effectively impeding crack growth, thereby contributing to enhanced mechanical performance and greater structural reliability.

Keywords: Q-P-T steels; crack initiation and propagation; phase-field; polycrystalline phase-field finite element modeling

Study and Analysis on the Effect of Stirring Rate on Quenching Cooling Effect

Zhu Hua, Han Bing, Xue JianMing

Nanjing Kerun Industrial Media Co., Ltd., China

E-mail address: 0358@njkerun.com (Zhu Hua)

Abstract: Currently, in heat treatment equipment, the most commonly used stirring method for quenching tanks is propeller stirring. The quenching medium generates impact through the high speed rotation of propeller blades, which can achieve a good stirring effect. A reasonably designed propeller agitator can make the quenching medium form turbulent flow, which is conducive to increasing the cooling capacity of the medium. However, there are few studies on the actual cooling effect of oil tank agitation in on-site heat treatment production. This experiment uses propeller stirring to analyze the influence on the actual cooling characteristics and the quenching effect of workpieces. The kinematic viscosity of the quenching oil at 40℃ is 61.59 mm^2/s, the maximum cooling rate is 85.2 ℃/s, and the on-site operating oil temperature is 90 ℃. The detection of the actual cooling characteristics of the oil in the tank is basically carried out with reference to the GB/T30823 standard. Cooling characteristic tests are conducted for the same type of quenching oil in the quenching oil tank of the same oil-quenching roller hearth furnace, at the same detection position, and under different stirring motor speeds. The test results show that when the stirring speed exceeds 600 r/min, the vapor film stage is sharply shortened and can no longer be detected by the cooling characteristic tester. In the convection stage, the increase in motor stirring speed intensifies the flow velocity in the corresponding area, reducing the cooling time to 200 ℃ by approximately 2.5 times. Meanwhile, standard cylindrical specimens with a length of 95 mm, a diameter of 18 mm, and a material of GCr15 are selected. Through 500 × microstructure detection, it can be observed that when the stirring speed is increased from 200 r/min to 800 r/min, the content of troostite significantly decreases, and the quenching effect is obviously improved. The research findings indicate the following: 1. The surface heat flux density of the workpiece is correlated with the stirring rate of the propeller. When the stirring rate is below 600 r/min, the surface heat flux density (qin) of the workpiece is greater than or equal to the minimum critical heat flux density (qcr1) required to form a vapor film. The quenching process sequentially undergoes four stages: transient boiling, vapor film, nucleate boiling, and convection. Due to the intensification of forced convection, the vapor film rupture time is shortened, the areas of the transient boiling (Stage I) and vapor film (Stage II) are reduced, the characteristic temperature increases accordingly, and the maximum cooling rate also rises. 2. When the stirring rate exceeds 800 r/min, the medium removes more heat from the workpiece surface, causing a decrease in the workpiece's qin. Consequently, the surface heat flux density (qin) becomes less than qcr1. In this case, the vapor film stage during quenching is extremely brief, and after entering the transient boiling (Stage I), the process transitions almost directly to the nucleate boiling (Stage Ⅲ). 3. During the quenching process, as the stirring rate of the quenching oil tank motor increases, the characteristic temperature in the high-temperature stage rises. Under high-speed stirring conditions, the characteristic temperature becomes nearly undetectable by the nickel chromium alloy probe, and the film rupture time is extremely short. 4. In the low-temperature stage, the convective heat transfer effect is significantly enhanced. The comprehensive cooling capacity of the quenching oil shows a clear upward trend, ultimately leading to an effective improvement in the actual quenching quality.

Keywords: quenching oil tank; stirring rate; cooling characteristics; heat flux density

Heat Treatment Simulation of SAE 1060 Steel Using Polyacrylate Aqueous Solution as Quenching Media

Sergio Gallegos-Cantú[1], D. Scott MacKenzie[1], Jihong Yao[2], Min Ma[2]

[1] Quaker Houghton, Norristown, Pennsylvania, USA

[2] Quaker Houghton, QingPu, China

E-mail address: sergio. gallegos@ quakerhoughton. com (Sergio Gallegos-Cantú)

Abstract: Simulations of heat treatment processes can help predict both mechanical and microstructural properties of metals. However, simulation results are dependent on process inputs such as, material, specimen shape, heating and cooling cycles and cooling curves characteristics, among others. Quenching fluids such as polymers are still under investigation and need to be characterized to improve the accuracy of the simulation results. This work aims to contribute to the optimization of the simulation process in the heat treatment industry. Cooling curves of a polyacrylate-based (ACR) polymer solution at a concentration of 9 and 12%, using an axial flow rate of 1. 30 L/min on an inmersion system and a fluid temperature of 45 ℃ were acquired and analyzed. Air quenching was also used to compare the polymer quenching conditions. The specimens made of SAE 1060 steel were designed for a laboratory-scale immersion quenching system; a 1 mm diameter thermocouple was introduced on the cross section to measure cooling curves at 2 mm from the bottom surface of the specimen. The results of the cooling curves were used as boundary conditions in the simulation. The commercial software JMatPro© (Version 15, Sente Software, Guildford, UK) was used to predict hardness and phases on the SAE 1060 steel. Finally, physical tests were performed on the specimens for the polymer and air quenching conditions to validate the results provided by the simulation.

An Artificial Neural Network-Based Parameter Identification Model of Spray Quenching for 7050 Aluminum Alloy

Guo Ruichao*, Zhao Haizhen

School of Aeronautical Engineering, Shandong Engineering Research Center of Aeronautical Materials and Devices, Shandong University of Aeronautics, China

E-mail address: ytgrc@163.com (Guo Ruichao)

E-mail address: zhaohaiz2000@163.com (Zhao Haizhen)

Abstract: Aluminum alloys are one kind of the most commonly used materials for aerospace structural parts due to its lightweight, high strength, and good formability. High-strength aluminum alloys have been increasingly used as the structural materials in the aerospace industry. Spray quenching is a critical heat treatment process for achieving high-quality production of aerospace aluminum thick plates. However, the inappropriate spray quenching will lead low mechanical properties and high residual stresses. Thus, it is urgent to optimize the spray quenching processes. An artificial neural network-based parametric identification design framework was proposed in present work. Through the spray quenching experiments, which were combined with FEM and quenching factor analysis method, the residual stresses and mechanical properties of 7050 aluminum alloy were analyzed under different spray conditions. Combining physics and simulation data with artificial neural network, the parametric identification model was constructed. The establishment of themodel was mainly divided into two steps. The first step was to establish a forward model based on neural networks using radial basis functions. Orthogonal experiments were used to generate a small sample of process parameter dataset. The cooling curves of the central layer of the material under different spraying parameters were collected by spray quenching experiments. The distribution of mechanical properties from the central layer to the boundary layer were obtained by quenching factor analysis method, And the distributions of residual stresses were determined by finite element simulation. Then, the obtained data were applied for the forward model training. A large number of virtual spray parameter combinations were generated by Latin hypercube sampling in the process parameter space, and the corresponding material residual stresses and mechanical properties were predicted by the trained forward model. Finally, an expanded data set was obtained. The second step was to establish a parameter identification model using artificial neural networks. Residual stresses and mechanical properties were served as input variables, while the output variable as the spraying parameters. The obtained expanded data set was used for training the parametric identification model, and an early stop mechanism was set. The dynamic learning rate adjustment and Bayesian optimization were used to optimize the model parameters. The obtained results show that the designed model has high accuracy in identifying the spraying parameters, achieving good generalization performance. The minimum mean relative error of the recognition results is 1.03%, which corresponds well with the expected results. The model verifies the effectiveness of the established parameter identification method. This paper provides a promising solution for the deformation control problem in the heat treatment process.

Keywords: artificial neural network; radial basis function network; spray quenching; heat treatment; aluminum alloys; coordinated regulation

A Novel Quenching Strategy for Enhanced Mechanical Behaviour Homogeneity of Large-Size Parts via Bainite Kinetics Acceleration

Tian Yu[a,b], Pan Tao[a], Tan Zhunli[b], Shi Zhongran[a], Zhang Min[b]

[a] Institute for Structural Steels, Central Iron and Steel Research Institute Co., Ltd., Beijing 100081, China

[b] School of Mechanical, Electronic and Control Engineering, Beijing Jiaotong University, Beijing 100044, China

Abstract: Quenching is a key process for many large-size parts in manufacturing. However, property difference between the surface and the center of the part after quenching is amplified as its part size increases because of phase transition difference. As a mesothermal phase, bainite is potential to reduce above difference through appropriate microstructure regulation. A novel quenching strategy combining the advantages of spraying schedule and phase transformation characteristic is proposed for large-size parts here to get enhanced mechanical behaviour homogeneity. In this novel strategy, an additional holding at 700 ℃ for 600 s was used before traditional quenching cooling process during spraying of wheel block part. The results after novel treatment shown that the differences/gaps in strength and hardness between the surface and the center was reduced, and simultaneously the strength and toughness matching of overall part was improved than 20%. The mechanism of as-treated process was expounded in detail by bainite transformation kinetics. This additional holding can accelerate subsequent bainite formation, increase bainite volume fraction, and result in better homogeneity. These bainite with refining substructure is good to advanced strength and toughness matching. The FE model systematically considering thermal, phase transformation based on phase kinetics, and property were performed and its positive influence with new strategy on large-size parts were verified. It is considered that water spraying only has a great effect on the rim of the wheel. Hence, two blocks with size of 100 mm length × 145 mm width × 100 mm depth were cut from the wheel (Fig. 1 (a)). The detailed heat treatment schedule is as follows, shown in Fig. 1 (b): Block 1: directly cooling the tread (surface) with spraying for 245 s and followed by air cooling to room temperature; Block 2: cooling the tread (surface) with spraying for 20 s, followed by holding at 700 ℃ for 600 s and then spraying for 225 s, after that air cooling to room temperature. Fig. 1 (c) shows a schematic illustration of the water spray system, including a circular water system with a pump (SUNSUN, HQB-2000, 24W, 220V) with 2000 L/h water flowing, that can be switched on and off, a platform for placing blocks on top of circular water, spraying nozzle. After heat treatment, mechanical and microstructural analysis was executed via samples whose related positions and profiles were extracted from the 10 mm, 30 mm depth of every block, and mean hardness values for nine points were tested every 10 mm depth from tread to rim on the cross-section (Fig. 1 (d)).

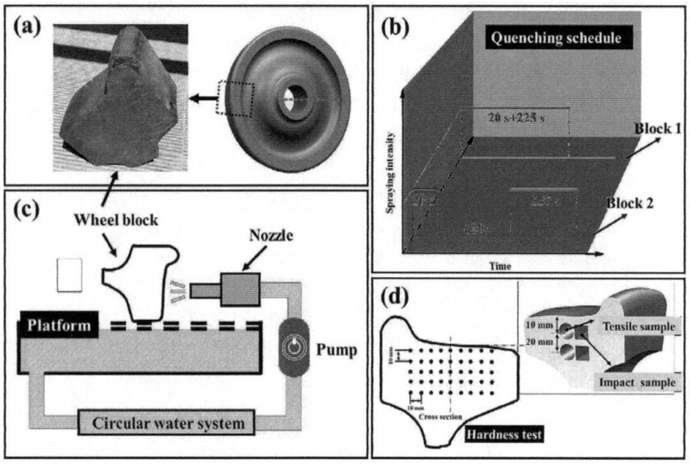

Fig. 1 (a) Test blocks cut from the typical large-size wheel, (b) quenching schedule realized by spraying vs time for two test blocks, (c) schematic illustration of the water spray system, and (d) position and profile of tensile and impact samples extracted from 10 mm and 30 mm depth of the block and tested hardness distribution

The mechanical properties at 10 mm, and 30 mm depth of the tread in two blocks are shown in Table 1. Block 1 has a tensile strength of 1356 ± 3 MPa and a yield strength of 1060 ± 8 MPa at 10 mm depth, while 1255 ± 8 MPa and 935 ± 7 MPa at 30 mm depth respectively, which is similar to Block 2. Two quenching schedules have little effect on the tensile properties and similar effects on total elongation with around 17.5%, on each block. But the difference in tensile strength at 10 mm, and 30 mm depth decreases from 101 MPa of Block 1 to 79 MPa of Block 2, and yield strength decreases from 125 MPa to 87 MPa. This indicates that the novelty quenching strategy can improve the tensile property homogeneity from the surface to the center. Meanwhile, the toughness of block is greatly improved after the strategy, while ensuring similar strength levels. The impact toughness obtained by this strategy also has better results, showing 23.5% enhancement at 10 mm depth from 85 ± 5 J to 105 ± 3 J and 25.8% enhancement at 30 mm depth from 62 ± 2 J to 78 ± 5 J. For strength and toughness matching, expressed as the product of tensile strength and impact toughness (unit: MPa · J), it improved 20% at 10 mm depth and 24% at 30 mm depth.

Table 1 Mechanical properties at 10 mm, 30 mm depth of the tread in two blocks

Sample		Tensile strength (MPa)	Yield Strength (MPa)	Total Elongation (%)	Impact toughness (J)
Block 1 (Direct cooling)	10	1356 ± 3	1060 ± 8	17.4 ± 0.1	85 ± 5
	30	1255 ± 8	935 ± 7	17.5 ± 0.1	62 ± 2
Difference/Gap		101	125	\	\
Block 2 (Novelty quenching)	10	1316 ± 2	1027 ± 4	17.6 ± 0.1	105 ± 3
	30	1237 ± 4	940 ± 5	17.6 ± 0.1	78 ± 5
Difference/Gap		79	87	\	\

Keywords: bainite acceleration; large-size part; incubation; surface quenching; homogeneity

Multiscale Gradient Nanostructure Design: Synergistic QPT and SMAT Processing for Enhanced Mechanical Properties of Plain Steels

Zeng Liyang[1], Chen Nailu[2], Zuo Xunwei[2*], Rong Yonghua[2], Lu Jian[1*]

[1] Department of Mechanical Engineering, City University of Hong Kong, China

[2] School of Materials Science and Engineering, Shanghai Jiao Tong University, China

E-mail address: jeepling@ sjtu. edu. cn (Zuo Xunwei)

E-mail address: jianlu@ cityu. edu. hk (Lu Jian)

Abstract: Since the introduction of the QuenchingPartitioningTempering (QPT) process in 2007, highcarbon QPT plain steels have demonstrated unprecedented simultaneous improvements in strength and ductility. However, the persistent challenge of overcoming the inherent trade-offs among strength, toughness, fatigue resistance, and cost efficiency remains. To address this, we have integrated the QPT-LE (local equilibrium) model with the Dislocations Across Martensite/Austenite Interface (DAMAI) effect to achieve precise control over phase fraction and carbon distribution in a Fe-0.69C-1.56Mn-1.51Si-0.052Nb (wt%) high-carbon Q-P-T steel. Subsequently, Surface Mechanical Attrition Treatment (SMAT) was applied to engineer a multiscale gradient nanostructure. Mechanical testing demonstrates that the Q-P-T + SMAT steel exhibits exceptional properties: an ultimate tensile strength of 1600 MPa coupled with over 30% elongation; a high-cycle fatigue limit of 820 MPa at 10^7 cycles, with a remarkable cost-performance ratio of 1653 MPa · kg/USD (14 times higher than conventional maraging steels); and a fourfold improvement in impact toughness, effectively overcoming the traditional strength-toughness dilemma. Mechanistic investigations reveal that the surface nanocrystalline layer and high dislocation density synergistically impede crack initiation and propagation, while the gradient compressive stress field combined with approximately 10% retained austenite significantly bolsters fatigue life and impact resistance. This novel multiscale gradient nanostructure design strategy opens new avenues for developing cost-effective, high-durability steels suitable for demanding applications in aerospace, automotive lightweighting, and dynamically loaded structures.

Keywords: Quenching-Partitioning-Tempering (QPT) steel; Surface Mechanical Attrition Treatment (SMAT); gradient compressive stress; gradient nanostructure; multiscale design

Influence of Quench Oil Vapor Film Formation Characteristics on Surface Non-Martensitic Organization

Zuo Yongping, Ding Lin, Tian Fang, Xia Baicheng
Nanjing Kerun lubricants Co. , Ltd. , Nanjing, China
E-mail address: zuoyp@ njkerun. com (Zuo Yongping)

Abstract: During the quenching process of gears, bearings and other critical components in heat treatment oil, non-martensitic structures often form on the surface, typically in the transition zone of the effective wall thickness of the workpiece, with a depth of approximately 10 to 30 micrometers, leading to a significant decrease in contact fatigue strength. The cause is closely related to the shock boiling stage and the steam film rupture and wetting behavior during the heat transfer process of the quenching oil. Specifically, when a high-temperature workpiece is immersed in quenching oil, there is a 0. 1 to 0. 2 second shock boiling stage, where the instantaneous contact between the workpiece surface and the oil medium causes a rapid cooling, partially consuming the incubation period for the transformation of austenite to martensite. Subsequently, a stable steam film stage is entered, and the rupture and wetting process of the steam film is significantly influenced by the geometry and size of the workpiece. In the transition zone of the wall thickness, the steam film rupture and wetting speed of the quenching oil is slower, prolonging the high-temperature residence time and promoting the formation of more non-martensitic structures. In engineering practice, by regulating the cooling performance of the quenching medium, such as optimizing the characteristic temperature of the quenching oil, the film-breaking efficiency can be enhanced, effectively shortening the steam film time and quickly entering the boiling heat transfer stage. This helps to avoid excessive consumption of the incubation period by shock boiling and film boiling, and significantly reduces the cooling time of the workpiece surface at the " nose" of the C-curve. In this paper, a variable cross-section cylindrical bar-shaped workpiece was designed, and high-definition camera technology was used to directly observe the differences in the steam film rupture and wetting speed in different cross-sectional areas during the quenching process. This wetting delay phenomenon further exacerbates the consumption of the incubation period and becomes the core mechanism for the formation of the non-martensitic layer on the surface. The local delay in the wetting speed directly leads to abnormal microstructure in the transition zone, highlighting the importance of the dynamic behavior of the quenching medium in controlling the microstructure.

Keywords: non-martensitic structures; quenching; shock boiling; heat treatment

Effect of Copper Loading on the Tribological and Antibacterial Properties of Arc-Sprayed Cu-TiN/TiO_2 Laminate Coatings

Wu Lin[1], Jia Chuanbao[1*], Yu Zhentao[2*]

[1] MOE Key Lab for Liquid-Solid Structure Evolution and Materials Processing, Institute of Materials Joining, Shandong University, China

[2] Institute of Advanced Wear & Corrosion Resistant and Functional Materials, Jinan University, China

E-mail address: jiachuanbao@ sdu. edu. cn (Jia Chuanbao)

E-mail address: yuzhentao@ jnu. edu. cn (Yu Zhentao)

Abstract: Structural steel is widely employed in marine engineering applications, such as sea-crossing bridges, offshore oil platforms, and ship hulls, due to its inherent high strength and impact resistance. However, the harsh marine environment presents a complex combination of mechanical, chemical, and biological degradation mechanisms, which significantly compromise the structural integrity of steel components and increase maintenance demands. To address these challenges, the development of multifunctional protective coatings with enhanced durability, antimicrobial activity, wear resistance, and corrosion resistance is essential. In this research, Cu-TiN/TiO_2 laminate-structured metal-ceramic composite coatings with varying Cu loadings (8. 3 wt. % to 100 wt. %) were fabricated via a supersonic arc spraying technique by modulating the diameter ratio of heterogeneous metal filaments. The effects of Cu loading on the microstructure, mechanical performance, tribological and corrosion behavior, as well as the antibacterial efficacy of the coatings were systematically investigated. The results demonstrate that the composite coatings exhibit a dense and well-bonded layered architecture, with a gradient distribution of hardness and Young's modulus corresponding to changes in Cu loading. The alternating spatial arrangement of ductile Cu and hard TiN/TiO_2 ceramic phases contributes to strong interfacial bonding and an optimized hard-toughness balance, resulting in superior mechanical strength and erosion-corrosion resistance. Additionally, the heterogeneous laminate structure effectively modulates the dissolution and sustained release of Cu ions in corrosive environments, endowing the coatings with robust and long-lasting antibacterial properties. Notably, coatings with Cu loading exceeding 18. 6 wt. % achieved complete (100%) bactericidal efficiency.

Keywords: arc spraying; metal-ceramic laminated-structured; Cu-TiN/TiO_2 coating; sterilization and antifouling coating; erosion-corrosion

Mechanism of Water Jet-Guided Laser Machining of Al/SiCp Metal Matrix Composites

Ai Tiancheng, Xu Dongdong*

School of Mechanical Engineering, Tongji University, Shanghai, China

E-mail address: atc@tongji.edu.cn (Ai Tiancheng)

E-mail address: Dongdong_xu@tongji.edu.cn (Xu Dongdong)

Abstract: Metal matrix composites (MMCs) have attracted considerable research interest for their applications in the aerospace, automotive, and semiconductor industries. However, MMCs present significant challenges in machining. Water jet-guided laser (WJGL) machining emerges as a promising hybrid technique, which is highly suitable for machining these difficult-to-process materials. Nevertheless, the fundamental mechanisms of WJGL and effective guidelines for process optimization remain inadequately understood. For example, the unique surface morphologies produced by the laser-coupled water jet, the distinctive mechanisms induced by the impact and cooling effects of high-pressure water jet, and the additional reactions generated by laser ablation in a water- and air-rich environment are yet to be fully elucidated. To address these issues, this work systematically investigates the WJGL machining performance on aluminum-based silicon carbide MMCs (Al/SiCp MMCs), focusing on geometric accuracy, surface morphology, and subsurface characteristics. The results demonstrate that the process achieves good geometric accuracy with minimal thermal damage. Roundness errors increase moderately with hole diameter and depth, while dimensional errors remain relatively stable, indicating distinct influencing mechanisms. The taper is well-controlled, with ΔD values below 20 μm for holes and −46 μm for micropillars. Surface morphology analysis confirms the absence of significant protrusions and burrs, highlighting that the water-assisted cooling in WJGL effectively suppresses thermal effects and debris accumulation, enabling the formation of clean, accurate holes with smooth edges. Surface morphology analysis reveals several distinctive features resulting from the combined effects of pulsed laser ablation and high-pressure water jet impact. WJGL machining produces ripple-like patterns with periodic spacing, water-jet-induced flow marks, and localized pits along the sidewalls. SEM observations confirm that vapor recoil, rapid solidification, and hydrodynamic forces contribute to the formation of resolidified protrusions and pores. Comprehensive surface chemical reaction analyses reveal that WJGL machining induces notable surface chemical modifications on Al/SiCp MMCs. Despite the cooling effect of the high-pressure water jet, localized high temperatures lead to oxidation of matrix elements (forming Al_2O_3, MgO, CuO), decomposition of SiC, and the formation of carbides such as Al_2C_3. Additionally, SiO_2 and various C-O compounds are formed due to reactions between free Si/C and water vapor. Furthermore, the subsurface analysis of WJGL-machined Al/SiCp MMCs reveals that the process induces distinct microstructural modifications with minimal thermal damage compared to conventional laser machining. A shallow heat-affected zone (~14 μm) is formed, consisting of a porous recast layer (~7 μm) and a transition layer with thermally softened and fractured SiC particles. The high-pressure water jet contributes to SiC pullout, particle fragmentation, and crack formation, while rapid cooling limits heat diffusion. Overall, these findings indicate that WJGL enables localized material modification with reduced subsurface damage, preserving composite integrity while shaping the microstructure through the combined effects of laser heating and hydrodynamic impact, and showing significant improvement over conventional machining.

Keywords: water jet-guided laser; metal matrix composites; Al/SiCp; surface integrity

Fast Reversion of Hydrophility-Superhydrophobicity on Textured Metal Surface by Electron Beam Irradiation

Wang Li, Wang Haidou

Institute of Surface/Interface Science and Technology, Key Laboratory of Superlight Material and Surface Technology of Ministry of Education, College of Materials Science and Chemical Engineering, Harbin Engineering University, Harbin, 150001, China

Abstract: Reversible hydrophilic-hydrophobic transitions on metal surfaces are challenging in surface-wetting treatment. Existing preparation methods have the disadvantages of requiring a long time, producing a metal oxide layer, and being difficult to implement. In this paper, a rapid reversible transitions between hydrophilicity and superhydrophobicity was achieved on the surface of 316 L stainless steel using laser etching combined with electron beam irradiation. The contents of C-C/C-H hydrophobic functional groups was found to be notably increased by low electron dose in the experiments. The contents of C-O-C, O-C = O and -OH hydrophilic functional groups decreased. The interaction of excited secondary electrons with precursor molecules generated free radical fragments that adsorbed onto surfaces and formed new chemical bonds, changing the content of the functional groups. Reversible hydrophilic and superhydrophobic transitions could be achieved by adjusting the electron dose. The contact angle continually changed from 91° to 160° with a beam current of 100 μA, to 17° with 300 μA, and back to 90° with cleaning. The entire procedure required only four minutes. Meanwhile, the superhydrophobic state with contact angle of 160° has a sliding angle of 6°, presenting excellent resistance to bacterial adhesion, UV irradiation, and delayed icing.

Keywords: pitted texture; electron beam irradiation; functional group; superhydrophobicity; reversible transition

The Effect of Laser Etching Process Parameters on Surface Quality of Copper Film Deposited by Magnetron Sputtering

Zhang Yingtao[1], Sun Haoran[1], Wang Gang[2]

[1] College of Mechanical & Electrical Engineering, Hohai University, Changzhou, 213200, China

[2] Beijing Key Lab of Precision/Ultra-precision Manufacturing Equipments and Control, Tsinghua University, Beijing, 100084, China

Abstract: The structures with copper (Cu) thin film deposited on polyimide (PI) is widely used for satellite antennas in the aerospace field. As an advanced flexible manufacturing method, nanosecond laser etching is the preferable technology for precision manufacturing of electromagnetic functional components. The etching accuracy of thin films is the key point for the production requirements. The Cu film with specific geometric shape need to be removed by the etching process without damaging the PI layer. In this paper, the Cu film with thickness 5 μm is prepared on the surface of PI with thickness 120 μm by magnetron sputtering. The 125 sets of laser etching experiments are conducted to study the effects of process parameters on etching morphology. The scanning speed is set between 200 mm/s and 1000 mm/s. The pulse width varies from 6 μs to 14 μs and the pulse frequency is 30-50 kHz. The three-dimensional morphology is observed through laser confocal microscopy to get the cross-sectional morphology after laser etching. The groove size after laser etching is measured to determine the effects of scanning speed, pulse width and pulse frequency on the etching depth, etching width, height of protrusions on both sides, and roughness of the bottom surface. The etching depth changes from 0 to 13 μm with the increase of energy acting on the Cu film. Meanwhile, Cu material solidifies on the surface during the laser etching process, leading to the height on both sides of the groove. According to the testing results, the optimized parameters is set as scanning speed 400 mm/s, pulse width 10, pulse frequency 40 kHz, The Cu film can be completely removed without damaging the substrate material PI.

Keywords: laser etching; Cu film; microscopy; process parameters; polyimide

Tailoring Atomic Interfaces in One-Dimensional van der Waals Heterostructures for Semiconductor Applications

Kang Lixing

CAS; Suzhou Institute of Nano-tech and Nano-bionics, China

E-mail address: lxkang2013@ sinano. ac. cn (Kang Lixing)

Abstract: Owing to their pronounced nanoscale interfacial coupling effects, one-dimensional (1D) van der Waals (vdW) heterostructures exhibit highly distinctive electronic and magnetic behaviors. Among them, carbon nanotube (CNT) -derived 1D atomic crystal heterojunctions represent a novel class of vdW systems, where atomic-level interface engineering becomes possible. CNTs provide an ideal nanoconfinement platform due to their exceptional chemical stability, precisely defined inner cavities on the sub-nanometer to few-nanometer scale, and cylindrical surface topologies, enabling fine control over nanoscale heterointerface formation, phase stabilization, and interfacial orbital interactions. Building on this principle, we have developed a general confinement-guided synthesis strategy to fabricate diverse 1D atomic chains and heterostructures within single-walled CNTs. The encapsulation not only drives the formation of metastable or previously inaccessible 1D phases but also induces strong interfacial charge redistribution between the guest atomic chains and the host CNTs. Using low-voltage aberration-corrected transmission electron microscopy (AC-TEM) and advanced spectroscopic analyses, we achieve atomic-resolution structural characterization and probe the nature of charge transfer and orbital hybridization at the heterointerface. Magnetic and transport measurements reveal that these 1D confined heterostructures exhibit emergent magnetic phases—such as spin-glass-like states, superparamagnetism, and enhanced ferromagnetic ordering—which are markedly distinct from their bulk counterparts. These findings underscore the critical role of dimensional reduction and interfacial confinement in tuning quantum functionalities and offer new opportunities for interface-driven applications in nanoelectronics, spintronics, and surface engineering.

Keywords: semiconductor; CNTs (carbon nanotubes); nanostructures

Effect of Laser Shock Strengthening on Microstructure and Properties of Cronidur 30 Steel

Chen Wenbo[1], He Tiantian[1*]

[1] Henan University of Science and Technology, Luoyang, China

[2] National United Engineering Laboratory for Advanced Bearing Tribology, Henan University of Science and Technology, Luoyang, China

E-mail address: 17637401656@163.com (Chen Wenbo)

E-mail address: tthe@haust.edu.cn (He Tiantian)

Abstract: Aero-engine main shaft bearings operating under extreme conditions frequently suffer from surface degradation such as scratching and fatigue spalling, demanding superior material performance. This study systematically investigates the effects of laser shock peening (LSP) on third-generation bearing steel Cronidur 30 through comprehensive microstructural and mechanical characterization using SEM, TEM, XRD, residual stress analysis, and microhardness testing. The results demonstrate that increasing pulse energy and impact frequency during LSP treatment leads to: (1) surface roughening (Ra 0.013-0.62 μm) accompanied by lattice distortion evidenced through (110) α peak broadening and high-angle shifting, along with the formation of a deep compressive residual stress layer (−913MPa, 1400μmdepth); (2) significant microstructural refinement featuring fragmented lamellar martensite and redistributed carbides (size reduced from 0.49 to 0.45 μm with increased number density); and (3) enhanced mechanical properties including substantial work hardening (hardness increased 23% to 903 HV) and extended hardened layer depth. These findings collectively validate LSP as an effective surface engineering approach for improving the durability and performance of advanced bearing steels in critical aerospace applications, where resistance to surface-initiated failures is paramount. The study establishes quantitative process-structure-property relationships that can guide the optimization of LSP parameters for Cronidur 30 components.

Keywords: Cronidur 30 bearing steel; laser shock peening; microstructure; mechanical property

Strength Improvement of Al-Mg-Sc Alloy by Powder-Wire Coaxial Laser Deposition

Liu Xianyue[1], Wang Gang[2*], Ji Longcen[3], Sun Haoran[3]

[1] Department of Mechanical and Energy Engineering, Southern University of Science and Technology, China

[2] Beijing Key Laboratory of Transformative High-end Manufacturing Equipment and Technology, Tsinghua University, China

[3] College of Mechanical & Electrical Engineering, Hohai University, China

E-mail address: 11930500@ mail. sustech. edu. cn (Liu Xianyue)

E-mail address: gwang@ tsinghua. edu. cn (Wang Gang)

Abstract: Aluminum alloy structural components are widely used in aerospace applications due to their high strength-to-weight ratio, which meets the critical performance requirements of aircraft and spacecraft. Coaxial laser wire directed energy deposition (Coaxial LW-DED) has attracted significant research interest in recent decades owing to its unique direction-independent characteristic compared to common lateral wire feeding process. In this process, the metal wire is fed coaxially through annular laser radiation, enabling enhanced geometric flexibility for repairing damaged components or fabricating complex structural parts. To improve the strength of laser-deposited Al-Mg-Sc alloy, nano-ZrO_2 particles was introduced through synchronous powder-wire feeding. The nano-ZrO_2 powder was coated on the surface of AlSi10Mg micro powder. The forty-layer thin-wall part was built by using Al-Mg-Sc wire as forming material and AlSi10Mg/nano-ZrO_2 powder as reinforcement. Different proportions of nano-ZrO_2 additions (5 wt%, 10 wt% and 20 wt%) and different powder feeding rate (0. 5 g/min and 1. 5 g/min) were employed to obtain the optimal mechanical properties of ZrO_2-modified Al-Mg-Sc alloy. The samples fabricated using higher proportions of ZrO_2 particles and higher powder feeding rate tend to have lower yield strength and lower elongation. The grain morphology was analysed by electron backscatter diffraction (EBSD). The microstructure of Al-Mg-Sc/nano-ZrO_2 composites was characterized using the scanning electron microscope (SEM) and transmission electron microscopy (TEM). The spherical nano-sized ZrO_2 reinforcements hinder the dislocation motion, resulting in the strengthening effect on laser-deposited Al-Mg-Sc alloy.

Keywords: Al-Mg-Sc alloy; coaxial laser wire directed energy deposition; synchronous powder-wire feeding; ZrO_2; strength

Study of Surface Modification based on Multi-Physics Synergy

Piao Zhongyu*, Zhou Zhenyu, Zheng Qiuyang

Zhejiang University of Technology, Hangzhou, China

E-mail address: piaozy@zjut.edu.cn (Piao Zhongyu)

E-mail address: zhouzy@zjut.edu.cn (Zhou Zhenyu)

Abstract: In response to the critical scientific challenges of harsh service conditions, superficial structural instability, and performance degradation in critical components of high-end equipment for aerospace and nuclear power applications, this study introduces a suite of surface modification strategies applicable to diverse material surfaces, including laser-deposited layers, thermal spray coatings, and as-cast substrates. A comprehensive theoretical and technological framework for surface modification, driven by multi-physics synergy, has been systematically established. This technology targets core bottlenecks such as surface defects, microstructural heterogeneity, and interfacial failure. From a multi-scale and multi-physics perspective, the methodology leverages the spatio-temporal synergy of ultrasonic, electric current, laser, and mechanical fields to facilitate the transformation of the surface layer from a rough, porous state to a dense, homogeneous one; the refinement of grain structures from the micro-scale to a nano-gradient architecture; and the evolution of the interfacial bond from mechanical interlocking to metallurgical integration via elemental interdiffusion. A pathway has been established for the concurrent enhancement of microstructural continuity, interfacial toughness, and a balanced residual stress state. The investigation systematically elucidates the intrinsic mechanisms governing crystallographic texture reconstruction, coupled plastic deformation, micro-defect healing, and the transition of residual stress from tensile to compressive states under the influence of multi-physics synergy. A multi-scale correlation linking microstructural evolution, interfacial behavior regulation, and macroscopic service performance enhancement has been constructed. The developed surface modification framework significantly improves the fatigue, fretting, wear, and corrosion resistance of critical components under extreme service conditions, providing a robust theoretical foundation and technical support for advancing the operational reliability and safety margins of high-performance equipment.

Keywords: Multi-physics synergy; surface modification; microstructure control; interfacial strengthening; residual stress regulation

A New Method for Preparing Mg-Al-layered Double Hydroxides coating using supercritical CO_2 fluid

Zhang Xiaochen

Heilongjiang Institute of Technology, China

E-mail address: zxc161616@126.com (Zhang Xiaochen)

Abstract: A new method for preparing Mg-Al-layered double hydroxides (Mg-Al-LDHs) coating using supercritical CO_2 fluid was developed. The morphologies, structures, compositions, and corrosion resistances of the surface coatings were investigated. Unlike traditional methods, the method using supercritical CO_2 fluid generated ultrafine nanocrystal coatings with an average size of 30 nm, and a thickness of ca. ~11 μm. The Mg-Al-LDHs coating was mainly composed of C, O, Mg, and Al. Corrosion resistance measurements show that the amount of hydrogen evolution for the nanocrystal LDHs coating was only 0.133 mL · cm^{-2}, thc polarisation resistance was 3174.9 Ω · cm^2, and the immersion test corrosion grade of 7G. Finally, the formation mechanism of the Mg-AlLDHs coating under supercritical CO_2 fluid was proposed.

Keywords: surface; corrosion resistances; Mg-AlLDHs coating

Keynote Speakers in Advanced Technology for Functional Surface Modification

Kang Lixing

E-mail address: lxkang2013@ sinano. ac. cn (Kang Lixing)

Abstract: Owing to their pronounced nanoscale interfacial coupling effects, one-dimensional (1D) van der Waals (vdW) heterostructures exhibit highly distinctive electronic and magnetic behaviors. Among them, carbon nanotube (CNT) -derived 1D atomic crystal heterojunctions represent a novel class of vdW systems, where atomic-level interface engineering becomes possible. CNTs provide an ideal nanoconfinement platform due to their exceptional chemical stability, precisely defined inner cavities on the sub-nanometer to few-nanometer scale, and cylindrical surface topologies, enabling fine control over nanoscale heterointerface formation, phase stabilization, and interfacial orbital interactions. Building on this principle, we have developed a general confinement-guided synthesis strategy to fabricate diverse 1D atomic chains and heterostructures within single-walled CNTs. The encapsulation not only drives the formation of metastable or previously inaccessible 1D phases but also induces strong interfacial charge redistribution between the guest atomic chains and the host CNTs. Using low-voltage aberration-corrected transmission electron microscopy (AC-TEM) and advanced spectroscopic analyses, we achieve atomic-resolution structural characterization and probe the nature of charge transfer and orbital hybridization at the heterointerface. Magnetic and transport measurements reveal that these 1D confined heterostructures exhibit emergent magnetic phases—such as spin-glass-like states, superparamagnetism, and enhanced ferromagnetic ordering—which are markedly distinct from their bulk counterparts. These findings underscore the critical role of dimensional reduction and interfacial confinement in tuning quantum functionalities and offer new opportunities for interface-driven applications in nanoelectronics, spintronics, and surface engineering.

Progress of the Research on Diffusion Barrier for Cr-coated Zr Alloys at High Temperature

Yang Jianqiao[1*], Zhao Fen[3], Yun Di[2]

[1] Key Laboratory of Thermo-Fluid Science and Engineering of MOE, School of Energy and Power Engineering, Xi'an Jiaotong University, Xi'an 710049, China

[2] School of Nuclear Science and Technology, Xi'an Jiaotong University, Xi'an 710049, China

E-mail address: jianqiao. yang@ aliyun. com (Yang Jianqiao)

Abstract: Currently, zirconium alloys are the primary materials used for nuclear fuel cladding due to their extremely low thermal neutron absorption cross-section and high melting point. However, the loss-of-coolant accident at the Fukushima nuclear power plant in Japan in 2011 highlighted a critical limitation of zirconium alloys: under high-temperature steam conditions during accidents, they oxidize rapidly and release hydrogen, potentially leading to hydrogen explosions. To address this drawback and enhance the safety of nuclear reactors, the development of Accident Tolerant Fuel (ATF) cladding materials has become a research focus in recent years. One important approach in ATF cladding development is applying protective coatings on zirconium alloy surfaces to improve their oxidation resistance. Various types of coating materials have been developed for this purpose, including pure metal coatings, MAX phase coatings, alloy coatings, and oxide coatings. Among these, pure chromium coatings have shown significant potential in improving the high-temperature oxidation resistance and mechanical strength of zirconium alloy claddings. Chromium coatings are also advantageous due to their simple processing methods and good economic feasibility. As a result, chromium-coated zirconium alloys are considered highly promising candidate materials and have become a research hotspot, with in-reactor testing already carried out in several countries. However, at elevated temperatures, interdiffusion between chromium in the coating and zirconium in the substrate presents a significant bottleneck for the further development of chromium-coated zirconium alloys. Studies have shown that in a 1200 ℃ steam environment, the oxidation and diffusion consumption rates of the chromium coating are nearly identical. Zirconium diffuses along chromium grain boundaries and reacts with inward-diffusing oxygen to formZrO_2, which leads to grain boundary coarsening and creates additional pathways for oxygen diffusion, thereby accelerating coating degradation. Moreover, under beyond-design-basis accident conditions, rapid and complete coating failure can occur within a short time due to the eutectic reaction between chromium and zirconium, which limits the maximum service temperature of chromium coatings to no higher than 1330 ℃. To address the Cr-Zr interdiffusion issue, this study developed interlayers based on metallic, ceramic, and high-entropy alloy systems. These interlayers are introduced as thin layers between the chromium coating and zirconium alloy substrate to suppress mutual diffusion, thereby enhancing the high-temperature stability under accident conditions and improving the coating-substrate adhesion. The main research results are summarized as follows: (1) Ceramic Coating: A Cr-CrN-Cr composite coating was successfully fabricated. After oxidation in high-temperature steam at 1200 ℃ for 1 hour, the weight gain of the coated sample was less than one-third of that of the Cr-CrN/TiSiN-Cr coated sample, four-fifths of that of the CrN-coated sample, and four-fifths of that observed in the pure Cr-coated sample under 1100 ℃ conditions—only one-seventh of the weight gain of the uncoated zirconium alloy. This indicates a significant improvement in high-temperature oxidation resistance. After 6 hours of steam oxidation at 1200 钨℃, the sample exhibited a parabolic weight gain behavior without signs of accelerated oxidation, suggesting the coating still maintained its protective function. Microstructural analysis revealed that, after oxidation at 1200 ℃, the structure from the outermost layer inward consisted of: Cr_2O_3, CrN, Cr_2N, ZrN, a mixed phase of α-Zr (O) and α-Zr (N), α-Zr (O), and the Zr alloy substrate. Nitrogen released from CrN decomposition formed [1 1 0] // [1 1 0] twin layers with the underlying Zr matrix, and numerous stacking faults were observed at the coating-substrate interface. (2) Metallic Coating: A Cr-Cr/Mo-Cr composite coating was successfully fabricated. High-temperature diffusion experiments at 1200 ℃ showed that Mo diffused into the Zr substrate to form a solid solution phase, and the CrMo interlayer disappeared after 30 minutes. Outward diffusion of Zr led to the formation of a $Zr(Cr_{1-x}Mo_x)_2$ phase within the CrMo interlayer. The thickness of this diffusion layer was governed by the diffusion rate of Zr and increased in a parabolic manner. TEM-EDS analysis revealed the formation of a Cr: Mo: Zr = 3: 2: 4 mixed phase, with high-resolution imaging showing orderly atomic arrangements of Zr and Cr. This structure exhibited a strong inhibitory effect on oxygen diffusion, with almost no oxygen detected. Under conditions of 1400 ℃ for 5 minutes, only small Kirkendall-type voids (approximately 20-50 nm in size) were observed, indicating that the addition of Mo effectively raised the eutectic temperature of the coating.

(3) High-Entropy Alloy Coating: A quaternary Cr-Mo-Zr-Fe mixed interlayer was successfully fabricated between the chromium coating and the zirconium alloy substrate. The composition of the interlayer was 17 wt% Cr, 16 wt% Fe, 29 wt% Nb, and 38 wt% Mo, with a calculated configurational entropy of 1. 38R, classifying it as a medium-entropy alloy (MEA). Under a 1200 ℃ argon atmosphere, noticeable elemental segregation occurred in the MEA layer after 2 minutes, and the interlayer disappeared after 5 minutes, with a continuous layer of voids forming at the coating-substrate interface. Microstructural analysis revealed that segregation of Mo in the MEA layer was accompanied by void formation. This segregation resulted from the relatively lower diffusion rate of Mo compared to Cr, Fe,

and Nb. The remaining MEA layer developed dislocation structures due to volume changes.

Keywords: high temperature; diffusion; cr coating; zirconium alloy; loca

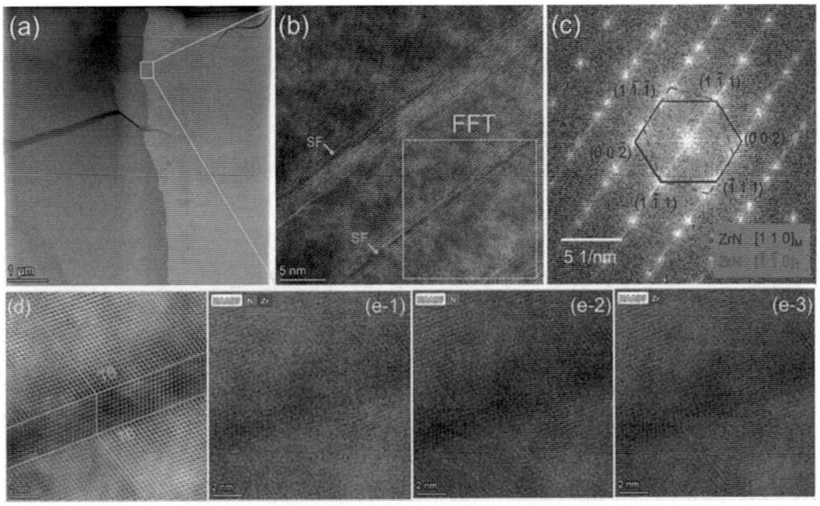

Fig. 1 Ceremic interdiffusion layer

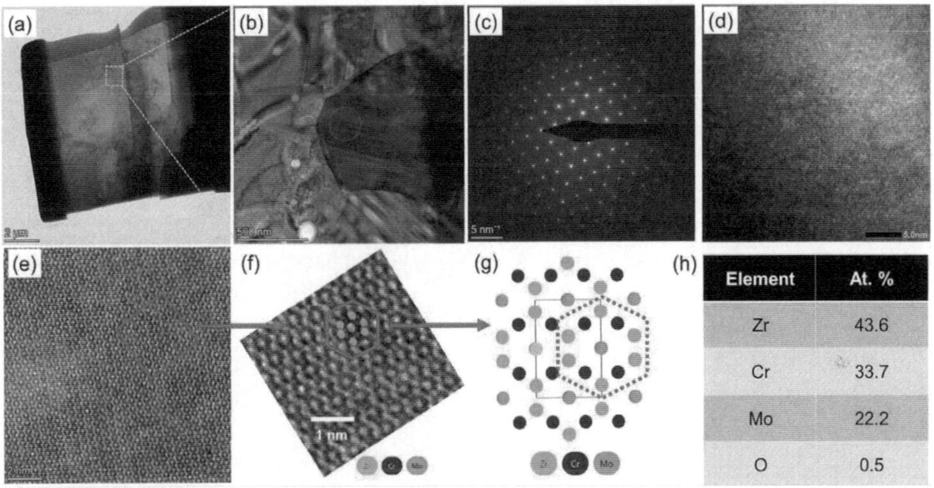

Fig. 2 Cr/Mo interdiffusion layer

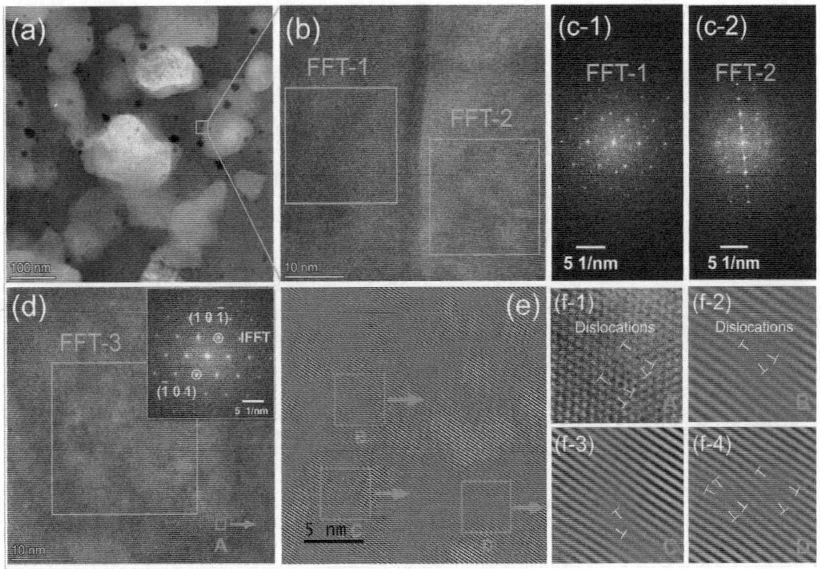

Fig. 3 High entropy interdiffusion layer

The First-Principles Study on the Effect of Irradiation on Oxygen Transport Properties in Zirconium Alloy Oxide Films

Xie Yaoping

National Key Laboratory of Nuclear Power Key Materials, School of Materials Science and Engineering, Shanghai University

E-mail address: xieyp@shu.edu.cn (Xie Yaoping)

Abstract: Zirconium alloys are universally employed as nuclear fuel cladding materials in pressurized water reactors (PWRs). The oxide film formed on the zirconium alloy surface acts as a barrier between the alloy matrix and the corrosive environment. Since zirconium alloy oxidation occurs at the oxide/metal interface, oxygen transport within the oxide film governs the corrosion rate of the alloy. To elucidate the corrosion mechanism of zirconium alloys, this study investigates the influence of irradiation on oxygen transport properties in zirconium alloy oxide films using first-principles calculations and the dual-end stochastic potential energy surface method.

Keywords: nuclear fuel cladding materials in pressurized water reactors; zirconium alloy

Electroshock Treatment Mechanism on Tensile Behavior of Laser Melting Deposited Ti55531 alloy

Xie Lechun

Wuhan University of Technology, China

E-mail address: xielechun@ whut. edu. cn (Xie Lechun)

Abstract: A novel method of electroshock treatment (EST) is introduced to modify the tensile properties of LMD Ti55531 alloy, and the optimizing mechanism is revealed. These results provide insight into a further improvement of the EST process to prepare alloys with better strength and toughness. The important results are highlighted as follows. The microstructure of LMD Ti55531 showed a layer structure with alternating short columnar and equiaxed grains, the α phase at the interface was refined and uniformly dispersed after 0. 06 s EST, and the α phase at the interface was transformed into β with EST increased to 0. 08 s. The elongation of LMD Ti55531 increases from 14. 5% to 17. 6% with 0. 06 s EST and the tensile strength increased from 826. 7 to 937. 8 MPa as well. The plasticity of LMD Ti55531 was obviously improved after EST, which corresponded to the variation of α-phase content and distribution. The effect mechanism of EST on tensile behavior of Ti55531 could be ascribed to the fine size α phase precipitated after EST with 0. 06 s. The fine size α phase showed a "pinning effect" on the slip line, which hindered and stop the movement of the slip line and resulted in the improvement of elongation. All results indicated that EST can optimize the microstructure of LMD Ti55531, and improve the plasticity without the loss of tensile strength. It is an effective and novel way to optimizing the tensile properties of LMD titanium alloys.

Keywords: electroshock treatment (EST); laser melting deposition (LMD); tensile property; Ti55531 alloy; slip lines

Residual Stress Release Mechanism and Theory on Post Weld Heat Treatment of Pressure Vessels

Gu Wenbin[1], Jiang Wenchun[1*]

[1] College of New Energy, China University of Petroleum (East China), Qingdao, 266580, China

E-mail address: jiangwenchun@126.com (Jiang Wenchun)

Abstract: Welding residual stress is the primary cause of stress corrosion cracking, creep failure, and fatigue damage in large pressure vessels. Heat treatment has become an effective approach to relieve such welding-induced residual stresses. Currently, the pressure vessel are becoming extremely large in diameter, thickness and length, Due to limitations in furnace size, special structures, in-service repair welding, etc., large pressure vessels often cannot undergo integral heat treatment and require local post-weld heat treatment (LPWHT). However, the applications of LPWHT may result in detrimental effects without proper process optimization. In this paper, the evolution of residual stresses during overall and local heat treatment considering creep effects is given for large pressure vessels. The results show that creep is a key factor in regulating residual stresses during the integral heat treatment. The release of welding residual stresses occurs mainly in the heating stage. When the heat treatment temperature reaches the creep temperature, the residual stress is rapidly released in a short period of time and remains largely unchanged during the holding stage. The residual stress evolution pattern during the heating and cooling stages of local heat treatment is quite different from the integral heat treatment. During the cooling stage, due to the cold-end constraints from unheated regions and the effect of axial temperature gradient, waist-contracting deformation occurs, generating significant secondary tensile stress on the inner wall, which leads to the ineffective relief of the inner wall stress. To address the above issues, this paper establishes an analytical solution for the stresses and deflections induced by local heat treatment of pressure vessels, and presents process criteria to approximate the residual stress relief effect of integral heat treatment. Furthermore, a primary-secondary local heat treatment method and theoretical model are proposed. Through theoretical derivation, the optimal process parameters for maximizing residual stress relief are determined, and a process design approach to induce compressive stress on the inner wall.

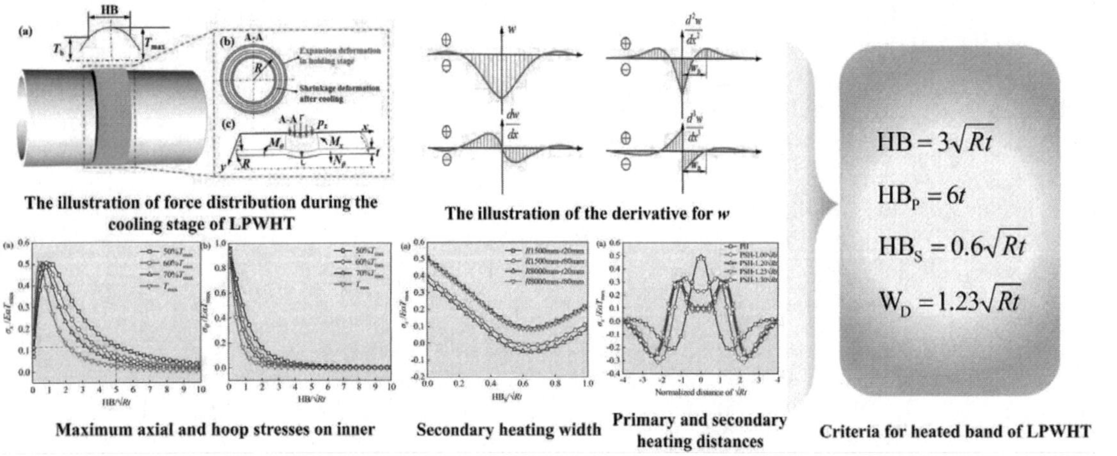

Process and results of the theory on local post weld heat treatment of ultra-large pressure vessels

Keywords: local post weld heat treatment; width of heated band; residual stress; primary plus secondary local post weld heat treatment

Enhanced Steam Oxidation Resistance of Zr-Sn Alloy via ZrO_2/CrAl Composite Coating for Accident-Tolerant Fuel Applications

Zhou Qian[a,b], Xu Chi[a], Xue Luwei[c,d], Wang Rongshan[c,d], Zhao Wanxiang[c,d], Chen Weiwei[b], Xue Wenbin[a*]

[a] Key Laboratory of Beam Technology of Ministry of Education, School of Physics and Astronomy, Beijing Normal University, Beijing 100875, China

[b] Department of Materials Science and Engineering, Beijing Institute of Technology, Beijing, 100081, China

[c] Suzhou Nuclear Power Institute, Suzhou 215004, China

[d] National Engineering Research Center for Nuclear Power Plant Safety & Reliability, Suzhou 215004, China

E-mail address: 17812162775@163.com (Zhou Qian)

E-mail address: xuewb@bnu.edu.cn (Xue Wenbin)

Abstract: Zirconium alloys serve as critical nuclear fuel cladding materials in commercial reactors due to their excellent mechanical properties and low neutron absorption cross-sections. However, their rapid oxidation in high-temperature steam during loss-of-coolant accidents (LOCAs) remains a significant safety concern, as it leads to hydrogen generation and potential explosion hazards. To address this challenge, we developed an advanced $ZrO_2/Cr_{94}Al_6$ composite coating on Zr-Sn alloy using plasma electrolytic oxidation and vacuum arc deposition techniques. The coating features a ZrO_2 interlayer that prevents Cr-Zr interdiffusion and galvanic corrosion, while the CrAl outer layer forms protective oxide scales. After 3600s in 1000-1200℃ steam, the coated samples exhibited mass gains of only 16.7-27.8% compared to uncoated alloy, with oxidation activation energy increasing from 220.41 to 404.78 kJ/mol. Microstructural analysis revealed the formation of a dual-layer oxide structure, with an outer Cr_2O_3 scale and an inner Al_2O_3 barrier that effectively blocks oxygen penetration. The coating demonstrated excellent stability across the temperature range, though different interfacial reaction products were observed at varying temperatures. These results demonstrate the significant potential of this composite coating system for enhancing the accident tolerance of nuclear fuel cladding materials in severe operating conditions.

Keywords: Zr-Sn alloy; ZrO_2/CrAl coating; high temperature steam oxidation; Al_2O_3 barrier layer; accident-tolerant fuel

Study on the Accelerated Oxidation Mechanism of Zr Alloys Induced by LiOH

Abstract: Eddy current testing is the most widely used method for measuring the thickness of cladding tubes in pressurized water reactors (PWRs) . During poolside inspection, an automated system is used to scan the fuel assembly with an eddy current probe. To ensure the accuracy of this method, a set of Zr alloy samples with known and uniform oxide thicknesses is required for calibration, which can be called the standard sample. After three fuel cycles, the oxide thickness of cladding tubes can exceed 100 钨 μm. Therefore, a batch of standard Zr alloy samples with oxide thicknesses up to 150 μm is needed. These standard samples must exhibit physical properties—such as electrical and magnetic conductivity—equivalent to those of actual PWR cladding tubes. Moreover, the oxide film on these samples should be durable and resistant to cracking under cyclic compressive stress generated during eddy current inspection. Additionally, the electrical conductivity of the Zr alloy substrate should match that of the original cladding material. In this study, a fabrication method for producing Zr alloy standard samples with controlled oxide thickness was developed. Corrosion tests were carried out in a static autoclave at 360 ℃ and 18.6 MPa—conditions commonly used for evaluating the corrosion behavior of Zr alloy claddings. A high lithium ion concentration (450 mg/L) was used to accelerate oxidation. The maximum corrosion duration was 316 hours. The tested alloy composition was Zr-1Nb. Following corrosion testing, cross-sectional analysis of the oxide layers was conducted using scanning electron microscopy (SEM) . A transmission electron microscopy (TEM) foil was also prepared near the oxide/metal (O/M) interface using focus ion beam (FIB) milling. The results revealed that the oxide film growth followed a nearly linear rate. Four circumferential zones were examined by SEM, showing uniform oxide thickness across the sample. No significant cracks, either vertical or horizontal, were observed in samples with oxide thickness below 100 μm. However, vertical cracks were present in samples exceeding 100 μm in thickness, initiating at the outer surface and extending approximately halfway through the oxide layer. TEM analysis showed that the oxide film consisted of nano-crystalline ZrO_2 grains, approximately 20-30 nm in diameter. Numerous micropores were observed in the TEM foils. In some regions, continuous cracks with approximately 1 钨 μm in length were found, oriented parallel to the O/M interface. Energy-dispersive X-ray spectroscopy (EDS) mapping was conducted to detect niobium segregation; however, no distinct segregation was observed. Selected area electron diffraction (SAED) revealed two sets of diffraction patterns in grains near the pores: one corresponding to ZrO_2, and the other to Nb oxides. This indicates that Nb oxide segregation occurred, and that these oxides were preferentially dissolved by LiOH, resulting in the formation of nanoscale pores and cracks. These pores served as diffusion pathways for Li^+ and O^{2-} ions, undermining the compactness of the ZrO_2 oxide layer and thereby accelerating oxidation. Furthermore, considering the Pilling-Bedworth ratio of ZrO_2 is 1.56, significant tensile stress is expected when the oxide film exceeds 100 钨 μm in thickness. The presence of nano-sized pores helped to relieve such stress, thereby preventing the formation of cracks parallel to the O/M interface.

Influence of Time and Loading on the Long-Term Structural Stabilities and Mechanical Behavers of Some Alloys Used in Nuclear Power Plants

Chai Guocai

Linkoping University, China

E-mail address: chaiguocai@ riam. com. cn (Chai Guocai)

Abstract: The requirements of the materials for the nuclear power plants are not only their physical, thermal, corrosion, and mechanical properties, but also their long-term structural stabilities during the service. So far, such information is very few, especially the influence of loading. This study will provide a review on the long-term structural stabilities and their effect on the mechanical behavers of some nuclear alloys such as Alloy 800H, 316L/LN and 304L used in the nuclear power plants with ageing time up to about 260 000 hours (about 30 years), and a discussion on the influences of the applied stresses. For Alloy 800H, the g' phase is the main precipitate at temperature below 700 ℃. They are small with a low growth rate. At higher temperatures, $M_{23}C_6$ and MX are the main precipitates. The stress applied will increase MX phase in the matrix, but decreases M23C6 at the grain boundaries, which is beneficial to the creep strength. For 316L/LN, the stress applied can greatly reduce precipitation of $M_{23}C_6$ phase. Few can be observed in the sample tested with a stress of 45 MPa at 700 ℃ for 240 131 hours. However, it increases amounts of sigma phase and G phase. Long-term ageing can also lead to a phase transformation from Laves phase to nano X phase, which causes pinning of moving dislocations and consequently formation of nano dislocation sub-cells or fine grains. This leads to an unexpected high long-term creep strength or life. Sigma phase, $M_{23}C_6$ and ETA phase are the main precipitates in 304 L steel. Similar to 316 L/LN, the stress applied reduces amount of $M_{23}C_6$, but increase Sigma phase and ETA phase. Plenty of small sigma phase particles precipitated at the moving dislocations or dislocation sub-cells that contribute to the creep resistance of the material. This work may increase the understanding or knowledge to choose these materials for the applications.

Keywords: age; structural stability; mechanical property; creep; precipitation

High-Performance Ti-Based Alloys for Nuclear Power: Design and Optimization of HEAs and Ti-O Systems

Cai Zeyun[1], Leng Xuesong[1*], Xie Guoqiang[2,3*]

[1] Research Institute of Physical Sciences in Special Environments, Harbin Institute of Technology (Shenzhen), Shenzhen 518055, China

[2] School of Materials Science and Engineering, and Institute of Materials Genome & Big Data, Harbin Institute of Technology (Shenzhen), Shenzhen 518055, China

[3] State Key Laboratory of Advanced Welding and Joining, Harbin Institute of Technology, Harbin 150001, China

E-mail address: lengxuesong@hit.edu.cn (Leng Xuesong)

E-mail address: xieguoqiang@hit.edu.cn (Xie Guoqiang)

Abstract: Structural materials used in nuclear power systems must operate reliably under extreme conditions, including high temperature, high pressure, and neutron irradiation. These environments impose strict requirements on mechanical strength, ductility, phase stability, and corrosion resistance, which are difficult to achieve simultaneously in conventional alloys. To address these challenges, our group has focused on two classes of Ti-based alloy systems—high-entropy alloys (HEAs) and titanium-oxygen (Ti-O) alloys—as candidates for next-generation nuclear structural materials. Ti-based HEAs composed of refractory elements (e.g., Ti, Zr, Nb, Ta, Hf) were synthesized by mechanical alloying followed by spark plasma sintering (SPS). NbTaTiZr alloys exhibited ultrafine dual-phase BCC microstructures and high compressive yield strength (up to 2172 MPa) with good ductility (> 12%). First-principles calculations were also employed to guide the design of TiZrHfNb HEAs, identifying element combinations with low electronegativity differences and negative formation enthalpies to favor solid solution stability. A lightweight AlFeNiTiV alloy featuring $L2_1$-ordered phase precipitation achieved >1500 MPa tensile strength and >15% ductility at a reduced density (~6.0 g/cm^3), offering an optimal strength-to-weight balance. In parallel, Ti-O alloys were developed using Ti/TiO_2 powder metallurgy and SPS. Interstitial oxygen content was precisely controlled between 0.3 and 2.1 wt.%, resulting in significant strength enhancement. The Ti-1.95O alloy reached a yield strength of 2026 MPa with retained compressive ductility (up to 9%), due to synergistic effects of solid solution strengthening and grain refinement. Together, these alloy systems provide promising pathways toward structural materials capable of meeting the rigorous demands of advanced nuclear reactors.

Evaporation Behavior of Nickel Based Alloys and Protection Layer Under Space Extreme Service Environments

Yang Wanhuan, Zou Jichun, Yang Wen, Zhong Weihua

Division of Reactor Engineering Technology Research, China Institute of Atomic Energy, Beijing 102413, China;

Abstract: Space nuclear reactors show great potential for application in deep space exploration and extraterrestrial base missions, where nickel-based alloys are critical candidate materials for key components. A systematic study was conducted on the evaporation behavior of nickel-based alloys in a thermal vacuum environment, revealing the evaporation mechanisms at grain boundaries and phase boundaries. Results demonstrate significant evaporation even at temperatures far below the melting point in ultra-high vacuum, with actual evaporation rates substantially exceeding theoretical predictions due to defects like grain boundaries. Nickel and chromium exhibit high volatility with their low melting points and high vapor pressures, while tungsten-rich precipitates resist evaporation. To address rapid evaporation, we developed Ni-W evaporation-resistant coatings through optimized composition, structure, and processing. These coatings significantly reduce substrate evaporation rates, providing effective protection. This research advances safety design for space nuclear reactors.

Keywords: space environments; nickel based alloys; evaporation behavior

Fabrication of High-Strength Tough Boron-Containing Hard Coatings and Performance Evaluation in High-Speed Dry Machining of Ti-6Al-4 V alloy

Cai Fei[1*], Tao LangLang[1,2], Zhang Shihong[1,2]

[1] Key Laboratory of Green Fabrication and Surface Technology of Advanced Metal Materials, Ministry of Education, Anhui University of Technology, China

[2] School of Materials Science and Engineering, Anhui University of Technology, Maanshan City, Anhui Province, PR China

E-mail address: caifei32@126.com (Cai Fei)

E-mail address: 13615675772@163.com (Tao LangLang)

Abstract: AlTiBN composite coatings with various B contents and bilayer structures were prepared by multi-arc ion plating method, and their microstructures, mechanical properties, impact fatigue resistance and cutting performance during high-speed dry machining of titanium alloys were systematically evaluated. As B content increased, the microstructure of the AlTiBN coatings showed progressive refinement, accompanied by a phase transformation from the single-phase *c*-(Al, Ti)N structure to a dual-phase structures consisting of *c*-(Al, Ti)N and *h*-AlN. The adhesion, fatigue resistance, and tribological performance also increased with higher boron contents. The impact fatigue resistance of AlTiBN coatings initially increased and then decreased with B contents. Furthermore, introducing of multilayer structures increased the adhesion and impact fatigue resistance of AlTiN/AlTiBN multilayer coatings compared with single AlTiBN coatings. High-speed dry turning tests of titanium alloy revealed that the tool life of AlTiBN-coated tools firstly increased and then decreased with increasing B contents at both cutting speeds of 80 m/min and 100 m/min. Among them, the B10 coating demonstrated superior cutting performance (2500 m at 80 m/min; 780 m at 100 m/min). The AlTiN/AlTiBN multilayer coatings also exhibited a similar trend with increasing number of layers, and the 5C-AlTiN/AlTiBN multilayer coating exhibited extended tool life (2800 m at 80 m/min; 1200 m at 100 m/min). The wear mechanisms of coated tools were not affected by the cutting speed, and the adhesive wear and oxidative wear were observed on the rake face, while abrasive wear, adhesive wear, and oxidative wear were identified on the flank face.

Keywords: boron-containing coating; tool coatings; impacting fatigue resistance; high-speed dry cutting performance

Interface Regulation of Zr-B-N/ZrO_2 Multilayer Composite Tool Coatings and the Transformation Mechanism for High-Efficient Milling of Hardened Steel

Wang Tiegang*

Tianjin University of technology and education, Tianjin, China

E-mail address: tgwang@ tute. edu. cn (Wang Tiegang)

Abstract: In the national strategic fields such as aerospace and defense industry, the efficient and precise processing of key hot-end components faces the problems of easily chipping and wearing of cutting tools, as well as short service life. A method of " reverse design of coating composition" was proposed. Using high-power pulsed magnetron sputtering and pulsed direct current magnetron sputtering combined technology, Zr-B-N/ZrO_2 multi-layer composite tool coatings were developed. A certain amount of reducing H_2 was mixed into the reactive N_2 to prepare high-purity Zr-B-N coatings. Then, ZrO_2 layers were periodically implanted as thermal barriers and chemical barriers to exert the synergistic effect of the two coating layers, achieving the controllable preparation of Zr-B-N/ZrO_2 multi-layer composite coatings. The phase composition of the coating was analyzed by X-ray diffraction (XRD), and the microstructure of the coating was characterized by scanning electron microscopy (SEM) and high-resolution transmission electron microscopy (HRTEM). The mechanical properties and tribological behavior of the coating were tested using nanoindentation instrument, scratch tester, and high-temperature friction testing machine. The cutting performance of the coating tool was evaluated by a vertical machining center. The results showed that the Zr-B-N/ZrO_2 multilayer composite coating was mainly composed of fcc-ZrN phase and t-ZrO_2 phase. When the modulation period of the Zr-B-N/ZrO_2 multilayer composite coating was 80 nm and the modulation ratio was 7: 1, the hardness of the coating reached the maximum value of 30. 2 GPa. Compared with the single-component Zr-B-N coating (23. 2 GPa) and the ZrO_2 coating (24. 2 GPa), the hardness of the Zr-B-N/ZrO_2 nano-multilayer coating increased by 30. 2% and 24. 8% respectively, and at this time, the wear rate of the coating reached the minimum value of 5.0×10^{-7} mm^3/ (N · mm). Using uncoated milling cutters and deposited and annealed Zr-B-N/ZrO_2 multi-layer composite coatings, continuous dry cutting of 45# quenched steel was performed. The cutting lifetimes were 6 minutes, 35 minutes and 39 minutes respectively. Compared with the uncoated milling cutters, the cutting lifetimes of the deposited and annealed coating milling cutters were increased by 5. 8 times and 6. 5 times respectively.

Keywords: coating interface control; Zr-B-N/ZrO_2 multilayer composite coating; mechanical properties; frictional behavior; cutting performance

Structure Design and Enhanced Properties of CrAlSiN Coatings

Fan Qixiang*, You Shilong, Wang Yuelin

Tianjin Key Laboratory of High-Performance Manufacturing Technology & Equipment,

Tianjin University of Technology and Education, Tianjin 300222, China

E-mail address: qxfan@ tute. edu. cn (Fan Qixiang)

Abstract: Bias voltage, a critical parameter applied to the substrate, plays a pivotal role in tuning ion energies and tailoring the microstructure and properties of coatings. An optimal bias voltage enhances coating densification, improving mechanical, tribological, and anti-corrosive properties. Increasing the bias voltage typically refines grain structure and increases coating density, as higher energy enhances surface adatom mobility, thereby improving hardness. However, excessively high bias voltages can introduce excessive residual stress and defects, potentially compromising adhesion strength. To address this, a bias-graded deposition method has been developed to optimize coating performance. To investigate the influence of bias voltage on CrAlSiN coatings, six coatings were deposited under constant or linearly increasing bias voltages. All coatings exhibited a single fcc- (Cr, Al) N phase, with minimal variation in chemical composition. As bias voltage increased, grain size and surface roughness decreased, resulting in higher hardness but also greater residual stress and reduced adhesion strength. Compared to coatings deposited at constant bias voltages (−160 V and −200 V), the bias-graded coatings showed a slight decrease in hardness but achieved 3-20% higher adhesion strength and 11-49% improved fracture toughness due to reduced residual stress. This enhancement stems from the grain size gradient along the growth direction, which mitigates stress concentration and inhibits crack propagation. In cutting tests, the bias-graded coatings demonstrated a 44% longer tool life compared to constant-bias coatings, with performance further improving at higher ultimate bias voltages. This superior performance is attributed to their enhanced adhesion strength and toughness, highlighting the effectiveness of the bias-graded approach in optimizing coating durability. The implementation of nanolayered architecture through precisely controlled alternating deposition ofCrAlN and CrSiN layers with varying modulation periods (106-283 nm) represents another effective approach for enhancing the comprehensive performance of CrAlSiN coatings. The multilayer interfaces create additional strengthening mechanisms that preferentially improve coating hardness and wear resistance compared to monolithic coatings. By precisely adjusting the alternating sputtering times of CrSi and CrAl targets, three distinct CrAlN/CrSiN nanolaminated coatings with different modulation periods were successfully fabricated. Extensive characterization revealed that the multilayered coatings consistently outperformed single-layer CrAlN and CrSiN coatings in both hardness and toughness while simultaneously reducing internal stresses. Progressive reduction of the modulation period from 283nm to 106nm yielded systematic improvements in coating properties: residual stress decreased substantially, while both hardness and toughness showed significant enhancement, reaching peak hardness of 34. 3 GPa at the optimal 106nm modulation period. Quantitative measurements documented that coating adhesion strength increased from 45. 08 N to 63. 73 N, while the wear rate decreased from 7.3×10^{-7} $mm^3/N \cdot m$ to 4.8×10^{-7} $mm^3/N \cdot m$ as the modulation period was reduced from 283 nm to 106 nm. The superior performance stems from multiple reinforcing mechanisms at the nanoscale interfaces, including stress field-induced dislocation blocking, crack path deflection, and additional energy dissipation channels. Furthermore, the multilayer design demonstrated excellent potential for industrial applications, with coated tools showing substantially extended service life, where the cutting tool longevity increased progressively as the modulation period decreased.

Keywords: PVD coatings; bias voltage; microstructure; mechanical properties; cutting properties

Deposition Mechanism and Defect Control of CrN/NbN Coatings with Excellent Tribocorrosion Performance

Liu Yongkang, Yang Ying*, Zhang Shihong*

Key Laboratory of Green Fabrication and Surface Technology of Advanced Metal Materials,

Ministry of Education, AnhuiUniversity of Technology, Ma'anshan, China

E-mail address: yangying@ahut.edu.cn (Yang Ying)

E-mail address: shzhang@ahut.edu.cn (Zhang Shihong)

Abstract: The rapid exploitation of marine resources in China has heightened the need for advanced marine engineering equipment and imposed more stringent requirements on the surface performance of its key components. CrN/NbN coatings, with their excellent corrosion and wear resistances, demonstrate potential for applications in marine service environments. In this study, CrN/NbN coatings were deposited on 45# steel substrates using arc ion plating technology. A multilayer/nanolayer design and ion etching process were implemented to reduce coating defect densities, thereby enhancing overall coating performance. Scanning electron microscopy analysis revealed that S2-S6 coatings exhibited fine columnar structures, with well-defined and cohesive sublayer interfaces in S2 and S3 multilayer coatings. XRD and TEM analyses confirmed that the primary phases of the coatings were CrN and NbN. HRTEM images of the S6 coating demonstrated a nanolayer structure with a modulation period of 8.9 nm, where CrN and NbN sublayer thicknesses were approximately 2.7 nm and 6.2 nm, respectively. A coherent interface was observed in the S6 coating, accompanied by the interdiffusion of Nb and Cr elements between the CrN and NbN sublayers. The fast Fourier transform image displayed streak-like features characteristic of stacking faults, as well as two sets of diffraction patterns indicative of coherent sublayer interfaces. Nanoindentation tests revealed that among the fabricated coatings, the S1 monolayer coating exhibited the lowest hardness of (21.8 ± 0.7) GPa, while the S4 coating demonstrated the highest hardness of (30.1 ± 1.4) GPa, attributed to its coherent interfaces and stacking faults. Ion etching had minimal impact on coating phases and mechanical properties. However, ion bombardment effectively interrupted the continuous growth of large particles, resulting in smoother surfaces and interfaces and thereby reducing surface defect proportions. The defect percentages for S3 and S5 coatings were (2.7 ± 0.19)% and (2.43 ± 0.49)%, respectively. These lower defect densities contributed to higher pore resistance (Rpo) and charge transfer resistance (Rct). As sublayer thickness decreased, the electrochemical and tribocorrosion performance of CrN/NbN coatings improved progressively, with the S6 sample achieving the lowest corrosive wear rate of 2.42×10^{-6} $mm^3/(N \cdot m)$. The tribocorrosion failure mechanism was preliminarily explored, identifying layer-by-layer peeling as the dominant failure mode. Compared to NbN monolayer coatings, CrN/NbN multilayer/nanolayer coatings exhibited superior mechanical properties and corrosion resistance due to interface blocking and reinforcing effects. Furthermore, the application of ion etching to CrN/NbN multilayer/nanolayer coatings enhanced their electrochemical corrosion and tribocorrosion properties by disrupting the growth of large defects.

Keywords: CrN/NbN; multilayer/nanolayer coatings; ion etching; tribocorrosion

Ni as a Substitution of Mn in steel: Microstructural Engineering for Broad Mechanical Property Tailoring

Hao Mei*, Zhu Kangying[3], Roxane Massion[2], Olivier Bouaziz[2], Lu Xing[1]

[1] School of Materials Science and Engineering, DalianJiaotong University, Dalian 116028, China

[2] Université de Lorraine, CNRS, Arts et Métiers Institute of Technology, LEM3, F-57000 Metz, France

[3] ArcelorMittal Research and Development Voie Romaine-BP30320, 57283Maizières-lès-Metz Cedex France

E-mail address: hao. mei_ fr@ foxmail. com (Hao Mei)

E-mail address: kangyin. zhu@ arcelormittal. com (Zhu Kangying)

Abstract: This research introduces a methodology for engineering a novel third-generation advanced high-strength steel (AHSS-Gen3Ni) through nickel (Ni) substitution for manganese (Mn), achieving comprehensive tuning of mechanical properties by integrating the " unisteel" concept. The three devised heat-treatment processes (P1-P3) facilitate meticulous regulation of the microstructure while compatible with simple cooling techniques: intercritical isothermal quenching (P1) generates a ferrite-martensite dual phase microstructure; isothermal treatment above the Ms point (P2) leads to A718 steel containing 8.2% retained austenite (RA); air-cooling with partitioning treatment (P3) induces the formation of bainite and 7.3% RA in A717 steel. This work presents innovative approaches to addressing manganese resource consumption in conventional medium-manganese steels, and its " composition-process-structure" regulation mechanism establishes a theoretical basis for lightweight design of automotive steels.

Keywords: advanced high-strength steel; nickel substitution for manganese; microstructure; process design

Simulation of Microstructure and Texture Evolutions During Cold Rolling and Annealing of Ferritic Stainless Steel

Zhang Chi, Zhang Liwen, Song Kangjie

School of Materials Science and Engineering, Dalian University of Technology, Dalian 116024, Liaoning, China

E-mail address: zhangchi@ dlut. edu. cn (Zhang Chi)

Abstract: The formability of ferritic stainless steel is strongly influenced by its grain size distribution and crystallographic texture. The microstructure and texture evolutions during cold rolling and annealing exhibit complex behaviors along with the initial microstructure, texture and processing parameters. In this work, an integrated crystal plasticity (CP) and cellular automaton (CA) model was developed to investigate the microstructure and texture evolutions of ferritic stainless steel during cold rolling and subsequent annealing. To address convergence issues under high deformation, a remeshing method compatible with CP was introduced to transfer inhomogeneous microstructural information to the undeformed configuration. The cold rolled microstructure, subgrain structure, texture and dislocation density was simulated using the CP model. Then the deformed polycrystalline structure was used as the initial microstructure for recrystallization simulation using the CA model. The CA framework incorporated nucleation criteria based on critical dislocation density and grain boundary migration dynamics governed by disorientation. The microstructure and texture evolutions of interested grains with typical initial orientations were tracked by the developed method. Furthermore, the effects of initial orientation, grain morphology and interaction between neighbor grains on the microstructure and texture evolutions were analyzed.

Enhanced Wear Resistance of Cold-Sprayed Cu/MoS_2 Composite Coatings for Steel Distributor Plate Applications

Hao Zishuo, Gong Chunzhi, Tian Xiubo, Liu Yuan, Zhou Changzhuang

State Key Laboratory of Precision Welding & Joining of Materials and Structures, Harbin Institute of Technology, Harbin 150001, P. R. China.

E-mail address: haozishuo0@163.com (Hao Zishuo)

E-mail address: chunzhigong@163.com (Gong Chunzhi)

Abstract: The life-span of plunger pumps is frequently limited by wear at the interface between the steel distributor plate and copper rotor pad. Molybdenum disulfide (MoS_2), renowned for its exceptional solid lubrication properties, offers significant potential for improving tribological performance. This study employed high-energy ball milling to synthesize nano-MoS_2-coated Cu powder, subsequently blended with micron-sized MoS_2 powder to form a hybrid micro-nano feedstock. Composite Cu/MoS_2 coatings were deposited onto 30CrMnSiA steel substrates, pre-coated with an Al bond layer, using the cold spray (CS) process. Process parameters, notably gas temperature and powder feed rate, were systematically varied. Due to significant differences in the deposition efficiency of the constituent powders, the resulting coatings exhibited a distinct architecture comprising a Cu-rich underlayer and an MoS_2-enriched top surface. Key findings include: The formation of vortex-dominated interfacial microstructures at the Cu-Al interface, enhancing mechanical interlocking; A substantial reduction in the average coefficient of friction from 0.77 (without lubrication) to 0.09, attributed to the lubricious MoS_2 surface layer under standardized pin-on-disk testing; Enhanced coating thickness and further reduced COF achieved at elevated gas temperatures and higher powder feed rates. The optimized cold spray processing window identified herein provides valuable guidance for fabricating durable, low-friction coatings for critical pump tribo-components.

Keywords: cold spraying; friction performance; composite Coatings; molybdenum disulfide; high-energy ball milling

Study on the Evolution Mechanism of Heat-Treated Residual Stresses in Welded Joints of X52 Oil and Gas Pipeline

Zhang Jin, Wang Huining*, Zhang Yiman

College of New Energy, China University ofPetroleum (East China), Qingdao, China

E-mail address: zhangjin151. slyt@ sinopec. com (Zhang Jin)

E-mail address: s23150047@ s. upc. edu. cn (Wang Huining)

Abstract: This work takes X52 oil and gas pipeline welded joints as the object, considers the effect of solid-state phase transformation, and uses numerical simulation to study the microstructure of the welding process and the residual stress evolution law before and after heat treatment. It is found that the ferrite phase transformation and pearlite phase transformation occurring during the welding cooling process have a large impact on the annular residual stress level, cooling compression produces a large range of annular residual compressive stress, the peak axial residual stress is not much affected by the presence or absence of phase transformation, but it is obvious that the top of the weld bead has a larger coverage of axial compressive stress after considering the phase transformation. The post-weld heat treatment has a certain relaxation effect on the residual stress, the axial residual stress is changed from a large compressive stress to a small tensile stress, the stress level at the weld decreases significantly, showing a low stress in the weld and a high distribution law of the base material; the circumferential residual stress maintains the original distribution of the stress, and the proportion of the stress release is about 50 per cent.

Keywords: X52 steel; welding residual stress; heat treatment; solid-state phase transition

Multiscale Thermal Management for Solidification Defect Control: Graded Cooling Design and Ceramic Core Thermal Redesign in Heavy-Duty Turbine Vane Casting

Chang Yu, Wang Yao*, Dong Jiasheng

Institute of Metal Research, Chinese Academy of Sciences, Shenyang, China

E-mail address: ychang21s@ imr. ac. cn (Chang Yu)

E-mail address: yaowang@ imr. ac. cn (Wang Yao)

Abstract: This study develops an integrated thermal management strategy to eliminate shrinkage porosity defects in large-sized gas turbine vane in investment casting through systematic process optimization from Process Ⅰ to Ⅲ. Combining numerical simulation and engineering-scale verification, the graded shell cooling strategy in Process Ⅱ establishes controlled sequential solidification, effectively resolving platform-airfoil junction shrinkage porosity through optimized transverse feeding efficiency. Concurrently, the strategic redesign of ceramic core thermal conductivity in Process III extends solidification windows for diaphragms with ultra-small thickness-to-length ratio (thickness/length < 0.018), achieving densification of the diaphragm by setting phase evolution sequences and maintaining critical melt feeding channel. The results of engineering-scale validation confirm the superiority of the optimized process Ⅲ, showing remarkable porosity reduction at the platform-airfoil junction and on the diaphragm. The verified integrated process optimization provides a replicable paradigm for manufacturing large-sized complex superalloy components.

Keywords: investment casting; nickel-based superalloy; shrinkage porosity; thermal gradient control; solidification simulation; gas turbine vane

Influence of Heat Treatment on the Microstructure and Mechanical Properties of WC-Reinforced Steel-Bonded Cemented Carbide

Zhu Rui, Zhail Yazhong, Li Jinpu, Che Hongyan

Special Steels Institute, Central Iron and Steel Research Institute Co. , Ltd. , Beijing, 100081, China

E-mail address: chehy. 2009@ tsinghua. org. cn (Che Hongyan)

Abstract: Steel-bonded cemented carbide is a type of wear-resistant material between cemented carbide and high-speed steel. Its properties can be regulated through forging and heat treatment. Usually, WC or TiC is used as the hard phase. Due to the better wettability of WC with ferrous-based materials, it has been more widely applied in scenarios requiring toughness and wear resistance. However, the mutual diffusion between WC and the matrix makes the heat treatment of steel-bonded cemented carbide more complex. This paper takes steel-bonded cemented carbide with 25% WC as the research material to systematically study the effects of different quenching temperatures (970 ℃, 1010 ℃, 1030 ℃) and cryogenic treatment at −80 ℃ on its microstructure and mechanical properties. Meanwhile, the comprehensive regulation of quenching and tempering temperatures was conducted to explore the synergistic effects of multiple parameters on the microstructure and properties of the material. The results show that with the increase of quenching temperature, the number and size of WC particles undergo significant changes, the material hardness continues to rise, and the bending strength and impact toughness both exhibit a " first decreasing then increasing" trend. For the 1010℃ quenched samples, cryogenic treatment (−80℃ × 1 h) of the sample quenched at 1010 ℃ can significantly improve the comprehensive properties of the material, with impact toughness increased by 28.63% and bending strength increased by 14.5%. Through the comprehensive regulation of quenching and tempering processes, the performance of the optimized material is excellent, with impact toughness of 16.5 g/cm^2, hardness close to 63 HRC, and bending strength approaching 4000 MPa. This study clarifies the correlation mechanism between the heat treatment process parameters and the microstructure-performance of WC-reinforced steel-bonded cemented carbide, providing theoretical support and data reference for the engineering application and process optimization of such material.

Keywords: WC reinforced steel-bonded cemented carbide; heat treatment; microstructure; mechanical properties

A Dual Phase Steel Structure with Synergetic Strength-ductility Enhancement: Heat Treatment and Deformation Mechanisms

Wu Junyin[1], Chen Ran[1], Shen Yao[1]

School of Materials Science and Technology, Shanghai Jiao Tong University, China

E-mail address: user_ wujunyin2017@ sjtu. edu. cn (Wu Junyin)

E-mail address: chenra@ sjtu. edu. cn (Chen Ran)

Abstract: Dual-phase (DP) steels are widely used in engineeringapplications, particularly automotive manufacturing, due to their cost-effectiveness and good formability. However, traditional approaches of increasing strength — typically by raising martensite content — often result in a significant loss in plasticity. Achieving a synergetic improvement in both strength and ductility while not raising the martensite content remains a challenge in the development of advanced DP steels. Inthis study, a novel fast-heating intercritical step-annealing (FA-ISA) process was developed to fabricate a dual-phase steel, featuring a unique " ferrite core + necklace martensite shell" (" core-necklace shell") phase distribution, from a homogeneous plate with a commercial DP780 composition (denoted as 780-FA) . While maintaining martensite volume fraction, this phase rearrangement led to an approximate increase of 100 MPa in tensile strength and a 3% gain in total elongation, corresponding to an enhancement of nearly 20% in the strength-ductility product. The formation of the " core-necklace shell" structure isrealized in four steps: (1) Rapid heating to increase the density of nucleation sites for austenites; (2) Fast intercritical annealing (FIA) to produce austenite along ferrite grain boundaries, stopping right at impingement of neighboring austenites; (3) Mid-temperature annealing (MA) to facilitate ferritization of austenite boundaries, thereby tuning the inter-austenite spacing (i. e. , the width of ferrite channels) . (4) Finally, quenching to transform the microstructure into the desired core-necklace shell configuration. μ-DIC & EBSD experiments, coupled with damage assessment, reveal that both the strengthening of 780-FA steel and ductility enhancement are attributed to enhanced mesoscale strain homogeneity. The enhanced homogeneity is manifested in several aspects: (1) increased fraction of medium-strain area and (2) reduced difference in the strains in the ferrite and austenite phases, (3) Even in the slip bands typifying strain concentrations, the bands of high strain concentration become more fragmented. These features are supposed to improve the strength and retard the damage nucleation and development, therefore leading to the synergetic improvement. The " core-necklace shell" phase distributionhelp to utilize the strengthening capacity of the hard phases and the deformation ability of the soft phases, to a larger extent simultaneously, and it holds promise for broader application in the development of dual-or multi-phase structural steels to enhance the strength-ductility synergy.

Keywords: dual-phase (DP) steels; structure regulation; deformation mechanism; rapid heat treatment

Evaluation of the Effect of Ultrasonic Impact on the Fatigue Properties of Q235 Steel Welded Joints Based on the Force-Magnetic Effect

Shan Zihao, Liu Bin*

Jiangsu University of Science and Technology, Zhenjiang, China

E-mail address: 231210601106@stu.just.edu.cn (Shan Zihao)

E-mail address: liubindely@163.com (Liu Bin)

Abstract: During the service of welded joints, the residual stress after welding and complex engineering load environment of welded joints will cause fatigue damage to the welded joints, thereby reducing the safety of the structure. Therefore, it is very important to quickly regulate the stress of the welded structure and detect early damage. This paper uses ultrasonic impact to add beneficial compressive stress to the surface of the sample to adjust the stress, and evaluates the stress redistribution of samples under different coverage rates during fatigue based on the force-magnetic effect. First, fatigue tests are carried out to determine the relationship between different stress ratios and different loads and fatigue life, and magnetic signals on the sample surface under different states are collected. Then, the reliability of online and offline acquisition signals was compared, and it was found that offline detected signals had better response to fatigue state. Secondly, the effect of ultrasonic impact coverage on fatigue life is determined through fatigue tests. Finally, the accuracy was verified through OM, SEM experiments.

Keywords: metal magnetic memory; ultrasonic impact treatment, fatigue damage; fatigue life

Gradient Microstructure-Mediated Superior Torsional Properties in Ti-55531 Alloy Based on High-Frequency Induction Heating

Huang Chaowen *

College of Materials and Metallurgy, Guizhou University, Guiyang, China

School of Mechanical & Aerospace Engineering, Nanyang Technological University, Singapore

E-mail address: cwhuang@ gzu. edu. cn (Huang Chaowen)

Abstract: This study explores a gradient microstructure design strategy for Ti-55531 alloy, utilizing high-frequency electromagnetic induction quenching (HFEIQ) to establish a surface-to-core microstructural gradient that optimizes torsional performance. Through synergistic integration of HFEIQ and aging treatments, we fabricated a hierarchical gradient microstructure characterized by nanoscale secondary α (α_s) lamellae at the surface to enhance strength, while retaining semi-equiaxed primary α (α_p) phases in the core to preserve ductility. The optimized gradient structure achieved remarkable mechanical improvements, with the HFEIQ-5. 6 s specimen demonstrating a maximum shear stress (τ_{max}) of 1158. 08 MPa—representing an 8. 75% enhancement over conventional bimodal microstructures—while retaining 9. 63% shear ductility, thereby achieving an exceptional strength-ductility balance. Microstructural analysis reveals that the gradient structure promotes dislocation accumulation at α_s/β_r (retained β matrix) interfaces and deformation twinning within α_s lamellae, enabling coordinated plastic deformation. These mechanisms, tailored by the gradient design, are critical for the enhanced performance. However, prolonged HFEIQ time (5. 7 s) leads to α_s coarsening, shifting the dominant deformation mechanism to dislocation slip and reducing twin formation, thereby diminishing the strengthening effect. This work demonstrates that precisely controlled gradient microstructures can optimize the mechanical response of Ti-55531 alloy, offering a promising pathway for advanced structural applications.

Keywords: ti-55531 alloy; high-frequency electromagnetic induction quenching; gradient microstructure; torsion properties; deformation mechanism

High Performance Alloy Design Guide by ICME and Advanced Characterization Techniques

Zhang Jingqi[1], Hou Ziyong[1*]

[1] School of Materials Science and Engineering, Xi'an Jiaotong University, China

[2] State Key Laboratory of Mechanical Transmission for Advanced Equipment, International Joint Laboratory for Light Alloys (MOE), Institute of Advanced Interdisciplinary Studies Chongqing University, China

E-mail address: jingqi.zhang@xjtu.edu.cn (Zhang Jingqi)

E-mail address: houzy@cqu.edu.cn (Hou Ziyong)

Abstract: Alloy design and heat treatment optimization are crucial methods for developing high-performance metallic alloys, such as steels and titanium, that achieve an exceptional balance of strength and toughness. The behavior of precipitation during heat treatments and solidification is fundamental in shaping the microstructure and enhancing the mechanical properties of these materials. This is driven by processes such as precipitation hardening and grain refinement. The conventional trial-and-error approach to material development can be prohibitively time-consuming and costly. In contrast, more efficient strategies, such as Integrated Computational Materials Engineering (ICME), are emerging as vital tools in the pursuit of innovative material design. In this study, we will utilize quantitative modeling, complemented by experimental validation through advanced characterization techniques, to advance the development of Fe-C-Cr steel and titanium alloys. Our goal is to precisely adjust their microstructure and mechanical properties for superior performance. Furthermore, we will explore the complexities of precipitation modeling, including nucleation, growth, and coarsening, while examining how input parameters derived from microstructural characterization can significantly impact simulation outcomes. This integrated approach not only illustrates a pathway toward high-efficiency alloy design but also lays the groundwork for groundbreaking advancements in material science.

Keywords: alloy design; heat treatment; ICME; microstructure; mechanical properties

Improving Hardness-Toughness of DC53 Cold Work Die Steel via Tailoring Isothermal Quenching Process

Liu Yuxiang, Wang Zhigang*

School of Materials Science and Engineering, Jiangxi University of Science and Technology, Ganzhou, China

E-mail address: wzgang2008cn@163.com (Wang Zhigang)

Abstract: Achieving a balance between hardness, wear resistance, and toughness remains a key challenge in developing advanced cold work die steels for precision components and molds. This study addresses this challenge by applying an optimized isothermal quenching process to modified DC53 steel (with W addition), targeting a refined microstructure of lower bainite and martensite. The results show that isothermal quenching induces a dual-phase (martensite + lower bainite) structure in DC53 steel, with the lower bainite fraction increasing with isothermal treatment duration—reaching 24%, 51%, and 67% after 3, 8, and 10 hours, respectively. The best mechanical properties are achieved at a near 1 : 1 martensite-to-bainite ratio. The optimal heat treatment parameters consist of austenitizing at 1040 ℃ for 20 min, followed by rapid quenching in a 300 ℃ molten salt bath for 8 hours. Compared to conventional high-temperature tempered DC53, the isothermally quenched steel maintains 61 HRC hardness while increasing unnotched impact toughness from 32.8 J/cm^2 to 41.3 J/cm^2, alongside lower friction coefficients and reduced wear rates. The main mechanisms of performance optimization of DC53 steel include: (1) The lower bainite acts to partition the original austenite grains, which produced an effect like fine-grain strengthening. (2) Lower bainite can disperse the stress at the original austenite grain boundaries, alleviate stress concentration, thereby enhance toughness. (3) Film-like retained austenite improves thermodynamic and mechanical stability, reducing wear-induced cracking. This study provides both experimental guidance for industrial isothermal quenching of DC53 die steel and deepens the fundamental understanding of the strengthening-toughening mechanisms in dual-phase microstructures.

Keywords: DC53 cold work die steel; isothermal quenching; lower bainite; mechanical properties; strengthening and toughening mechanisms

Effects of Pre-strain Deep Cryogenic Aging on the Mechanical and Corrosion Properties of Al-Cu-Li Alloys

Chen Junfeng[1*], Liu Yan[1], Chi Haitao[2]

[1] School of Materials Science and Engineering, Fuzhou University, Beijing, China

[2] Fujian Xiangxin Light alloys manufacture Co., Ltd., Fuzhou, China

E-mail address: chts@chts.org.cn (Chen Junfeng)

E-mail address: getotten@gmail.com (Liu Yan)

Abstract: In order to improve the mechanical properties and corrosion resistance of Al-Cu-Li alloy, this work develops a novel strain heat-treatment technology. Surprisingly, Al-Cu-Li alloys treated by 4% pre-strain and 48-hour deep cryogenic aging exhibits excellent mechanical properties, whose yield strength, tensile strength, and elongation are 573.7 MPa, 593.6 MPa, and 14.25%, respectively. Furthermore, the corrosion resistance of sample treated by the above pre-strain deep cryogenic aging process is two times as much as traditional T6 aged. This study deeply investigates the influence of various precipitates in Al-Cu-Li alloys on its performance through microstructure analysis. The results show that the introduction of pre-strain inhibits the precipitates of T1 and δ′ phases at grain boundaries, and prevents the formation and widening of precipitate-free zones. At the same time, it reduces the potential difference between grain boundaries and matrix, and enhances the corrosion resistance of Al-Cu-Li alloys. Deep cryogenic can inhibit recover and maintain deformation dislocation, so that a large number of defects produced by pre-strain promote the nucleation and diffusion of precipitates in the matrix, resulting in tiny and uniformly distributed precipitates. Therefore, the novel pre-strain deep cryogenic aging not only endues Al-Cu-Li alloys excellent performance, but also offer a new a new strategy for controlling precipitation.

Keywords: 2A97 Al-Cu-Li alloys; pre-strain deep cryogenic aging; precipitate; mechanical property; corrosion property

Precision-controlled Technology and Application of High-Strength and High-Toughness Multiphase Microstructure for High-End Equipment Bearings

Qian Dongsheng, Lu Yunbin, Wang Feng*

Wuhan University of Technology, Wuhan, China

E-mail address: qiands@ whut. edu. cn (Qian Dongsheng)

E-mail address: wangfengwut@ whut. edu. cn (Wang Feng)

Abstract: Bearings are essential components of high-end equipment including high-speed machine tools, aerospace, and new energy vehicles, directly affecting the operational performance and service life of the equipment. Domestically produced bearings exhibit shorter service lives and inferior reliability, leading to prolonged dependence on imported products, which severely restricts the independent development of high-end equipment industry. The study in response to the significant demand for high-performance manufacturing technology research and application of bearings in the independent development of high-end equipment in China. developed a precision-controlled technology to achieve a high-strength and high-toughness multiphase microstructure in bearings. The genetic evolution mechanism of multiphase heat treatment microstructure of bearings was revealed. The formation, strengthening, and toughening mechanisms of martensite-bainite multiphase microstructures was clarified. The mapping relationship between heat treatment conditions, martensite/bainite microstructure states and mechanical properties was established. The optimal heat treatment conditions for a high-strength and high-toughness martensite-bainite multiphase microstructure was determined. The multiphase heat treatment method that precisely controls the volume fraction, morphology, and size of bainite was developed. The resulting GCr15 bearing steel exhibited a tensile strength of 2215 MPa and an impact energy (AKU) of 20 J/cm^2, representing the highest strength-toughness combination reported in current literature. The developed technology has found extensive application in high-end industrial machines, aerospace, and new energy vehicles bearings, demonstrating significant industrial value.

Keywords: multiphase heat treatment; martensite-bainite microstructures; bearing; mechanical properties

Effect of Y Precursors on the Synthesis of Cu-Y_2O_3 by Mechanical Alloying and Spark Plasma Sintering

Liu Jiaqin, Xu Jie, Ma Bing, Wu Yuceng*

[1] Beijing University of Chemical Technology, Beijing, China

[2] Hefei University of Technology, Hefei, China

E-mail address: jqliu@ buct. edu. cn (Liu Jiaqin)

E-mail address: ycwu@ hfut. edu. cn (Wu Yuceng)

Abstract: This work prepared the Cu-Y_2O_3 alloy through mechanical alloying and spark plasma sintering. Based on the in-situ reaction principle, Y salts were used as a precursor of Y_2O_3 dispersoid instead of pure Y powder, and the influence of different Y salts on the microstructure, mechanical, and physical properties of the copper matrix was researched. The results indicate that Y salts can effectively prevent the cold welding effect of pure Y metal powder, thereby improving the homogeneity and refinement effect of alloying. Among different Y salts samples, $Y(NO_3)_3 \cdot 6H2O$ and $Y_2(SO_4)_3 \cdot 8H_2O$ may cause varying degrees of contamination to the copper matrix, affecting the material properties. In comparison, $Y_2(CO_3)_3$ has the advantage of low decomposition temperature and the decomposition product CO_2 does not react with the copper matrix. It successfully refined the average particle size of Y_2O_3 from 8.79 ± 3.73 nm to 5.48 ± 2.82 nm, increased the number density from $2.37 \times 10^{21}/m^3$ to $7.46 \times 10^{21}/m^3$, and obtained a tensile strength of 291 MPa and an elongation of 19.8% at room temperature, with a thermal conductivity of 322 W/(m · K) at 400 ℃ for the Cu-Y_2O_3 alloy. This validates the feasibility of combining Y salts with mechanical alloying. The preparation of high-strength, high-conductivity Cu alloys holds significant reference value.

Keywords: Cu-Y_2O_3; mechanical alloying; spark plasma sintering; Y salts; heat sink materials

Effect of a Modified Quenching on the Uniformity of Microstructures and Fatigue Properties of 52100 Bearing Steel

Li Yonghan, Liu Tengyuan, Jiang Zhonghua, Wang Pei*, Li Dianzhong*

Shenyang National Laboratory for Materials Science, Institute of Metal Research, CAS, China

E-mail address: pwang@ imr. ac. cn (Wang Pei)

E-mail address: dzli@ imr. ac. cn (Li Dianzhong)

Abstract: In order to improve the uniformity of microstructures, and thus enhance the mechanical properties, especially the fatigue properties and impact toughness, of 52100 bearing steels, modified quenching heat treatments with different austenitizing processes have been designed. The results show that 740 ℃ is the optimal holding temperature (approaching Ac 1 = 750 ℃) to improve the L_{10} fatigue life by ~3 times and impact toughness by 37% after quenching and tempering. Microstructure observations reveal that 740 ℃ holding shortens the nucleation incubation period of austenitization phase transformation, and inhibits the overfast growth of austenite in some regions, which leads to simultaneous nucleation and growth of austenite during austenitizing, and homogeneous dissolution of carbides in different austenite grains. The changes in microstructure evolution improve the microstructure uniformity (carbide and grain size) after quenching. Uniformly distributed carbides and grain size effectively reduce stress concentration and retard crack propagation during fatigue and impact loading, thus resulting in excellent mechanical properties.

Keywords: austenitizing process; carbides distribution; fatigue life; impact toughness; 52100 steels

Tuning Heterostructure for Enhanced Strength-Ductility Synergy in Mg-Matrix Composites via Heat Treatment and Severe Plastic Deformation

Xu Qiong[1*], Li Yuhua[2], Ma Aibin[3]

[1] School of Materials and Chemical Engineering, AnhuiJianzhu University, China

[2] School of Mechanical and Electrical Engineering, Anhui Water Conservancy Technical College, China

[3] College of Materials Science and Engineering, Hohai University, China

E-mail address: 15251856392@163. com (Xu Qiong)

E-mail address: sduhua@126. com (Li Yuhua)

Abstract: In this study, a magnesium matrix composite (SiCp/AZ91) with heterostructure was fabricated using heat treatment combined with severe plastic deformation. The microstructural evolution and mechanical properties of the composite were investigated. The results indicate that the SiCp particles exhibited excellent dispersion. After the combined processing, a heterogeneous microstructure consisting of fine β-$Mg_{17}Al_{12}$ phases and a high fraction of refined grains formed in the composite matrix. The area fraction of fine grains (<5 μm) was 68. 4%, while that of coarse grains ($\geqslant 5$ μm) was 31. 6%. The ultimate tensile strength (UTS) significantly increased to 426 MPa, and the yield tensile strength reached 372 MPa. Furthermore, the heterostructured composite achieved a ductility of 6. 8%. This non-uniform grain structure resulted from the synergistic effects of fine SiCp particles and dispersed β-$Mg_{17}Al_{12}$ precipitates, which induced heterogeneous recrystallization and inhibited grain growth. The findings of this study provide a novel strategy for developing magnesium matrix composites with heterogeneous particle structures capable of optimizing both ductility and strength.

Keywords: magnesium matrix composite; solid solution; severe plastic deformation; heterogeneous microstructure; mechanical properties

Enhanced Built-in Electric Field and Photoelectric Properties in Self-polarized $BiFeO_3$ Film

Li Ze[1*], Liu Jingshun[1], Fei Weidong[2]

[1] Inner Mongolia University of Technology, Hohhot, China

[2] Harbin Institute of Technology, Harbin, China

E-mail address: ngdzeli@163.com (Li Ze)

E-mail address: jingshun_liu@163.com (Liu Jingshun)

Abstract: It is significant for the design and application of photoelectric effect of thin films to introduce the built-in electric field of vertical thin films through self-polarization. In this work, the self-polarized $BiFeO_3$ films are prepared with sol-gel method. The built-in electric field and photoelectric effect are enhanced with the methods of designing double layer films and Zn^{2+}-Ti^{4+} ions co-doping. Among which, the self-polarized $BiFeO_3$ film under low power density illumination achieves an open circuit voltage and short circuit current of 0.46 V and 0.6 $\mu A/cm^2$, respectively. The $BiFeO_3/ZnTiO_3$ double layer film under ultraviolet illumination exhibits a responsivity and detectivity of 7.4 mA/W and 7.5×10^{11} Jones, respectively. Additionally, Zn^{2+}-Ti^{4+} ions co-doping can enhance the built-in electric field strength and self-polarization stability of $BiFeO_3$ film. The results provide references for the regulation of polarization behavior and the improvement of photoelectric performance in ferroelectric materials.

Keywords: self-polarized $BiFeO_3$ film; ion doping; self-powered photodetection; built-in electric field

Nanoscale Co-precipitation Strengthened High-entropy Alloys

Jiao Zengbao*

Hong Kong Polytechnic University, Hong Kong, China

E-mail address: zb. jiao@ polyu. edu. hk (Jiao Zengbao)

Abstract: Nanoscale co-precipitation-strengthened high-entropy alloys can leverage the synergistic effects of different precipitate phases in terms of strengthening, plastic deformation, and work hardening, thereby overcoming the limitations of single-phase precipitation and achieving an excellent balance of strength and toughness at ultra-high strength levels. This study found that adding Nb to FCC high-entropy alloys can promote the composite precipitation of coherent L12 nano-phases and incoherent Laves nano-phases. The coherent L12 nano-phase can be sheared by dislocations, facilitating long-range dislocation glide in the alloy, while the incoherent Laves phase can effectively hinder dislocation motion, leading to significant dislocation pile-up at its interface and enhancing the alloy's work hardening capability during the early stages of plastic deformation. The study also found that adding Nb to eutectic high-entropy alloys not only promotes the precipitation of coherent L12 nano-phases in the FCC matrix, but also facilitates the composite precipitation of α' nano-phases and Laves nano-phases in the B2 matrix, forming a complex microstructure with a dual-phase matrix and multiple nano-precipitates. This high-entropy alloy exhibits a yield strength of 900 MPa and 50% ductility at 700 ℃, demonstrating good high-temperature mechanical properties.

Keywords: high-entropy alloy; nanoscale precipitate; mechanical properties; alloy design

Research on the Collaborative Regulation of Microstructure Pattern on Multiple Service Properties of High-performance Aluminum alloys

Liu Danyang, Gan Kefu, Li Ruidi, Li Jinfeng*

School of Materials Science and Engineering, Central South University, Changsha, Hunan 410083

E-mail address: liudanyang@ csu. edu. cn (Liu Danyang)

E-mail address: lijinfeng@ csu. edu. cn (Li Jinfeng)

Abstract: High-performanceAluminum alloys possess outstanding comprehensive performance in various applications, which meet the current demands of integrated engineering scenarios. The research progress in this field has attracted significant attention from many high-tech sectors such as aviation, aerospace, marine, and rail transportation. As a typical representative of high-performance alloys, Al-Li alloys are renowned for their superior comprehensive performance and are currently the focus of extensive research. The main characteristics of Al-Li alloys lie in the diversity of their age-hardening precipitates, including δ′, θ′, T1, and S′, along with minor phases such as T2, TB, δ, and Ω. The state of these precipitates is complex and variable, and is easily influenced by external factors such as temperature and stress. The grain structure morphology of Al-Li alloys is also susceptible to external conditions, presenting complex features that differ from those of conventional high-performance Aluminum alloys. The combination of various precipitate phases and grain structure characteristics is highly correlated with the multi-performance of the alloy under specific (micro) alloying conditions, profoundly affecting the actual service performance of the alloy. The report is based on the research foundation of the multi-performance and microstructure regulation of Al-Li alloys that our research group has summarized over many years, and it mentions the effective research practice of constructing the correlation of multi-performance combinations under the regulation of composition design and microstructure patterns of Al-Li alloys.

Keywords: aluminum alloy; microalloying; multi-performance regulation; phase characteristics

High-entropy Nitride Coatings for Marine Applications: Microstructural Evolution and Electrochemical Properties

Yang Wei[1], Wang Zhenyu[2], Wang Aiying[2*]

[1] Faculty of Mechanical Engineering & Mechanics, Ningbo University, Ningbo, China

[2] Key Laboratory of Advanced Marine Materials, Ningbo Institute of Materials Technology and Engineering, Chinese Academy of Sciences, Ningbo, China

E-mail address: yangwei2@nbu.edu.cn (Yang Wei)

E-mail address: aywang@nimte.ac.cn (Wang Aiying)

Abstract: High-entropy materials possess high hardness and strong wear resistance, yet the key bottleneck for their practical applications is the poor corrosion resistance in harsh marine environments. In this work, $(MoNbTaTiZr)_{1-x}N_x$ ($x = 0$-0.47) high-entropy nitride (HEN) coatings were fabricated using a hybrid direct-current magnetron sputtering technique. This study focused on the effect of nitrogen content on the microstructure, mechanical properties, and electrochemical performance. Results indicated that the as-deposited coatings exhibited a typical body-centered cubic (BCC) structure when nitrogen-free ($x = 0$). At $x = 0.17$, the structure comprised an amorphous matrix with embedded face-centered cubic (FCC) nanocrystallites. Further increasing x to the range of 0.35-0.47 resulted in the formation of a polycrystalline FCC phase. Compared to the MoNbTaTiZr metallic coating, the nitrogen-containing coatings achieved significantly higher hardness (13.7-32.4 GPa) along with superior tolerance to both elastic and plastic deformation. Furthermore, these N-containing coatings demonstrated low corrosion current densities ($\sim 10^{-8}$-10^{-7} A/cm^2) and high electrochemical impedance ($\sim 10^6\ \Omega \cdot cm^2$) in 3.5 wt.% NaCl solution, indicating exceptional corrosion resistance. The enhanced electrochemical performance is attributed to the spontaneous formation of protective passive layers on the coating surface. These layers primarily consist of multi-elemental oxides with inherent chemical stability. Notably, the $(MoNbTaTiZr)_{0.83}N_{0.17}$ coating displayed the highest hardness (32.4 ± 2.6 GPa) and H/E ratio (0.09), coupled with remarkable corrosion resistance, suggesting its strong potential for demanding applications requiring both superior wear and corrosion resistance.

Keywords: High-entropy nitride coating; magnetron sputtering; microstructure; mechanical property; corrosion behavior

The Influence of β-stabilizing Elements on the Microstructure Evolution and Mechanical Properties of the Zr-containing TiAl Alloy

Guo Yingfei *

College of Materials and Metallurgy, Guizhou University, Guiyang, China

E-mail address: yfguo@ gzu. edu. cn (Guo Yingfei)

Abstract: This study systematically investigates the influence of β-stabilizing elements (Cr, Mn, and Fe) on the microstructure evolution and mechanical properties of the Ti-46Al-2Zr alloy. The results indicate that the incorporation of β-stabilizing elements facilitates the nucleation of the primary β phase during solidification. Due to the balance between the growth of primary β and α phases, the addition of Cr and Mn significantly reduces both the grain size and lamellar spacing, while the introduction of Fe primarily results in a refinement of lamellar spacing alone. Moreover, the accumulation of β-stabilizing elements in the residual liquid phase contributes to the formation of the B2 phase, while the addition of Fe is also associated with the emergence of the honeycomb τ_3 phase. The inclusion of β-stabilizing elements effectively enhances the room-temperature mechanical properties of the Zr-containing TiAl alloy, with Cr addition demonstrating the highest compressive strength and Mn addition resulting in the greatest fracture strain. The significant improvement in compressive strength from Cr addition is attributed to fine grain strengthening, lamellar strengthening, and solid solution strengthening. Conversely, the toughening mechanism induced by Mn addition involves the refinement of grains and lamellae, a reduction in the c/a ratio, and the formation of nanotwins. The effect of Fe addition on lamellar refinement is the least favorable, resulting in the weakest strength improvement. However, the increase in the content of the blocky γ phase leads to higher plasticity compared to Cr addition.

Keywords: uphill process; quenching; heat treatment; aluminum alloys; tensile test; mechanical properties. TiAl alloys; Zr alloying; β-stabilizing elements; microstructure evolution; mechanical properties; strengthening and toughening mechanisms

Preparation and Interface Design of High Strength Be/Al Composites

Yang Wenshu, Ju Boyu

School of Materials Science and Engineering, Harbin Institute of Technology, Harbin 150001, China

E-mail address: yws001003@163.com (Yang Wenshu)

E-mail address: jby@hit.edu.cn (Ju Boyu)

Abstract: Due to its excellent specific strength and specific stiffness properties, Be-Al alloy (also known as Be/Al composite) is a key material for the lightweight of new generation aerospace equipment. However, the strength performance of the existing beryllium-aluminum materials has been low for a long time, which is due to the defects of beryllium powder itself and the low bonding strength of beryllium-aluminum interface. In view of the above problems, a new manufacturing technology route of beryllium aluminum material (pressure infiltration technology) was designed in this work. The aluminum liquid was infiltrated into the microcracks of beryllium powder under high temperature and high pressure, and the defects of beryllium powder were repaired in situ. The strength performance and strengthening efficiency of the material were higher than those of the materials prepared by traditional precision casting and powder metallurgy. On this basis, based on the first-principles calculation and experimental verification, the influence mechanism of the addition of typical alloying elements such as Cu, Mg and Si in Al on the bonding mode and bonding strength of beryllium-aluminum interface was studied. It was found that the addition of Cu element increased the interfacial adhesion work between Be/Al and the bonding trend, which was beneficial to the improvement of the bonding strength of beryllium-aluminum interface, thus improving the strength and plasticity of the material.

Keywords: Be-Al alloy; matrix alloying; interface regulation; strong-plasticity mechanism